World Forests

Volume 23

World Forests Description

As forests stay high on the global political agenda, and forest-related industries diversify, cutting edge research into the issues facing forests has become more and more transdisciplinary. With this is mind, Springer's World Forests series has been established to provide a key forum for research-based syntheses of globally relevant issues on the interrelations between forests, society and the environment.

The series is intended for a wide range of readers including national and international entities concerned with forest, environmental and related policy issues; advanced students and researchers; business professionals, non-governmental organizations and the environmental and economic media.

Volumes published in the series will include both multidisciplinary studies with a broad range of coverage, as well as more focused in-depth analyses of a particular issue in the forest and related sectors. Themes range from globalization processes and international policies to comparative analyses of regions and countries.

More information about this series at http://www.springer.com/series/6679

Timo Särkkä · Miquel Gutiérrez-Poch
Mark Kuhlberg
Editors

Technological Transformation in the Global Pulp and Paper Industry 1800–2018

Comparative Perspectives

Springer

Editors
Timo Särkkä
Department of History and Ethnology
University of Jyväskylä
Jyväskylä, Finland

Mark Kuhlberg
Department of History
Laurentian University
Sudbury, ON, Canada

Miquel Gutiérrez-Poch
Department of Economic History, Institutions, Policy and World Economy, The Research Centre in Economics and Economic History 'Antoni de Capmany'
University of Barcelona
Barcelona, Spain

ISSN 0785-8388 ISSN 1566-0427 (electronic)
World Forests
ISBN 978-3-030-06945-2 ISBN 978-3-319-94962-8 (eBook)
https://doi.org/10.1007/978-3-319-94962-8

This Springer imprint is published by the registered company Springer Nature Switzerland AG
The registered company address is: Gewerbestrasse 11, 6330 Cham, Switzerland

Preface

This book is the product of a joint effort by 15 authors representing ten different universities around the world, including institutions in Finland, Spain, Canada, Sweden, Norway, Portugal, Australia and New Zealand. The book originated in a seminar organised at the Department of History and Ethnology, University of Jyväskylä, Finland, on 21–22 October 2013. The focus of the seminar was to create new collaborative networks among scholars interested in the history of the pulp and paper industry. It was followed by a seminar held at the Department of Economic History, Institutions, Policy and World Economy, University of Barcelona, Spain, on 25–26 November 2015. The purpose of that gathering was to discuss and develop theories, methods and approaches in dialogue with the authors of the book. Dr. Jussi Uusivuori, Research Professor, at the Natural Resources Institute Finland, acted as the invited commentator of the seminar. First chapter drafts were presented at the 1st World Congress on Business History/20th Congress of the European Business History Association in Bergen, Norway, on 25–27 August 2016.

The seminars and congresses were followed by an intensive research period, and during this time, two of the editors visited their colleagues' institutions to work on this project. Dr. Mark Kuhlberg was a visiting professor for two and a half months (3 April–15 June 2017) at the Department of History and Ethnology, University of Jyväskylä, and Dr. Timo Särkkä was a visiting researcher for a month (6 November–8 December 2017) at the Department of Economic History, Institutions, Policy and World Economy, University of Barcelona. For an international research project, researcher mobility is a necessary and vital element to its success. It is important that the members of the research group are able to work together intensively and in close collaboration with each other. The purposes of the visits were to strengthen research cooperation with the members of the research team, prepare the book manuscript, conduct research work, advance career development, add a more international and critical element to the national research systems and provide new insights for the research. This, we hope, will be manifest in the quality of the book.

The seminars, conferences and researcher visits were supported by the Academy of Finland (grant numbers 267720; 298453), Department of History and Ethnology, University of Jyväskylä, and The Research Centre in Economics and Economic History 'Antoni de Capmany', Department of Economic History, Institutions, Policy and World Economy, University of Barcelona. The editors would like to thank the funding bodies, the hosting institutions, the authors, the publisher and the series editor for the faith displayed in their work.

Jyväskylä, Finland — Timo Särkkä
Barcelona, Spain — Miquel Gutiérrez-Poch
Sudbury, Canada — Mark Kuhlberg

Contents

Contributors

Raimo Alén Laboratory of Applied Chemistry, Department of Chemistry, University of Jyväskylä, Jyväskylä, Finland

Ann-Kristin Bergquist Department of Geography and Economic History, Umeå University, Umeå, Sweden

Amélia Branco GHES/CSG, ISEG-Lisbon School of Economics and Management, Universidade de Lisboa, Lisboa, Portugal

Gordon Dadswell School of History and Philosophy of Science, University of Melbourne, Melbourne, VIC, Australia

Miquel Gutiérrez-Poch Department of Economic History, Institutions, Policy and World Economy, The Research Centre in Economics and Economic History 'Antoni de Capmany', University of Barcelona, Barcelona, Spain

Niklas Jensen-Eriksen Department of Philosophy, History and Art Studies, University of Helsinki, Helsinki, Finland

Mark Kuhlberg History Department, Laurentian University, Sudbury, ON, Canada

Juha-Antti Lamberg Department of History and Ethnology, University of Jyväskylä, Jyväskylä, Finland

Pedro Neves GHES/CSG, ISEG-Lisbon School of Economics and Management, Universidade de Lisboa, Lisboa, Portugal

Panu Nykänen Department of History and Ethnology, University of Jyväskylä, Jyväskylä, Finland

Jari Ojala Department of History and Ethnology, University of Jyväskylä, Jyväskylä, Finland

Michael Roche School of People, Environment and Planning, Massey University, Palmerston North, New Zealand

Timo Särkkä Department of History and Ethnology, University of Jyväskylä, Jyväskylä, Finland

Kristina Söderholm Department of Business Administration, Technology and Social Sciences, Luleå University of Technology, Luleå, Sweden

Stig Tenold Department of Economics, NHH - Norwegian School of Economics, Bergen, Norway

Jussi Uusivuori Natural Resources Institute Finland, Helsinki, Finland

Chapter 1
Technological Transformation in the Global Pulp and Paper Industry: Introduction

Timo Särkkä, Miquel Gutiérrez-Poch and Mark Kuhlberg

1.1 Research Setting

This contributed volume endeavours to analyse the past 200 years of technological transformation in the global pulp and paper industry from a comparative perspective. The study is motivated by the realisation that using comparative methods is a highly revealing way of exposing the complexities the modern pulp and paper technologies have undergone in the past and of analysing today's business environment with changing market dynamics and consumer behaviour. Methodologically, the study combines transnational, national and regional level analyses with micro level case studies by focusing on the development of a single industry—pulp and paper. The research concentrates on the various historical trajectories of the manufacture of pulp and paper and the technologies related to them. It covers the entire history of technology transfer in the global pulp and paper industry from raw materials to mill management, and from transportation to environmental issues. As a result, the book is arguably the most comprehensive historical analysis of papermaking technology available.

The investigation is directly linked to *The Evolution of Global Paper Industry 1800–2050: A Comparative Analysis* (Lamberg et al. 2012), which revealed several

T. Särkkä (✉)
Department of History and Ethnology, University of Jyväskylä, Jyväskylä, Finland
e-mail: timo.sarkka@jyu.fi

M. Gutiérrez-Poch
Department of Economic History, Institutions, Policy and World Economy, The Research Centre in Economics and Economic History 'Antoni de Capmany', University of Barcelona, Barcelona, Spain
e-mail: mgutierrez@ub.edu

M. Kuhlberg
History Department, Laurentian University, Sudbury, ON, Canada
e-mail: mkuhlberg@laurentian.ca

T. Särkkä et al. (eds.), *Technological Transformation in the Global Pulp and Paper Industry 1800–2018*, World Forests 23, https://doi.org/10.1007/978-3-319-94962-8_1

impetuses for an in-depth study of technological transformation in the global pulp and paper industry. The present volume is primarily intended as a basic introduction to the history of papermaking technology and it is aimed at the same category of readers as *The Evolution of the Global Paper Industry*; it is also geared toward students and teachers as course material at the secondary and tertiary levels and as a handbook for professionals working either in industry or research centres. The content could be best utilised for raising important questions among the more advanced graduate audiences by thoroughly familiarising them with the characteristics of the technological transformation of the global pulp and paper industry. It caters to graduate audiences in four intersected but conventionally separate fields of study, each having its distinct but interlinked interest and relationship within the pulp and paper industry: forestry, business, technical sciences, and history.

The primary idea behind the present book was to write a comprehensive volume that covers all the important areas related to the history of papermaking technology. Furthermore, it was believed that the need for a volume of this type is very real and continuing, and that updated research on changes that have been taking place in some of the key areas during recent years are useful for purposes of teaching and research. However, it is also obvious that this book pays less attention to certain topics and regions, which have already been covered in *The Evolution ofGlobal Paper Industry* (Lamberg et al. 2012). For this reason, in these cases a complementary use of both volumes is recommended. The world's pulp and paper industry is becoming increasingly global and with a speed unforeseen in the past. Thus covering what is termed here the global pulp and paper industry runs the risk of becoming outdated as the geographical orientation of the industry is constantly shifting. For instance, from the present point of view the emerging pulp and paper industry regions in Latin America could arguably be considered worthy of investigation. The present study is not intended to deny their importance. Instead, it outlines the development processes of the technological transformation of the industry from a historical perspective, and concentrates on phenomena and regions that are considered to be revealing for the topic as a whole.

1.2 The History of Papermaking Technology

The research starts with a comprehensive historical analysis of the evolution of cellulose chemistry. Paper can be defined as being an aqueous deposit of any plant fibres and other materials in the form of a web or sheet. It is manufactured from a filament product, which has been obtained by mechanically or chemically separating plant fibres from each other. In papermaking the liberated fibres are brought together again on the paper machine, where they interact with each other in the presence of chemicals charged to the liquid stock. As paper consists of plant fibres, paper partakes not only of the chemical nature of the fibre, but its physical nature as well. For instance, cotton rags formed an excellent papermaking material for the reason that a single cotton fibre is immensely strong: it is capable of supporting enormous weight in comparison with its thickness. Therefore, the characteristics of finished products can always be traced to the form, size, and chemical behaviour of the ultimate fibre itself. For this reason, any investigation into the history of papermaking technology inevitably encompasses the history of cellulose chemistry as well (Chap. 2; see also Alén 2007, pp. 18, 20; Beadle 1908).

By the judicious choice of raw material, and also by modifying both the mechanical and chemical treatments to which it is subjected, divergence in the qualities of paper can be achieved. The processes that are used to convert fibrous feedstocks into a mass of liberated fibres by dissolving the components (mainly lignin) that binds the cellulosic fibres together are collectively called "pulping". The purpose of pulping is to separate the fibres from the plant and render them suitable for papermaking. These conversions can be accomplished either mechanically (i.e., by means of mechanical beating) (Sundholm 1999, p. 17) or chemically (i.e., by means of chemicals) (Gullichsen 2000, p. 14) or by combining these two types of treatment. In fibres with low yields of cellulose that consume a large volume of chemicals, the cost of chemical treatment per ton of finished pulp is often so high as to prohibit the material from industrial use, regardless of whether the resulting pulp is of excellent quality. These factors need to be carefully considered in estimating the value of any raw material for the purpose of manufacture as several case studies (Chaps. 9–11) in this volume vividly demonstrate.

Until very recently, paper offered the most convenient, cheapest and democratic medium for communication in the form of newsprint, and printing and writing papers. In the 2000s, the rise of electronic media and information technology has slowly but what appears to be rather irreversibly eroded the role paper once played as the main medium for communication. During roughly the same period, a significant increase in the demand for packaging (e.g. liner, fluting, boxboard), hygienic, health care and other specialised end uses of paper have taken place. The expanded diversity of output can be determined from total global production. In 2016, wrapping and packaging paper grades accounted for more than half (57%) of total global production (409 million tonnes). Newsprint, printing and writing paper grades constituted approximately a third (31%) of the total, followed by household and sanitary paper grades (8%) (FAO 2016). Significantly, however, for most of the research period investigated in this study, by far the dominant use of paper was for newsprint, printing or writing. Naturally, then, the fundamentals that have defined technology transfer until very recently have mainly been connected to communication in one way or another.

One of the fundamentals that has defined technology transfer in the global pulp and paper industry is connected to the technical development of the paper machine itself (Chaps. 2 and 8). The Fourdrinier machine represented a straightforward mechanisation of what was formerly done by hand. In principle, it performs the same sequence of actions as in handmade papermaking, but does so at a much faster rate. Since its introduction more than 200 years ago, the maximum speed (m/min) and width (m) of the machines have increased dramatically, thus contributing—arguably more than any other factor—to the increase in production capacity. During the period from 1900 to 2005, for instance, the maximum speed of the machine rose ten folded (from 200 to 2000 m/min), and the maximum width of the machine grew from circa three metres to eleven. During roughly the same period, global production of paper grew from less than 10 million tonnes (in 1900) to 409 million tonnes (in 2016) (Diesen 2007, p. 11; FAO 2016). Without improvements in technology—whether related to papermaking engineering, cellulose chemistry, energy efficiency or transport—such a dramatic increase in production capacity would not have been possible.

Besides paper production capacity, global production of fibre furnish (i.e. wood-based pulp, non-wood-based pulp and recycled fibres) is an important measure of technology transfer. Paper is manufactured from plants containing cellulosic fibres, and they can be planted or grow naturally under favourable conditions of climate and soil. Today, the typical plants used for paper manufacture are coniferous trees, such as pine (*Pinus* spp.), spruce (*Picea* spp.), fir (*Abies* spp.) and hemlock (*Tsuga* spp.), and deciduous trees, especially eucalypts. In 2016, global production of fibre furnish amounted to 415 million tonnes. From this total refuse materials (e.g. recycled paper) represented 54% (217 million tonnes) of global fibre furnish, with the rest (46%) coming from virgin forest resources and various fibrous, non-wood feedstocks together with industrial forest plantations for the production of pulpwood (FAO 2016).

Prior to the mid-nineteenth century paper was manufactured almost exclusively from non-wood feedstocks. Grasses and straws were among the oldest raw materials used by the papermakers in East Asia. Paper was first introduced in China as a writing material to replace the use of the wooden tablets (i.e. flat pieces of wood on which records have been written in *sumi*, the traditional medium for writing in East Asia using an ink composed principally of soot and binders). Wood-block printing technology emerged during the Tang Dynasty (618–907). The expansion of the imperial dynasty to Korea brought the Chinese in contact with Japan, where some papermakers still use the paper mulberry as raw material to make *washi*, the Japanese paper (Chap. 2).

The domestication of important papermaking plants such as cotton and flax and their use for fabrics were important milestones in the history of papermaking. Gradually both the Arabs and the Greco-Roman world became familiar with the secrets of the cotton fibre and printing technology. The use of the Arabic and Latin alphabet gave the technological advantage to the Arabs and the Europeans over the Chinese, the Koreans and the Japanese, who used Classical Chinese characters (i.e. the relatively simple character sets facilitated the introduction of printing technologies that used types casted from metal as opposed to blocks of wood engravings). The immensely strong cotton and flaxen fibres in the form of rags from fabrics and technological innovations connected with printing ensured since the late fifteenth century the establishment of paper mills and printing presses first in the Old World, and then following European colonisation in the New World (Chap. 2).

The use of rags as the primary papermaking material started to show symptoms of saturation in the early-Victorian Britain—the first "journalising" society in the world (i.e. the mass media can be interpreted as the ideological environment of the early-Victorian society). The extension of education and literature, and the increased literacy and heightened social consciousness directly increased demand for paper. Furthermore, the mechanisation of the industry indirectly gave people and institutions more reason to need paper. The rise of the popular press and technological improvements in lithographic printing allowed for the mass production of penny and halfpenny newspapers, journals, magazines, reviews and cheap editions of books; they thus came within the reach of the very poorest members of society. The more economical methods of manufacturing an exponentially greater output to meet larger demand led to the quest for new raw materials. Two separate developments took place: the introduction of new processing treatments to some of the old papermaking materials (e.g. esparto and bamboo) (Chap. 11) and the introduction of a range of new technologies connected with the utilisation of coniferous wood-fibres, the use of which prevailed.

The utilisation of wood-fibres for making paper created the foundation for an industrial sector—the pulp and paper industry—in countries endowed with abundant coniferous forests, which served as its raw material, and hydro-electric power for energy; two prime examples are Finland (Chap. 3) and Canada (Chap. 7). From a historical perspective, wood-based fibres entered paper manufacture relatively late, circa 150 years ago, but the impact of their utilisation has been an important economic determinant for the geographical orientation of the industry. Previously, the availability of rags influenced the location of paper mills since rags from fabrics were not typically available in abundance. Hence, the early paper mills tended to spring up at a fairly short distance from large urban centres, which were both the biggest markets for paper and the centres of rag supplies, or cotton and flax mills, which were other major sources of these refuse materials. From the mid-nineteenth century onwards, mills began to spring up in regions remote from economic centres—on the shores of lakes and rivers, where it was possible to obtain the raw material needed from the forests. Where to locate production remained the crucial question until the latter part of the twentieth century. Thereafter the technological developments in seaborne transportation, and the subsequent dramatic reduction in the cost of shipping goods, started to erode the importance of distance *vis-à-vis* fibre sources and markets. The expansion of overseas trade highlighted the importance of proximity to good harbours so that the raw materials and finished goods did not have to be transhipped far from ocean-going vessels (Chap. 12).

The practically unlimited access to wood-based fibre resources some of the major producers had enjoyed since the mid-nineteenth century became an even greater advantage during and immediately after the First World War, which caused a violent disruption in the global paper trade. During the conflict, a shortage of shipping caused trade between the pulp producing countries and the rest of the world to collapse. The British Empire and Commonwealth countries, with the single exception of Canada, (Chap. 7) relied heavily on softwood imports. In the UK, paper makers imported practically their entire softwood supply (Chap. 11). Apart from Canada, only in New Zealand were there extensive stands of coniferous woods and this was the case only because the country had embarked on a vigorous afforestation programme of species that had been transplanted from California (Chap. 9). In Australia, coniferous species were limited to certain areas of the vast continent, rendering the exploitation of softwood resources economically unviable. In their absence, Australian papermakers adopted technologies that were aimed at pulping hardwoods, especially the indigenous eucalypt species (Chap. 11). In Europe, Portugal (Chap. 6) was among the first countries in which eucalypts were established in plantations for the production of pulpwood.

With little doubt, environmental regulation has occupied one of the main impetuses for technology transfer over the past fifty years. It was during the 1970s that effluent loadings and sulphur emissions from the pulp and paper industry were for the first time subjected to considerable scrutiny. Thereafter, emissions have fallen significantly in many technologically advanced pulp and paper producing countries. The reasons for the decrease can be found in changes in production technology and in the finished products themselves. For example, in many countries the calcium-based sulphite mills, which cause high emissions, have been closed down. In the case of pulp and paper products, major global upheavals such as the oil shocks of the 1970s have had a major impact on environmental protection. The crisis led to the introduction of numerous measures that resulted in improvements in energy efficiency and reductions in mill water consumption and emissions into the environment. Regulatory bodies also played a role in precipitating this progress by stepping up their work towards greening the pulp and paper industry. They introduced new and tightened existing legislation, particularly in relation to effluents. The main economic instruments available for the authorities were compensation procedures for defraying the cost of environmental protection and measures related to taxation. Lifting the turnover tax on environmental protection investments can be mentioned as an example. Furthermore, nongovernmental organisations have put pressure on producers for lower emissions and a more sustainable use of forest resources (Chap. 4; see also Hynninen 1998). From the mid-1980s onwards, the rise of green consumerism in some major paper consuming countries such as Germany and the UK functioned as an additional incentive for the transfer of environmentally driven technology. More recently, environmentally sustainable products using wood and forestry residues as well as other forms of biomass have gained considerable attention (Chaps. 2 and 4).

As a final point, it can be said that an appreciation of the surrounding institutional environment is of paramount importance in analysing the evolution of the global pulp and paper industry. As the case studies highlight the institutional environment differs considerably from one country to another (Chap. 5). Modern papermaking technologies would not have been realised without dedicated individuals, educational institutes, research laboratories, mills and workshops, (Chaps. 3, 7–11) and even political bodies that have created incentives—and in some illustrative cases even fetters—on adopting technological innovations (Chap. 7). Sometimes institutional instability per se has hindered the technology transfer (Chap. 8). Thus, while papermaking technology encompass the research work done in the field of cellulose chemistry, it also involves a function of changes in the market demand, availability and supply of capital, energy resources, raw materials knowledge and technology, the surrounding institutional framework and organisational solutions. Together these variables create what can be termed the main drivers of technology transfer in the global pulp and paper industry.

1.3 The Case Studies

In compiling the volume, it was deemed best to employ a simple organisational framework that grouped the 11 case studies (Chaps. 2–12) in three parts (Parts I–III) that were followed by a conclusion (Chap. 13). It is hoped that this straightforward approach will help the reader trace connections between the different case studies and better understand the research setting as a whole. Part I, "Research and Development", focuses on the evolution of cellulose chemistry, research and development of papermaking engineering and environmentally driven technology transfer. In Chap. 2, "Manufacturing Cellulosic Fibres for Making Paper: A Historical Perspective," Raimo Alén traces the long evolution of the development of pulp and paper technology from the inception of paper manufacture in China about 2000 years ago up to its recent developments. The analysis focuses on the last two centuries, during which period the manufacture of cellulosic fibres has been closely integrated with the growth of our fundamental knowledge of chemistry in general and the enhancement of our knowledge of the possibilities of chemical processing of cellulosic fibre in particular. The more recent developments deal mainly with the quest for new cellulose-based products and their possible applications as well as environmental concerns. The chapter represents arguably the most comprehensive and detailed, and yet relatively condensed, account of the history of how cellulosic fibres have been manufactured into a plethora of paper products over the last two millennia.

The tradition of research and development in the paper industry has historically been divided into two separate branches of research and development. One consists of chemistry, namely studying the possibilities of chemical processing of fibre. The other part of the history of research and development is mechanical engineering, and its roots can be traced back to the construction and use of paper machines. In Chap. 3, "Research and Development in the Finnish Wood Processing and Paper Industry, c. 1850–1990," Panu Nykänen investigates the tradition of research and development in the context of the Finnish wood processing and papermaking industries. In addressing its subject, the chapter is mainly concerned with the operation of Finnish paper mills from the 1850s onwards when the utilisation of coniferous wood as the new raw material for making paper revolutionised the entire industry. The chapter also addresses the related subjects of technical research, formal higher education and work-related, practical learning in Finland.

Environmental concerns are the focus in Chap. 4 "The Greening of the Pulp and Paper Industry: Sweden in Comparative Perspective," by Ann-Kristin Bergquist and Kristina Söderholm. It analyses the environmentally driven technology transfer in the pulp and paper industry by focusing on Sweden, which has pioneered parts of this transition. The chapter illustrates that the overall transition towards cleaner and more energy efficient production technologies in Sweden was the result of long-term incremental development, which gained momentum with the environmental awakening in the 1960s and was followed by the rise of green consumerism in the 1980s. More recently, the burning issue of climate change has been a major impetus to environmentally driven technology transfer, for instance in the development of cellulose-based products such as biofuels.

Part II "Regulations and Institutions" assesses the role of regulatory institutions in technology driven transfer in the global pulp and paper industry, within regional, national and transnational organisational frameworks. In Chap. 5 "Varieties of State Aid and Technological Development: Government Support to the Pulp and Paper Industry, the 1970s to the 1990s," Jari Ojala, Niklas Jensen-Eriksen and Juha-Antti Lamberg analyse government support for the pulp and paper industry in the Organisation for Economic Co-operation and Development (OECD) countries. The chapter reports that the favourable regulatory environment tended to exist in such major pulp and paper producing countries as Finland and Sweden, in which pulp and paper companies had very strong bargaining power over the government. On the contrary, in those OECD countries in which the pulp and paper industry's share of total manufacturing was marginal, bargaining power and consequently direct government support tended to remain rather limited.

Interestingly, in Chap. 6 "From Backward to Modern: The Adoption of Technology by the Pulp Industry in Portugal, 1891–2015," Amélia Branco and Pedro Neves find that in Portugal, where the pulp and paper industry remained technologically backward prior to the 1950s, government support was the main boost to the pulp industry. The main explanation can be found in the development of eucalyptus for plantation purposes and pulping hardwoods, which emerged as a major field of manufacturing in Portugal after the Second World War. The analysis highlights the roles played by the availability of knowledge, capital, and raw materials on the one hand and demand characteristics on the other in technologically driven transfer in the Portuguese pulp industry. Taking a longitudinal perspective on the Portuguese pulp industry, the chapter sheds light on the international economic integration of a peripheral country that suffered from a poor endowment of capital and natural resources.

In Chap. 7 "Natural Potential, Artificial Restraint: The Dryden Paper Company and the Fetters on Adopting Technological Innovation in a Canadian Pulp and Paper Sector, 1900–1950," Mark Kuhlberg analyses the issue of government support on national, provincial and organisational levels in Canada, one of the world's foremost pulpwood suppliers. The analysis focuses on the case of a pulp and paper mill in Dryden, northern Ontario, and highlights that the presence of abundant natural resources does not alone guarantee a technologically driven transfer in pulp and paper industry. The case of the Dryden Paper Company, which left finding economies of scale in the industry to others, is a reminder that the first mover in an industry does not always dominate it in the long run. Another significant conclusion of the chapter is that institutional support mechanisms and demand characteristics remain as important as abundant natural backings in analysing the main drivers in technology transfer.

The industrialisation process in the nineteenth century was based on the transfer of technology from pioneering countries to countries that had not yet participated in the process. Southern Europe in general and Spain in particular were very late in experiencing this modernisation wave. In Chap. 8 "The Endless Sheet: Technology Transfer and the Papermaking Industry in Spain, 1800–1936," Miquel Gutiérrez-

Poch explores the technology transfer in Spain, whereby foreign technology was the main driver in achieving the transfer in the papermaking sector. The chapter maintains that the institutional framework and its stability—into which the new technology is adapted—can accelerate or restrain the impact of the new technology. Therefore, the setting into which the new technology is received is a crucial consideration that affects the technology transfer. It is also influenced by numerous other factors, including the geographical concentration of the activity (i.e. industrial district) and simply the existence of an industrial base. In regard to papermaking in Spain, it was fundamental to have an active mechanical engineering sector that was able to adapt the new technology to local conditions.

Part III "Local Innovations and Global Markets" pays attention to the role of global upheavals and demand characteristics in analysing the birth of the pulp and paper industry in three peripheral regions of the global economy, namely New Zealand, Australia and India. In Chap. 9 "Technology Transfer and Local Innovation: Pulp and Paper Manufacturing in New Zealand, c.1860 to c.1960," Michael Roche traces the birth of a wood-based pulp and paper industry in New Zealand from its modest beginnings until the 1960s, by which time there were pulp and paper mills in operation serving local and export markets. In New Zealand the birth of a viable pulp and paper industry was fostered through a vigorous afforestation programme of *Pinus radiate*—a Californian coniferous species previously untried for papermaking, technical assistance from North American and Scandinavian countries, and a group of dedicated researchers working for both the Forest Service and private companies in New Zealand.

In place of conifers, pulping indigenous hardwoods, specifically the eucalypts, formed the basis for the emergence of a national pulp and paper industry in Australia. In Chap. 10 "Making Paper in Australia: Developing the Technology to Create a National Industry, 1818–1928," Gordon Dadswell traces the evolution of the pulp and paper industry in Australia, in which the comparative inaccessibility and the tyranny of distance had rendered the exploitation of the country's wood resources economically unviable until 1914. The need for establishing an Australian pulp and paper industry became imperative following the end of the First World War. The armistice resulted in a serious shortage of imported wood pulp, with the outcome that manufacturers in Australia were forced to switch to indigenous raw materials instead. The initial investigations for pulping indigenous hardwoods were undertaken in Perth, Western Australia after the war. The final stage of paper making in Australia shifted from the laboratory to commercial production in 1924, when the first Australian mill using indigenous hardwoods went into production in Tasmania.

Another remarkable story of pulping indigenous raw materials in place of conifers is told in Chap. 11 "The Quest for Raw Materials in the British Paper Trade: The Development of the Bamboo Pulp and Paper Industry in British India up to 1939," by Timo Särkkä. Bamboo—the fastest growing plant on earth—was introduced as the raw material for papermaking in India at the beginning of twentieth century. The impetus for this development was the global nature of the First World War, which led to the collapse of shipping between Britain and some of its colonies. The increased dependence on wood pulp, the likelihood of a pulp famine, and the consequent increase in the price for imported wood pulp were the means for drawing attention to the possibility of making commercial volumes of pulp by utilising Indian grasses in general and bamboo in particular. The technology was developed under British auspices, but was later adopted by Indian paper producers in response to the rising cost of imported wood pulp.

The final case study, Chap. 12 "Creating Global Markets: Seaborne Trade in Pulp and Paper Products over the Last 400 Years," by Jari Ojala and Stig Tenold, brings the very issue—where to locate production—to the fore of analysis by adopting a longitudinal perspective on the global seaborne trade in pulp and paper products. In contrast to the situation prevailing today, globally operating multinational companies were rarities in the world's pulp and paper industry prior to about the 1950s. Nationally and regionally conditioned technological and organisational solutions survived into the post-1945 era, although they soon went into a rapid decline. The development of technological solutions for dramatically improving the efficiency of ocean-going transportation were crucial for the high-volume, bulk products—both in terms of the raw material and end product—that define the pulp and paper industry. The chapter concludes that the declining cost of sea transport has been a necessary condition for the growth of the global pulp and paper industry into new regions of the world, ranging from South-East Asia to Latin America, which have traditionally been remote from the world's established economic heartlands.

Chapter 13 "Technological Transformation in the Global Pulp and Paper Industry: Concluding Remarks," by Mark Kuhlberg, Timo Särkkä, and Jussi Uusivuori summarises the theoretical framework presented in the introductory chapter and the key empirical findings of the case studies.

References

Alén R (2007) Introduction to papermaking. In: Alén R (ed) Papermaking chemistry, 2nd edn. Finnish Paper Engineers' Association, Helsinki, pp 16–26

Beadle C (1908) Chapters on papermaking. In: Comprising a series of lectures delivered on behalf of the battersea polytechnic institute in 1902, vol I. Crosby Lockwood and Son, London

Diesen M (ed) (2007) Economics of the pulp and paper industry. papermaking science and technology. Book 1. Finnish Paper Engineers' Association/Paperi ja Puu Oy, Helsinki

Food and Agriculture Organization of the United Nations (FAO) (2016) Global forest products facts and figures 2016. http://www.fao.org/3/I7034EN/i7034en.pdf. Accessed 21 Mar 2018

Gullichsen J (2000) Introduction. In: Gullichsen J, Fogelholm C-J (eds) Chemical pulping. Papermaking science and technology. Book 6A. Finnish Paper Engineers' Association/TAPPI, Helsinki, pp 14–16

Hynninen P (ed) (1998) Environmental control. Papermaking science and technology. Book 19. Finnish Paper Engineers' Association/TAPPI, Helsinki

Lamberg J-A, Ojala J, Peltoniemi M, Särkkä T (eds) (2012) The evolution of global paper industry 1800–2050. A comparative analysis. Springer, Dordrecht

Sundholm J (1999) What is mechanical pulping? In: Sundholm J (ed) Mechanical pulping. Papermaking science and technology. Book 5. Finnish Paper Engineers' Association/TAPPI, Helsinki, pp 17–21

Part I
Research and Development

Chapter 2
Manufacturing Cellulosic Fibres for Making Paper: A Historical Perspective

Raimo Alén

2.1 The Origins of Turning Cellulosic Materials into Paper

The use of cellulose-containing fibrous feedstocks for papermaking was known in early Far Eastern and Near Eastern cultures, from whence this knowledge gradually spread to Europe (Atchison and McGovern 1983; Whistler et al. 1984; Lindberg 2000; Alén 2007, 2018). It is known that many such materials have long fulfilled human needs for clothing and housing. Additionally, organised societies have recorded their thoughts and ideas on myriad materials, some whose existence is ephemeral (e.g., sand and leaves) and others that are far more permanent (e.g., stone, clay tablets, bone objects, leather, fabrics and walls of caves as well as surfaces of tree bark, wood and metals or their alloys). Hence, the physical form of cellulosic fibre (collectively called "paper") developed in China about 2000 years ago has been said to perhaps be humanity's most important invention. Paper has been made in many ways through the ages, and all the ancient methods are still being applied. The most common writing materials have been, besides cellulosic fibre, papyrus and parchment.

Papyrus (the ancient Egyptian word "pa-per-ah" means "pharaoh's own") was made from veneers of the stem of a reed-type plant (a marsh grass, *Cyperus papyrus*) flourishing in the Nile River valley and delta in ancient times, and it was widely used to make rugs, boats and sandals. Thin veneers, as wide as possible, were first soaked and then pounded into flat "basic sheets" that were then joined with either the plant's own "glue" or wheat starch. The final product was an even, 20 m long strip with a maximum width of about 30 cm and it was strong and flexible enough to be easily rolled into a "book". The writing surface was treated with pumice to prevent the ink from spreading and to cause it to adsorb properly on the surface. Papyrus rolls have been found since the First Dynasty of ancient Egypt (3100–2900 BC); the most

R. Alén (✉)
Laboratory of Applied Chemistry, Department of Chemistry,
University of Jyväskylä, Jyväskylä, Finland
e-mail: raimo.j.alen@jyu.fi

T. Särkkä et al. (eds.), *Technological Transformation in the Global Pulp and Paper Industry 1800–2018*, World Forests 23, https://doi.org/10.1007/978-3-319-94962-8_2

recent ones date from ca. 1000 AD. Papyrus survives well in dry climates, like the desert, but it absorbs moisture in humid climates and becomes moldy. Papyrus rolls that have not been exposed to oxygen or mold or those partially charred in fires have survived to the present time and tens of thousands of them can be seen in leading museums.

Initially, papyrus was mainly produced in Egypt and from there it spread—in spite of its "secret" method of preparation—as the preferred writing material of antiquity around the Mediterranean and nearby regions to the rest of Europe and Asia Minor. While papyrus as such cannot be fully included within the concept of paper, the word "paper", for example, and its numerous varieties in European languages, derive from its original name. During the period of Roman Caesars, the consumption of papyrus rolls was so large that their use had to be regulated. The production of papyrus decreased in the 800s and nearly ended in the 900s. The main reason for this was probably the introduction of a substitute, parchment, and gradually also actual "modern" paper. Moreover, the papyrus plant disappeared from large areas of Egypt at the same time.

Thus, paper was traditionally made from a filament product obtained by taking plant fibres that have been chemically or mechanically separated from each other. In the oldest method of making paper in China, a fixed amount of a solution of suspended fibres was poured into a mold that was partially submerged in water. The mold consisted of a wooden frame with a coarsely woven cloth stretched across it. The water drained through the cloth, leaving a thin and uniform fibre web on top, and it could be separated and dried. Into this basic fibre material could be mixed a wide range of additives (today mineral fillers and pigments, functional chemicals and process chemicals are added) for various purposes. Although the papermaking process has come a long way from its earliest days, and there have been many inventions that have affected the quality of paper products, papermaking is still in essence a "rather simple process". By the same token, the modern manufacture of paper embraces a wide range of technologies and fields of science, specifically surface and colloidal, organic and inorganic chemistry and even microbiology.

The actual art of papermaking apparently did not arise accidentally. According to the documents of the Chinese ruler of the time, Emperor Ho-Ti of the Han Dynasty (206 BC–220 AD), Ts'ai Lun, a court officer and scholar who entered the service in 75, introduced in 105 to the public the art of making use of a variety of raw materials to create fibre products for different purposes. However, it is also said that paper materials dated from 73 to 49 BC have been found in China. Under Ho-Ti's systematic leadership, modern papermaking started and paper material evolved into a high-quality product; several mentions exist about its use during the following centuries. The Chinese used phloem fibres from the bark of the paper mulberry tree (silk tree, *Broussonetia papyrifera*) and white mulberry tree (*Morus alba*) and fibres from ramie (*Boehmeria nivea*), and many other fibre materials, such as old rags and fish nets, which were all usually treated with potash and refined (i.e., macerated) until each filament was completely separate.

In China papermaking evolved into a significant home industry. Old Chinese paper found an unbelievable array of uses and was clearly of better quality than Egyptian papyrus. As a novelty, toilet paper was produced for the first time for the Emperor's court in 589, paper money used by the banks for receipts to the merchants ("flying money" that flew with the wind) in 807, playing cards in 969, casings for fireworks, banknotes for general use in ca. 1000 and paper lanterns in ca. 1100. In addition, the Chinese, since the 800s, began printing houses using wooden slabs that produced major printing works, such as the "Holy Books of the Buddha" that appeared in 972. The Chinese started using the fibres of rattan palm (*Calamus rotang*) as raw material in papermaking in the 900s, and it became the final improvement of the method. Only after this advance did paper finally replace the traditional use of bamboo strips in making "bound books".

The Chinese tried for a long time to keep their art of papermaking secret to protect their monopoly. However, it spread gradually, first to Korea in the 300s or 400s and then to Japan in ca. 610. The Japanese used the phloem layers of the mulberry tree and local mitsumata (*Edgeworthia papyrifera*) and gambi (*Wickstroemia canhescens*) bushes as raw materials. They also learned to make new products, such as silk paper, which became famous due to its excellent strength. On the other hand, the Arabs, who had created a powerful army during the Turkestan war, captured in the Battle of Talas (today it is in Kyrgyzstan) in 751 Chinese prisoners of war who knew the art of papermaking and printing. With the prisoners' labor, the city of Samarkand in Uzbekistan in Central Asia (southeast of Lake Aral) became an important centre of these new arts. Although it is likely that paper was already being produced there, the Arabs were quick to realise the advantages of paper materials made from rags, flax and hemp over traditional papyrus and parchment.

Thereafter, these papermaking skills spread expeditiously into other centres of the Arab empire (Baghdad 793, Damascus and Cairo 900, Fèz 1100, Xàtiva 1139 and Córdoba 1150), thus migrating from Asia first into North Africa (Egypt and Morocco) and then to Spain. Caliph Harun ar-Rašid (786–809), also known for his sagas, largely caused this development, as he understood the value of this knowledge obtained as war booty and decreed that papermaking was to be a state monopoly. This substantially enhanced the already powerful Arabic culture in its respect for the written word (ca. 750–1260). The Arabs further developed the techniques of rag grinding and paper sizing. They made coloured papers for various purposes, for example, blue for sorrow, yellow for riches and red for happiness and nobility.

The art of papermaking thus arrived in Europe, on one hand, from Morocco with Moor merchants and, on the other hand, with the Venetian explorer Marco Polo (1254–1324), who probably spent 17 years in the service of the Mongol emperor Kublai Khan (1215–1294) in Peking. Polo, a member of an esteemed line of merchants and aristocrats, started his journey in 1271 along the Silk Road through the enormous Mongol empire with a retinue including his father and uncle, who had travelled in the East earlier. It is reported that Polo returned to Venice in 1295, but he was taken prisoner in 1298 when Venice and Genoa were at war. He probably prepared with the help of a fellow prisoner (Rustichello da Pisa) an accurate and extensive travelogue known as the book, *Il Millione*.

Although Arabian paper, among others, was in common use in many countries for a while, there was a clear trend to start independent production. The first "paper mills" were established, for example, in Spain (Xàtiva, Valencia, in 1139), France (1189), Italy (Fabriano, 1220), Germany (1390), Switzerland (1432), Austria (1463), Poland (1473), England (1490), Sweden (1565), Denmark (1573), Russia (1576), Holland (1586), Finland (1667), North America (1690) and Norway (1695), in relatively rapid succession.

A short description of the development of coated papers illustrates the early origins of papermaking chemistry and paper converting: pigment-coated papers typically consist of a base paper covered by a mixture containing at least a binder and an inert pigment. An early Chinese document dated 450 AD reported starch sizing and gypsum surface treatments to improve its properties. During the Tang Dynasty between 618 and 907 AD, paper was coated with white mineral powders and wax, which filled the cavities between fibres to increase water repellency and smoothness for fine calligraphy. Furthermore, by the eighth century, Arab cultures modified paper with talc, gypsum or chalk, which could also be mixed with rice starch to coat the paper for increased whiteness. During the Middle Ages, paper was coated with white pigments to provide an appropriate surface for metal-point drawings. After this period, there were a seemingly infinite number of instances in which papers were coated for different purposes, and the nineteenth and twentieth centuries in particular fostered the development of new coating application techniques. Modern coatings are composed not only of pigments and binders (typically 90–94% pigment and 6–10% binder), but also additives and water. The most important pigments include clay, calcium carbonate and silicates, whereas the principal binders are either hydrophilic water-soluble colloids (e.g., starch and protein) or resins (or latexes) and resin emulsions in an aqueous medium.

2.2 The Basics of Industrial Applications of Cellulose

Cellulose is the world's most abundant natural biopolymer and this fibrous material is distributed in its natural form throughout the plant kingdom (Sjöström 1993; Alén 2000a, 2011, 2018). Almost all cellulose is formed by photosynthesis ("plant cellulose"), but there are also microbial extracellular carbohydrates, such as "biocellulose" or "bacterial cellulose" that are synthesised by various bacteria. The content of cellulose in plant materials and technical products (mainly containing cellulose, hemicelluloses and lignin) varies depending on its origin (Table 2.1).

Besides cellulosic wood fibres, there is archeological evidence that cotton, consisting of nearly pure cellulose, was used in India to make cloth and yarn in ca. 3000 BC, although this use may have started in India and elsewhere as early as 6000 BC. By 1500 BC, a centre of cotton manufacture had arisen in India, from where the knowledge spread to Persia (Iran), China and Japan. According to the writings of the Greek historian Herodotus and Roman Gaius Pliny the Elder, people of their time were familiar with cotton fibres. As indicated in Table 2.1 there are numerous other plant-based fibres, such as flax, manila, esparto, hemp, jute and sisal, and gradually the art of preparing these fibres into materials for cloth was learned.

Table 2.1 Examples of chemical composition of cellulose-containing materials (as a percentage of the dry solids)

Material	Cellulose	Hemicelluloses	Lignin
Softwoods	40–45	25–30	25–30
Hardwoods	40–45	30–35	20–25
Fibre plants (e.g. flax, cotton, hemp and sisal)	70–95	5–25	<5
Natural-growing reeds and agricultural residues	25–45	25–50	10–30
Chemical wood pulps	65–80	20–30	<5

Source Alén 2011

Elucidation of the chemical origin of cellulose dates back nearly two centuries. Over the course of 1837–1842, the French botanist Anselme Payen (1795–1871), an open-minded experimenter, noted that most plant materials contain a relatively resistant component with essentially the same elemental composition. This fibrous material isolated as a residue in the nitric acid treatment of plant materials was labelled "cellulose" according to his suggestion, and this name was confirmed in 1839 by the French Academy. The incrusting material of cellulose was isolated and named "lignin" (in Latin, "lignum" means "wood") in 1865 by Franz Eilhard Schulze (1840–1921). This term was probably already in use in a similar context in 1819 by Augustin Pyramus de Candolle (1778–1841).

A new area of cellulose chemistry started almost 50 years ago when non-aqueous solvent systems (e.g., *N*-methylmorpholine *N*-oxide) for cellulose were discovered. Solvents for cellulose are central to the more effective preparation of cellulose derivatives, but they are also needed in laboratory work. Those conventional solvents cannot be recovered and reused. Hence, novel solvents for cellulose have been systematically sought for industrial purposes, especially for the rayon and cellophane industries (i.e., for preparing regenerated cellulose fibres).

Cellulose derivatives, mainly based on the modification of dissolving pulps typically with high cellulose contents (80–90%), can be divided into cellulose esters and cellulose ethers. However, cellulose was used as a starting material for technically important derivatives even before its polymeric nature was fully understood. Already in the 1800s, it was possible, in the chemical industry, to produce nearly pure cellulose for use in the manufacture of cellulose derivatives with different properties. For example, Henri Braconnot (1780–1855) first prepared the oldest cellulose derivative of commercial importance, cellulose nitrate or nitrocellulose ("smokeless powder"), as early as 1832. The early history of this inorganic ester is related to the militaries of most European nations during the second half of the nineteenth century. In 1846, Christian Friedrich Schönbein (1799–1868) developed the preferred method of highly nitrated cellulose ("cotton powder" or "gun cotton") and in 1863 Frederick Abel (1827–1902) developed a method of safely handling cellulose nitrate, making possible its use as an explosive. Later in 1868, John Wesley Hyatt (1837–1920) discovered semisynthetic thermoplastic "celluloid", by combining cellulose nitrate ("collodion") with camphor and a minor amount of other plasticisers.

This discovery was made during a public competition which had as its objective finding substitute materials for ivory in producing billiard balls. Thus, cellulose nitrate gradually became the progenitor of numerous industries, including the production of explosives, plastics, lacquers, protective coatings, photographic films and cements. The synthesis of cellulose nitrate formed the basis for the first industrial process that produced "artificial silk".

The largest part of the cellulose-based artificial fibre, "viscose rayon", the first semisynthetic fibre product, is manufactured by the so-called "viscose process" originally invented in 1892 by Charles Frederick Cross (1855–1935), Edward John Bevan (1856–1921) and Clayton Beadle (1899–1976). This product was commercialised in 1894 for textile purposes. Today, rayon is used in many clothing articles; it may be the most useful synthetic fibre or filament to human beings. The corresponding production of regenerated cellulose from cellulose nitrate was already realised by Hilaire de Chardonnet (1839–1924) in 1885 (full-scale production began in 1899). This invention can even be traced back to 1850, when Joseph Wilson Swan (1828–1914) made electric lamp filaments by extrusion, and to 1855, when Georges Audemars prepared nitrocellulose-based artificial silk by a method impractical for commercial use. Due to the high flammability of this nitrogen-containing artificial silk, it was quickly taken off the market. Chardonnet's artificial silk was exhibited at the Paris Exhibition of 1899 and it won the Grand Prix.

During the 1890s, a development was stimulated not only by scientific curiosity but also by a more practical aim to transform using modern chemistry water-insoluble cellulose into a dissolved state. The ultimate aim was to prepare artificial silk from an endless cellulose thread. Among the long list of produced cellulose derivatives of great importance, cellulose acetate was first available in the 1930s. This cellulose ester was the first ester derivative of cellulose; it was first produced in 1865 by Paul Schützenberger (1829–1897).

2.3 Chemical Pulping

2.3.1 General Aspects

Paper and its often multi-layered stiffer variety, cardboard or paperboard (collectively called "board"), are today among our most important and versatile products that serve us in all aspects of our lives (Atchison and McGovern 1983; Alén 2007, 2018). The distinction between paper and board has been rather vague, especially in the past when products were named according to their purported use rather than their physical properties. The world's annual production of paper and board (including large amounts of additives and fillers) totals about 400 million tons, and the demand is expected to grow to nearly 500 million tons by 2020.

Pulping refers to the different processes that are used to convert wood or other fibrous feedstocks into a mass of liberated fibres by dissolving the lignin that binds the

Table 2.2 Commercial pulping methods

Method	Yield/% of wood
Chemical pulping	35–60
Kraft, polysulphide kraft and pre-hydrolysis kraft	
Soda-anthraquinone (AQ)	
Acid sulphite, bisulphite and alkaline sulphite-AQ	
Multistage sulphite	
Semichemical pulping	65–85
Neutral sulphite semichemical (NSSC)	
Soda	
Chemimechanical pulping	80–90
Chemithermomechanical (CTMP)	
Chemigroundwood (CGWP)	
Mechanical pulping	91–98
Thermomechanical (TMP)	
Refiner mechanical (RMP)	
Stone groundwood (SGWP)	
Pressure groundwood (PGWP)	

Source Alén (2011)

cellulosic fibres together; the process is thus called "delignification" (Ritman 1968; Sjöström 1993; Sundholm 1999; Alén 2000b; Gullichsen and Fogelholm 2000). This conversion can be accomplished either chemically or mechanically or by combining these two treatments. The term "pulp" is collectively used for chemical, semichemical, chemimechanical and mechanical pulps. Although pulps are mainly used for papermaking, some pulps (dissolving pulps) are processed into cellulose derivatives (cellulose esters and ethers) and regenerated celluloses, for example, viscose or rayon.

Table 2.2 gives a broad classification of commercial pulping processes and their yields. The average yield of chemical pulp is in the range of 45–55%. Yields of the products in dissolving pulping (i.e., pulps from acidic sulphite, multistage sulphite and pre-hydrolysis kraft methods) are generally 35–40%. Chemical pulping accounts for 70% of the total worldwide production; about 90% of chemical pulps (about 130 million tons) are currently produced in the dominant kraft (sulphate) process. The term "high-yield pulp" is often used for different types of lignin-rich pulps (mainly from neutral sulphite pulping) that require mechanical defibration. The annual production of high-yield pulps is about 45 million tons. Additionally, about 150 million tons of recovered fibres (deinked pulp, DIP) are used for papermaking, and one major trend in the industry is the growing use of recycled paper.

There are several major reasons why kraft pulping is dominant. The pulp it produces has excellent strength properties and yet it can use low quality wood as feedstock. In addition, it involves a well-established and efficient recovery of cooking chemicals, energy and by-products (i.e., crude turpentine and tall oil soap). In this conventional process, delignification takes place at elevated temperature (160–170 °C) and pressure (7 bar) under strong alkaline conditions with an aqueous solution of sodium hydroxide and sodium sulphide. Due to the lack of selectivity in kraft pulping, roughly one half of the raw wood substance that is used degrades and dissolves into the cooking liquor ("black liquor"). It thus contains, besides degraded lignin fragments, a substantial amount of carbohydrate-derived materials (mainly aliphatic carboxylic acids). The fundamental degradation phenomenon of wood carbohydrates was noticed in 1885, and it was named after its observers, Cornelis Adriaan Lobry van Troostenburg de Bruyn (1857–1904) and Willem Alberda van Ekenstein (1858–1937), and referred to as the "Lobry de Bruyn-van Ekenstein transformation" or the "Lobry de Bruyn-Alberda van Ekenstein transformation".

The main by-products of softwood kraft pulping are, besides black liquor, crude sulphate turpentine and tall oil soap. The availability of these by-products in a mill strongly depends on the wood species used, the method and time of storing logs and chips and the growth conditions of the trees; even among extractives-rich pine species there is significant variation in the volume of by-products. After recovery of most extractives-based compounds, the remaining black liquor mainly contains, in addition to inorganic substances, lignin, carbohydrate degradation products and residual extractives.

Crude turpentine is recovered from the digester relief condensates. The average yield of crude turpentine of pine species is between 5 and 10 kg per ton of pulp and somewhat lower for spruce species. Crude turpentine is purified in a distillation process, whereby impurities, such as sulphur-containing compounds, are removed. Main distillation fractions from turpentine have been used as paint thinners, varnishes and lacquers, and as rubber solvents and reclaiming agents. Today, they are used for making different products for the chemical industry, for pharmaceutical purposes (liniments) and perfumery.

Tall oil soap is removed from black liquor during its evaporation process and its chemically bound sodium is liberated by adding sulphuric acid to yield crude tall oil (CTO). The average yield of CTO is in the range of 30–50 kg per ton of pulp, corresponding to 50–70% of its initial volume in the raw material that is used for pulping. CTO is normally purified and fractionated by vacuum distillation. The main fractions include light oil, fatty acids, rosin (resin acids) and pitch residue. Of these fractions, only some parts of the light oil and pitch residue are generally used for combustion, other fractions and parts are utilised for versatile chemical purposes.

Unlike the alkaline kraft process, the sulphate process covers the whole range of pH, ranging from acidic conditions (pH 1–2, acid sulphite cooking) to alkaline conditions (pH 9–13, alkaline sulphite-AQ cooking) (Table 2.2). This variation causes a wide variance in pulp yields and properties; the pulps extend from dissolving pulps for chemical end uses to high-yield NSSC grades used mainly for packaging purposes. Hence, the active sulphur-containing species in the sulphite process also typically vary and similarly, the active base (cations) used depends on the pH range. As in the kraft cooking, in the acid sulphite and bisulphite cooking roughly one half of

Table 2.3 Typical chemical composition of wood and non-wood feedstocks used for pulping (% of the feedstock dry solids)

Component	Wood	Non-wood
Carbohydrates	65–80	50–80
Cellulose	40–45	30–45
Hemicelluloses	25–35	20–35
Lignin	20–30	10–25
Extractives	2–5	5–15
Proteins	<0.5	5–10
Inorganics	0.1–1	0.5–10

Source Alén (2011)

the wood substance degrades and dissolves into the cooking liquor (spent liquor) that contains degraded lignin fragments (lignosulphonates) and a considerable amount of various carbohydrate-derived materials. Sulphite pulping has decreased considerably during recent decades, but it is still important in certain countries and for making certain qualities of pulp. Its advantages over the kraft process include the production of brighter unbleached pulps, higher yields at a given removal of lignin, fewer odour problems and lower costs.

Today, wood is the predominant source of cellulosic fibre for pulp and paper manufacture, representing about 90% of the total, with the rest coming from various fibrous, non-wood feedstocks. The annual production of non-wood-derived pulps is about 20 million tons. Table 2.3 shows a general comparison between the chemical composition of wood and fibrous non-wood feedstocks utilised in pulping. One of the most harmful properties of non-wood lignocellulosics in pulp production is a high content of inorganics, mainly silica (typically, 0.5–7% of the feedstock dry solids). The main organic components in non-wood feedstocks are the same as those in wood feedstocks; they are found in varying amounts depending on species (genetic differences), presence of special tissues within individual plants and growing conditions. Because non-wood feedstocks are available in various forms, their physical and mechanical properties also vary. They typically include agricultural residues (e.g., sugarcane bagasse, sorghum corn stalks, cotton stalks, rice straw and cereal straws), natural-growing plants (e.g., bamboo, esparto, sabai, elephant grass and reeds), and non-wood crops (e.g., jute, ramie, hemp, kenaf, flax tow, abaca, sisal and cotton linter) grown primarily for their fibre content.

The most traditional delignification methods for non-wood feedstocks are soda and soda-AQ processes, although various sulphite and kraft processes are used as well. Pulping of non-wood as well as wood raw materials with organic solvents ("organosolv methods"), an approach dating back to the beginning of the 1930s (Theodor N. Kleinert and Kurt von Tayenthal), was not seriously considered for practical applications until the 1980s. The idea was to use basically "lignin solvents" for this purpose, but the systems were systematically investigated with a view to develop new industrial pulping processes.

The main driving force for developing organosolv pulping has shifted from energy-related considerations to the possibility of operating sulphur-free, less polluting and more economical (small-size) pulp mills with a simplified chemical recovery system, as well as improved recovery and upgrading of by-product lignin and hemicelluloses. These mainly acidic processes can provide solutions to problems of conventional non-wood pulping, including those related to chemical recovery of silica. The ambitious goals have given rise to many variations of organosolv pulping with organic solvents, mainly simple alcohols and organic acids. In general, the processes have been intensively studied but only a few commercial-scale operations have emerged. Additionally, most of the relevant solvents have been used in the presence of water; the introduction of suitable catalysts as the third component of the cooking liquor has enabled the production of pulps with a satisfactory degree of delignification, yield and strength. Organosolv methods, seen as potential examples of biorefinery concepts, may attract increased interest. As the ultimate aim is not only the effective production of chemical pulp but also the utilisation of by-products (lignin and carbohydrates) without their significant degradation, organosolv methods may play a significant role in the future.

2.3.2 Early Approaches

This section briefly outlines the main historical stages of chemical pulp production, especially in the 1800s, when the decisive technological developments occurred (Sjöström 1993; Alén 2011, 2018). In the West, in order to obtain a uniform product, paper was made primarily from pulped flax rags until 1843–1844. During this period and unknown to each other, Friedrich Gottlob Keller (1816–1895 in Saxony, Germany) and Charles Fenerty (1821–1892 in Nova Scotia, Canada) produced the raw material for papermaking from pulped wood into which rag pulp was added to increase its strength. Then in the early 1850s, Heinrich Voelter (1817–1887) and Johan Matthäus Voith (1803–1874), who bought the patent from Keller in 1846 in Heidenheim, Germany, presented the principles of a grinder suitable for large-scale production of groundwood pulp. After entering commercial production, the grinder won a gold medal in 1854 in the First General German Industrial Exhibition in Munich. It also evoked great interest at the Paris Exposition the following year. Consequently, the industrial production of pulp from wood using this groundwood process developed quickly in Scandinavia and Germany in the 1860s; the first pulp mill was founded in 1861 in the city of Poix in Belgium. At the Exposition Universelle of 1867 in Paris, it was possible to exhibit a modern pulp mill that even showed techniques for sorting, reject control and drainage.

In the manufacture of traditional rag (or cotton) paper made from cotton linterns or normally cotton rags, the collected and sorted material was reduced in size by cutting it into smaller pieces and removing buttons, pins and other foreign matter. This material was then treated with a lye solution, which removed fats and parts of dyes. Thereafter, the rags were allowed to "rot" through the retting or fermentation process at a warm temperature (the process was followed by observing the formation of mold), which caused the fibre strands to swell and become suitably brittle to enable the grinding of the material into a watery pulp. This process was first done by hand, but later stampers driven by water wheels, windmills or horses were used. The ground, watery pulp was drained to form paper sheets, the dried sheets were dipped into a solution of animal glue or starch to improve their properties and finally, they were hung on lines to dry.

The delignification from wood, the obvious and most common source of raw fibre material, proved surprisingly difficult and time consuming. In Europe at the turn of the eighteenth century, paper pulp was first made by cooking straw (its chemical defibration is easier than that of wood) in an alkaline, aqueous solution in an open vessel. Only in the early 1850s was a more efficient method (by A. Miller) of using a closed vessel under pressure introduced, enabling the use of higher temperatures. According to a patent (by Charles Watt and Hugh Burgess) granted in 1854, in addition to straw and other plant fibre pulps, wood raw material could also be cooked in an even stronger solution of sodium hydroxide at a higher temperature (150 °C). In this case, the losses from the cooking chemical were replaced with soda (sodium carbonate), whence the process was called "soda pulping"; the first factory was founded in the United States in 1854. Its patent also included chlorine-based bleaching of the product. The principles of "general chlorine bleaching" were already established nearly 100 years earlier in 1744, when Carl Wilhelm Scheele (1742–1786) noticed the bleaching effect of the gaseous matter (containing chlorine and oxygen) he had prepared. Soon thereafter, Claude Louis Berthollet (1748–1822) used mixed chlorine gas to bleach rags of poor quality for use as raw material for making white paper, a development that also enabled the use of coloured rags. It took another 30 years before Humphry Davy (1778–1829) realised that chlorine was an element, for which he proposed the name "chloride" (in Greek "light green").

Asahel Knowlton Eaton (1822–1906) patented in 1870 a method for replacing the sodium losses in alkaline cooking with sodium sulphate ("sulphate cooking process"). This method is normally known as "kraft cooking process" or "kraft process" and named according to the high strength of the resulting product (in German, "Kraft" means "strength"). Finally, based on the same idea, Carl Ferdinand Dahl presented in 1884 a practical process of employing wood as raw material, which was the decisive impetus toward the rise of the modern kraft industry. This quickly led to large-scale production; the first factory was founded in Sweden in 1885. The burning and causticising of the cooking chemicals (the sodium carbonate formed in the burning is converted to "caustic soda", sodium hydroxide, with calcium hydroxide) regenerates the active cooking chemicals into a mixture of sodium hydroxide and sodium sulphide formed from sodium sulphate in the reduction zone of the furnace. This method did not utilise a completely new idea because it had been observed already in the early 1800s that the addition of sulphur and sodium sulphide clearly accelerated the delignification of straw under alkaline conditions.

As shown in the sub-section entitled, "Sect. 2.2 The Basics of Industrial Applications of Cellulose", from 1837–1842 Anselme Payen laid the foundations for the other main line of development, acid cooking. He prepared remarkably pure cellulose by treating wood and other plant materials with acidified ammonia and concluded that cellulose formed a uniform basic component of plant cells. In terms of commercial production, Benjamin C. Tilghman patented in 1867 a method whereby the defibration of wood took place in a pressurised system of aqueous calcium hydrogen sulphite and sulphur dioxide (sulphite cooking). This process was not immediately realised on a large scale. When the world's first sulphite pulp factory began operating in Sweden in 1874, Carl Daniel Ekman (1845–1904) conducted pivotal research that realised the practical execution of large-scale processes. Alexandre Mitscherlich (1836–1918) also devised a similar, large-scale production method independently of Ekman. Previously, in 1851 Peter Claussen had proposed the possibility of making cellulosic fibre suitable for papermaking from wood with the help of sodium hydroxide and sulphur dioxide (this formed the precursor for neutral sulphite cooking). Several so-called "high-yield cooking" methods were proposed later in the 1880s, but their costs and products were found not to be very competitive in the following decades.

2.3.3 *Recovery of Kraft Cooking Chemicals*

In conventional kraft pulping, "white liquor" is used for cooking the chips; it is an aqueous solution that contains mainly active cooking chemicals, sodium hydroxide and sodium sulphide (Alén 2000b; Gullichsen and Fogelholm 2000; Finnish Recovery Boiler Committee 2004; Vakkilainen 2007). After cooking (or digestion) in a pressurised digester at 160–170 °C, the spent cooking liquor (black liquor) is separated from the pulp by washing and concentrated to a 65% solids (in a modern mill over 80%) content in multiple-effect evaporators. It is then combusted in the recovery furnace to recover the cooking chemicals and to generate energy. Combustion of black liquor in the recovery furnace produces an inorganic smelt of sodium carbonate and sodium sulphide. This smelt is then dissolved in water to form "green liquor", which is reacted in the causticising stage with calcium hydroxide to regenerate the original white liquor. Due to incomplete conversion reactions (about 90%) in the recovery cycle, white liquor also contains some dead-load inactive cooking chemicals.

The recovery boilers producing bioenergy have developed considerably in the past 80 years, culminating with units that are among the largest biofuel-fired boilers in the world. Early kraft pulp mills commonly discharged black liquor into local rivers, and it is toxic to aquatic life and fouls the water. G.H. Tomlinson invented the first industrial recovery boiler in 1934 and it was a distinct milestone in the advancement of the kraft process. This was also the first unit that had water walls (i.e., a unit with a completely water-cooled furnace) and where all combustion stages of sprayed black liquor droplets occurred in a single vessel. Typical features of the original recovery furnace have remained unchanged to this day, although many technical improvements to this technology have also emerged.

Kraft pulping would not be economically viable without chemical recovery. On the other hand, the recovery of chemicals also contributes to the capital intensity of the kraft process; in a modern pulp mill one third of the capital cost is attributed to the recovery process. As a result, significant effort has been spent on optimising the combustion of black liquor together with improving the energy efficiency of pulp and paper mills. Moreover, there has also been a major push to improve significantly the efficiency of the production of energy and chemical recovery cycle in kraft pulping by an attractive concept comprising gasification of black liquor; this technique has the potential to achieve higher overall energy efficiency than the conventional recovery furnace. However, many of these techniques still face big challenges and have not yet been realised on a commercial scale.

2.3.4 Recent Trends in Kraft Pulping

Since the 1950s, the kraft process has become the dominant production method for producing chemical pulps in the world (Alén 2000b; Gullichsen and Fogelholm 2000; Whiteman 2005; Lamberg et al. 2012). The fundamental concept behind the kraft pulping has remained the same, namely the removal of lignin from feedstock materials with the simultaneous liberation of fibres by the same chemicals. During its evolution of about 160 years, however, the kraft process has changed in many ways, and the technologies involved have undergone some significant developments, thereby giving birth to several process modifications. Additionally, the latest technology itself has been developed in order to allow for different options in terms of raw materials and pulp applications. In particular, technologies involved in chemimechanical pulping of wood have undergone some remarkable developments, and these have led to the production of new pulps with improved properties.

Excessive carbohydrate dissolution and degradation results in low pulp yield from wood in conventional kraft pulping. One approach to addressing this issue has been to apply the polysulphide kraft pulping to stabilise carbohydrates against alkali. The "actual" modified kraft cooking technique was developed in the early 1980s and this technique has become popular in many mills. The principle behind it is to charge the cooking chemicals more evenly throughout the cook. The reason for leveling out the alkali concentration (e.g., alkali charge) during delignification is mainly to obtain a selective delignification and it leads to lower contents of residual lignin in pulp without increasing the loss of carbohydrates. Several new variations of kraft pulping based on the principle of alkali concentration profiling have emerged. These include continuous flow processes, such as Modified Continuous Cooking (MCC), Extended Modified Continuous Cooking (EMCC), Isothermal Cooking (ITC) and LoSolids Cooking (LSC) as well as displacement batch processes, such as Rapid Displacement Heating (RDH), SuperBatch Cooking (SB) and White Liquor Impregnation (WLI).

One of the most significant improvements dealing also essentially with kraft pulping has been so-called "oxygen delignification" that has expanded very rapidly since the late 1960s (it was commercialised in the 1970s) and is today a well-established technique used in plants around the world (Sjöström 1993; Alén 2000b). This technique can be defined as a method in which a substantial fraction of the residual lignin in kraft pulp can be advantageously removed by oxygen and alkali prior to bleaching with more expensive chemicals. Hence, it is an environmentally friendly way of decreasing the formation of harmful bleaching effluents and an economical way of reducing the amount of expensive bleaching chemicals.

Carbohydrates and their derivatives have large potential as raw materials for making useful low- and high-molar-mass products. The primary approach of utilising wood and other carbohydrate-containing biomasses is their straightforward degradation into fermentable sugars; it can be accomplished by suitable pre-treatment followed by acid or enzymatic hydrolysis (Alén 2011, 2015). The hemicelluloses in wood are more readily hydrolysable by acids than cellulose, and their removal also enhances the reactivity of cellulose in the residual solids. The majority of hemicelluloses, along with some lignin, are dissolved into cooking liquor during chemical pulping. Thus, one potential approach is to separate a prominent part of hemicelluloses prior to delignification instead of recovering their degradation products from the spent cooking liquors. This is actually not a recent idea, as industrial pre-hydrolysis of hemicelluloses (mainly xylan), especially prior to hardwood kraft pulping, has been used for many years in the production of dissolving pulp. The process was first used commercially in Germany during the 1940s in two-stage, kraft pulping processes.

Especially in the Northern Hemisphere, the pulp and paper industry faces major challenges and needs new products with added value in order to remain competitive. Hence, integration of various pre-treatment stages prior to pulping is an attractive possibility. In developing concepts for the potential recovery of dissolved organics during alkaline pulping, certain limiting factors, including both technical and economic factors, should be considered. In all cases, the main product is cellulosic fibre; its strength properties must be maintained without interfering with the recovery of cooking chemicals. The extractives must also be efficiently separated. These prerequisites practically dictate that only partial recovery of the dissolved material is possible. The general aim is to maximise the recovery of carbohydrate-derived material with low heating value, while minimising the recovery of lignin-derived material with high heating value. It would also be advantageous if sulphur-free by-products could be produced by applying straightforward separation techniques. Such pre-hydrolysis processing of wood chips, belonging to the so-called "integrated forest biorefinery concept", have been investigated under a variety of conditions from several points of view.

As mentioned above, kraft black liquor as such is the most important by-product of kraft pulping and the combustion of this heterogeneous biofuel is a significant source of energy (Alén 2000b, 2015; Gullichsen and Fogelholm 2000; Vakkilainen 2007). However, black liquor has several combined features that make its combustion different from other fuels. For example, its high contents of inorganic material (25–40%) and water (15–35%) decrease the heating value (12–15 MJ/kg dry solids) of the black liquor to a clearly lower level than that of other common industrial fuels. Hence, another modern biorefinery type concept of kraft pulping has been to recover value-added organic solids dissolved during delignification from black liquor. In a modern kraft process, when surplus energy from the burning of black liquor is available, partial recovery of dissolved organic material (<20%) is becoming more attractive and can be realised without interfering with the recovery of cooking chemicals.

The partial removal of lignin from black liquor has a long history and this technique is used in the cases where the overloaded recovery furnace is the process bottleneck. Particularly during the last decade the separation technique based on the precipitation of lignin has been developed. In addition to lignin, large amounts of aliphatic carboxylic acids are formed in the kraft pulp process and their partial recovery is an interesting alternative to using them as fuel. The basic idea behind this approach is that about two thirds of the total heat produced by the liquor originates from lignin and only one third stems from the remaining constituents, mainly aliphatic acids.

As a general trend, the products of chemical pulping have become more diversified (Alén 2018). It is already indicated in the chapter entitled "Sect. 2.3.1 General Aspects" that a relatively small amount of chemical pulps (about 3%, corresponding to less than about 4 million tons annually) with flexible fibres from dissolving pulping are used as a source of purified cellulose (dissolving pulp, cellulose content typically 80–90%) for producing mainly cellulose derivatives (e.g., cellulose esters and ethers) and regenerated celluloses (e.g., rayon fibre, annual production about 3 million tons worldwide) used mainly for the production of threads (textile fibres) and foils. Furthermore, the production of microfibrillated cellulose (MFC), microcrystalline cellulose (MCC) and nanocrystalline cellulose (NCC) offer increasing possibilities to utilise the microfibrillated and crystallised structure of cellulose.

2.4 Bleaching of Pulps

Bleaching can basically be defined as a chemical process applied to chemical and mechanical pulps to increase their brightness (Dence and Reeve 1996; Alén 2000b; Hart and Rudie 2012). Thus, to reach an acceptable brightness level, bleaching is performed either by removing the residual lignin in chemical pulps ("delignifying bleaching" or "lignin-removing bleaching") or by converting (decolouring) chromophoric groups of mechanical pulps without loss of substance ("lignin-preserving bleaching" or "lignin-retaining bleaching"). In both cases, however, the obvious prerequisite is that no significant losses in pulp strength occur. Delignifying bleaching is performed in a "bleaching sequence" comprising a series of "bleaching stages" usually with intermediate washing between stages. A multistage process is necessary because it is not possible to achieve sufficient decolourisation of chemical pulp by the action of any chemical in a single stage. Depending on the case, a wide range of chemicals and bleaching systems are available. Additionally, chemical pulps are normally further purified in the oxygen delignification and the following bleaching stage where, besides the rather selective removal of residual lignin, for example, the residual extractives are also destroyed. The bleached, lignin-free chemical pulps are used as raw materials in the manufacture of paper and related products. In general, they still contain (20–30%) hemicelluloses necessary for suitable strength properties.

The current trend toward closed process-water circulation aims at a drastic decrease in the wastewater load (Alén 2000b). A totally effluent-free (TEF) mill represents the ultimate objective in pulp production. In this respect, to avoid environmental and corrosion problems caused by chlorine-containing compounds, traditionally used for more than a century, the application of oxygen-based chemicals (e.g., oxygen, ozone, hydrogen peroxide and peracids) in a totally chlorine-free (TCF) bleaching has gradually offered a potential alternative process. The proportion of elemental chlorine-free (ECF) bleached pulp is increasing in the world's bleached pulp production, representing today more than 50%; the corresponding TCF production is about 5%. Furthermore, ECF is acknowledged as a core component of best available technology (BAT) for the production of pulp, paper and board. Thus, in addition to the progress in the cooking phase, major developments have also taken place, especially in bleaching pulp products.

Early notable approaches (the "pre-chlorine era") performed by the ancient Gauls to whiten vegetable fibres were accomplished by the action of sunlight upon fibres moistened with an alkaline solution of wood-based ash (Lorås 1980; Singh 1991; Torén and Blanc 1997). Between this alkali treatment and washing, the goods were exposed on grassy meadows to the action of the sun (known as "grass bleaching"). The entire process was repeated several times prior to final treatment with lactic acid derived from sour milk. During the eighteenth century, this rather tedious and time-consuming practice reached its zenith around Haarlem in Holland and slightly later in Scotland where Francis Home (1719–1813) in 1756 proposed the use of dilute sulphuric acid in place of the traditional sour milk; this finding reduced the time of souring to about 4% of that formerly required. However, at that time, paper was manufactured from rags. Since white paper was obtained only from materials derived from white textiles, it was shown to be too difficult and expensive to bleach enough rags to meet the increasing demand for bleached textile fibres for paper production. In general, cotton and linen cloth have been bleached since ancient times. Additionally, it is known that Japanese papermakers, for example, bleached fibres by soaking them in water drawn from high mountain streams containing ozone formed by thunder storms.

As shown in the chapter "Sect. 2.3.2 Early Approaches", a major breakthrough was achieved a few years later after Francis Homes' innovation in 1774 when Carl Wilhelm Scheele noticed the bleaching effect of the gaseous chlorine-containing matter ("oxygenated muriatic acid"). This revolutionary innovation started the "chlorine era" and soon thereafter, Claude Louis Berthollet commercialised the chlorine-based method. However, due to the liberation of chlorine gas, a neutral water solution of chlorine and later its alkaline solution ("eau de Javel") were both unstable and difficult to transport and handle. Several years later in 1799, Charles Tennant (1768–1833), probably in close cooperation with Charles Macintosh (1766–1843), began to produce bleaching powder by passing chlorine into a mixture of lime (calcium hydroxide) and water. This product, calcium hypochlorite, had the advantage of being cheaper than the one generally used at the time because it substituted lime for potash (potassium salts) and an entirely dry bleaching powder also solved the transportation difficulties. Hypochlorite became the dominant bleaching agent for textiles and also later for wood-based pulps. In the 1920s, technologies that allowed for the use of gaseous chlorine began to displace hypochlorite in the first bleaching stage of multistage bleaching systems. This was possible primarily because bleaching equipment was now manufactured out of stainless steel, thus minimising the corrosive effects of the chlorine gas. The typical bleaching sequences (H refers to the hypochlorite stage, E to alkaline extraction and C to chlorination) were in the 1880s H, in the 1890s H-E-H and in the 1930s C-E-H or C-E-H-E-H. Typically, multistage bleaching was rather successful for sulphite pulps, but conditions necessary to obtain very bright kraft pulps resulted in depolymerisation of cellulose and loss in the fibre strength.

Compared to chlorine gas, chlorine dioxide (the bleaching stage D) reacts less aggressively on the cellulose and improvements in its manufacture were developed in the 1940s. It was first synthesized in 1811 by Humphry Davy, but it was not until the 1920s that its commercial bleaching properties were investigated. This development led, for example, in the 1950s to the five-stage bleaching sequence (C-E-D-E-D) which allowed the production of high-brightness products from kraft pulp with minimal loss of pulp strength. In recent decades, a variety of different bleaching sequences have been tailored for different purposes. Besides chlorine dioxide, the use of oxygen-based chemicals has significantly increased. Additionally, bleaching technologies have also been enormously improved. Today, the lignin-preserving bleaching of mechanical pulps with sodium hydroxide and sodium dithionite is also rather effective. However, throughout the history of bleaching, the issues of its effectiveness and its impact on product quality (brightness and strength) and its cost have been important factors.

2.5 Development of Industrial Papermaking

2.5.1 Brief History of Paper Machines

The actual "machinisation" of papermaking can be said to have started in the 1670s in Holland, where a grinder called the "Hollander beater" or the "Hollander" was developed to replace stamping machines, which had previously been used to disintegrate rags and beat pulp (Atchison and McGovern 1983; Sundholm 1999; Lindberg 2000; Alén 2018). This machine clearly expedited the production of pulp and improved its quality. The reign of rag pulp ended with the rapidly increasing demand for paper in the 1870s, when pulp makers learned to produce mechanical pulp from wood on an industrial scale. On the other hand, in 1789, along with the rapidly increasing number of newspapers during the French Revolution, a shortage of raw materials caused the price of rags to rise, rendering papermaking unprofitable. The development work of

Friedrich Keller described above followed the trends of the times. It was primarily based on the sharp observation of René-Antoine Ferchault de Réaumur (1683–1757) in 1719 that wasps (*Vespa vulgaris*) made very fine, grey and paper-like material for the walls of their nests from fibres of ordinary trees with their saliva and mastication. Inspired by this same observation, it took Bavarian Jacob Christian Shäffer (1718–1790) until 1765 to describe a non-commercial process for producing fibre material suitable for papermaking from sawdust and other woodworking residues.

The production could be "theoretically increased" only after the final principle of a paper machine was invented. The credit for it goes to Nicholas-Louis Robert (1761–1828) in France (the patent was awarded in 1799), preceded by close cooperation between Robert and Didot Saint-Lèger, whom Robert hired to work in his factory in Essonnes—close to Paris—in 1793. This was the first hand-operated paper machine (the typical width of French wall paper was 64 cm, length 260 cm and the machine speed 9 m/min). It only contained a belt of wire cloth and no dryer section, which meant that the paper sheets still had to be lifted off to be hung over a rope or wooden rod to air dry.

The paper machine based on Robert's patent remained on an experimental level for some time. The next step was taken by the English banker brothers, Henry Fourdriner (1766–1854) and Sealy Fourdrinier, together with John Gamble (the brother-in-law of Didot Saint-Léger). After acquiring the patent in question in 1803, they built in their machine workshop with Bryan Donkin (under John Hall) a modified paper machine—one that could be seen as an improved version of Robert's machine (width 120 cm and it included a wet press). With this machine, it was possible to produce paper continuously, and it was granted a new patent in 1807. Regarding production, a machine of this type was still rather small, and it soon proved uneconomical. However, the name of the brothers has survived in papermaking until this day; a traditional longitudinal wire papermaking machine is still called "a Fourdrinier machine" (or, more commonly, the forming section of a paper machine is called a "fourdrinier").

The introduction of surface sizing was another important breakthrough in papermaking from plant fibres. It enabled the production of a smooth surface on which the printing ink could be controlled. Papers made in the West were surface sized, before the introduction of large-scale starch sizing of machine papers in the early 1800s, with animal glues (gelatin sizing) originated by the Italians by boiling bones and skins (the Arabs used traditional starch sizing). However, the gelatin sizing method was an expensive and tedious process. Hence, in 1807, Moritz Friedrich Illig (1777–1845) in Germany published a sizing method based on rosin and alum. By the mid-1800s, this technique was in use across the world; it substantially improved the quality of printing and wrapping paper.

2.5.2 Current Papermaking

In spite of its lack of commercial success, the first continuously operating paper machine initiated rapid technical development that has continued until this day; the largest width of the web is now over 11 m and the driving speeds, especially in the production of LWC ("light-weight-coated") and newsprint, are at best about 2000 m/min and are being continuously improved (Alén 2018). The pilot machines used in product development can currently be driven at about 2400 m/min. The time interval from the invention of the paper machine to its wide industrial use was about 30 years with no significant differences emerging between paper and cardboard machines. However, the methods of papermaking are developing continuously. One interesting innovation being investigated is the so-called "foam forming", which reduces the consumption of water, energy and fibre.

Modern papermaking is a very dynamic process with a high-driving speed and all the individual phenomena occur rapidly (Alén 2007, 2018). Hence, to obtain acceptable retention (i.e., the retention of fibres and possible additives on the paper machine wire), suitable chemicals should be used. In addition to these chemicals, a wide range of other synthetic and natural additives are used for a variety of different proportions. All the chemicals are added (with the exception of fillers, generally 10–55% of the product) in proportions of up to 5% of the furnish (usually around 1%), and because of their relatively high cost in comparison to fibre and fillers, they also often represent a significant share of the total raw material cost. Papermaking additives can be divided into the following three groups: (i) mineral fillers and pigments, (ii) functional chemicals (directly affect paper quality and its other properties) and (iii) process chemicals (optimise the production process by improving "runnability" and deposit control or by reducing steam consumption). In spite of being used sparingly, chemical additives offer significant advantages in terms of productivity, quality, cost effectiveness and the environment and the intense development of even more effective additives is in progress.

The papermaking potential of pulps basically originates from the quality of the initial fibres in the raw material, but it is also largely determined by the fibres produced in various delignification or defibration operations. In general, paper can be considered to be a semisynthetic product originally based on cellulosic fibres, although its properties are basically adjusted as shown above with various additives. It is clear that the papermaking process has developed enormously during its 200-year history, but the basic phenomena of paper formation are still the same. Today, the products comprise a wide range of paper grades used, for example, for newspapers, magazines and books as well as for copying and laser or digital printing.

The industry faces several major challenges today (Whiteman 2005). The development of the modern electronic media, for example, has greatly reduced the demand for certain types of paper (e.g., newsprint) in certain parts of the world (e.g., North America). Another notable challenge is associated with the application of recyclable plastics, which may have an impact on cellulosic fibre-based packaging materials. On the other hand, the paper industry has numerous possible opportunities for the development of new products and the utilisation of new fibre sources, such as those based on non-wood resources. Moreover, today the demand for and production of packaging materials as well as tissue papers is increasing.

Intense international competition in the paper markets requires ever greater emphasis on the development of new products and production efficiency, including the effective use of raw materials and energy. Over the last decades there have been major technological improvements in the papermaking processes. These changes have clearly affected the end-product quality and process performance. At the same time, environmental impacts of the whole production chain are becoming important and have led to breakthroughs, such as a reduction in the consumption of fresh water by many mills. One significant factor has also been the application of sophisticated measuring and associated control systems as well as the improvement of pulp properties by means of more precise quality grading of the wood raw material.

2.6 Concluding Remarks

Several industries have been striving, especially during the last few decades, to promote a gradual shift toward a more efficient utilisation of different carbon dioxide-neutral lignocellulosic raw materials while simultaneously decreasing the use of fossil resources (Alén 2011, 2018). There are different opinions about how long we can continue to utilise our fossil fuels. These reserves are certainly limited and we need not only to develop new ways of producing energy but also to find alternative ways of manufacturing important chemicals and other products. The pulp and paper industry is an important branch of the global economy and its overall strategy to increase the large-scale utilisation of wood feedstocks has the potential to help significantly in changing over to an economy that is more focused on the better use of renewable, carbon resources ("biomass"). Currently, petroleum provides about 40% of the world's total energy consumption and about 90% of its vehicular fuel needs. For these recognised reasons and those related to environmental issues, industrial activities are occurring in a new era, the Green Industrial Revolution.

During most of our history, our survival was almost exclusively based on renewable resources; as indicated above, this changed during the first part of the nineteenth century. Production of organic chemicals and other products from fossil resources started as coal-based thermochemistry, coal carbonisation, about 150 years ago. Petroleum-based industrial chemistry followed some 60 years ago, leading to an enormous increase in the number of products. As we move toward a world of greater diversity and balance with the natural cycles of various materials, it is still important to learn the lessons from the oil refining and petrochemical industries.

The biorefinery concept can be defined as a process of fractionating and/or converting biomass, a carbon dioxide-neutral feedstock, in an eco-friendly way through advanced technologies into solid, liquid and gaseous bioproducts. The main objective is to maximise the value of the product while minimising the production of waste. In the simplest case, a biorefinery utilises only one feedstock with a single process that results in a single major product. This principle is analogous to that of petrorefineries utilising fossil resources, but biorefineries use a wider range of feedstocks and process technologies. It should be pointed out that the established conventional kraft process roughly separates, in much the same way as the new biomass processing technology ("reactive fractionation"), most lignin (dissolved into black liquor), extractives and a significant proportion of hemicelluloses (mainly in the form of black liquor-soluble aliphatic carboxylic acids) from cellulosic fibre suitable for versatile utilisation in paper and board. Hence, it can be expected that in the near future that there will be a major trend toward full-scale chemical pulping, and that it will represent, through an increasing diversification, one of the key processes which is capable of laying a natural foundation for an integrated biorefinery with a wide spectrum of novel by-products. Additionally, the current information society with many other factors of importance results in remarkable challenges for the pulp and paper industry.

Currently, only slight modifications to the kraft fibre production process have been made to enhance the recovery of energy and chemicals. One obvious reason for this "limited" progress is that the pulp industry is capital-intensive and the equipment, once installed, must have a long life-time. Therefore, sweeping changes of any significance in mill processes are even in most cases not likely.

References

Alén R (2000a) Structure and chemical composition of wood. In: Stenius P (ed) Forest products chemistry. Fapet Oy, Helsinki, pp 11–57
Alén R (2000b) Basic chemistry of wood delignification. In: Stenius P (ed) Forest products chemistry. Fapet Oy, Helsinki, pp 58–104
Alén R (ed) (2007) Papermaking chemistry, 2nd edn. Finnish Paper Engineers' Association, Helsinki
Alén R (ed) (2011) Biorefining of forest resources. Finnish Paper Engineers' Association, Helsinki
Alén R (2015) Pulp mills and wood-based biorefineries. In: Pandey A, Höfer R, Taherzadeh M, Nampoothiri KM, Larroche C (eds) Industrial biorefineries & white biotechnology. Elsevier, Amsterdam, pp 91–126
Alén R (2018) Carbohydrate chemistry—fundamentals and applications. World Scientific Publishing, Singapore
Atchison JE, McGovern JN (1983) History of paper and the importance of non-wood plant fibers. In: Kocurek MJ, Stevens CFB (eds) Pulp and paper manufacture. Properties of fibrous raw materials and their preparation for pulping, vol 1, 3rd edn. The Joint Textbook Committee of the Paper Industry, TAPPI & CPPA, pp 154–156
Dence CW, Reeve DE (eds) (1996) Pulp bleaching—principles and practice. TAPPI Press, Atlanta, GA
Finnish Recovery Boiler Committee (2004) 40 years recovery boiler co-operation in Finland. In: Proceedings of 40th anniversary international recovery boiler conference. Porvoo, Finland, 12–14 May 2004
Gullichsen J, Fogelholm C-J (eds) (2000) Chemical pulping, 6A and 6B. Fapet Oy, Helsinki
Hart PW, Rudie AW (eds) (2012) The bleaching of pulp, 5th edn. TAPPI Press, Norcross, GA
Lamberg J-A, Ojala J, Peltoniemi M, Särkkä T (eds) (2012) The evolution of global paper industry 1800–2050. A comparative analysis, Springer, Dordrecht
Lindberg NJ (2000) History of papermaking. In: Paulapuro H (ed) Papermaking, Part 1, stock preparation and wet end. Fapet Oy, Helsinki, pp 56–71
Lorås V (1980) Bleaching. In: Casey JP (ed) Pulp and paper—chemistry and chemical technology. vol I, 3rd edn. Wiley, New York, NY, pp 633–764
Ritman EL (1968) History of sulphite pulping. Appita, vol 21(5), pp xvii–xxi
Singh RP (1991) Historical and introductory. In: Singh RP (ed) The bleaching of pulp, 3rd edn. TAPPI Press, Atlanta, GA

Sjöström E (1993) Wood chemistry—fundamentals and applications, 2nd edn. Academic Press, San Diego, CA
Sundholm J (1999) History of mechanical pulping. In: Sundholm J (ed) Mechanical pulping. Fapet Oy, Helsinki, pp 22–33
Torén K, Blanc PD (1997) The history of pulp and paper bleaching: respiratory-health effects. Lancet 349:1316–1318
Vakkilainen EK (2007) Kraft recovery boilers—principles and practice, 2nd printing. Helsinki University of Technology, Energy Engineering and Environmental Protection, Espoo
Whistler RL, BeMiller JN, Paschall EF (eds) (1984) Starch—chemistry and technology, 2nd edn. Academic Press, New York
Whiteman A (2005) Recent trends and developments in global markets for pulp and paper. In: Proceedings of Paperex'2005—international technical conference on pulp and paper industry, New Delhi, 3–5 Dec 2005

Chapter 3
Research and Development in the Finnish Wood Processing and Paper Industry, c. 1850–1990

Panu Nykänen

3.1 Two Cumulative Masses of Knowledge

The tradition of research and development in the paper industry is divided into two separate branches. One consists of technical chemistry, studying the possibilities of chemical wood processing and wood chemistry. The other is part of mechanical engineering, and its roots are in the construction and use of paper machines. The difference between these two traditions is widely acknowledged,[1] and the practical differences between them result from their dissimilar relationships to basic scientific, technical and applied research.

The aim of this chapter[2] is to describe the general history of research and development work in the Finnish pulp and paper industry from the mid-nineteenth century until roughly 1990, a period that witnessed it undergo drastic changes because of the altered political environment in Europe and general globalization. In addressing its subject, this chapter is mainly concerned with the operation of paper mills, but it also addresses the related subjects of technical research, formal higher education and work-related, practical learning. In addition, it briefly discusses the challenges inherent in researching the development of technology because of its research traditions.

[1] For example, H. F. Rance, who wrote a presentation that was read by J. Carey at the fiftieth anniversary meeting of the Finnish Paper Engineers' Association in Helsinki in 1964, clearly delineated the fundamental distinctions between the two traditions.

[2] The author would like to thank the following persons for the invaluable co-operation they provided in conducting the research for this chapter: University Archivist Erin Dix, Lawrence University, Wisconsin, Jukka Kilpeläinen, DI, MBA, Helsinki, Sampsa Kaataja, Ph.D., Karjaa, University Archivist Sani Kivelä, Aalto University, Espoo, Archivist Ari Sirén, UPM Central Archives, Valkeakoski.

P. Nykänen (✉)
Department of History and Ethnology, University of Jyväskylä, Jyväskylä, Finland
e-mail: panu.nykanen@ttatv.fi

T. Särkkä et al. (eds.), *Technological Transformation in the Global Pulp and Paper Industry 1800–2018*, World Forests 23, https://doi.org/10.1007/978-3-319-94962-8_3

As disciplines studied at the technical universities, the two branches (i.e., chemical and mechanical) had separate curricula at the Helsinki University of Technology (HUT, now part of the Aalto University) up to the 1940s. The first professorships in wood chemistry were established in the early 1920s. The research at that time was focused on turpentine chemistry. In contrast, paper-machine technology was a part of general machine technology. As an indicator of the change in the needs of technology for industry, the department of forest products technology was founded in 1941. The rapprochement of the chemical and mechanical branches of research occurred a few decades later. The first students of paper engineering to change their major from machine shop technology did so in 1943, and the initial research papers that focused on paper machine technology were finished in 1946.[3]

As a result of a long-lasting discussion, professorships in wood processing machinery were established from the late 1960s in Finnish universities in an effort to address the industry's special needs.[4] This chapter deals with the background of this development, and the aim of the study is to reveal the impetus behind the organizational changes that occurred.

The nature of technology has been under discussion roughly since the First Industrial Revolution in the eighteenth century. Around the same time the sciences started to develop rapidly in universities in Western Europe and a special branch of industrial and technical education was created alongside the traditional scientific universities. Thus, two different masses of knowledge emerged, namely science and technology.

The line separating these 'mirror image twins' (Layton 1971), science and technology, is blurry. As de Solla Price has aptly stated, science is papyrocentric and technology is papyrophobic. Science is based on conducting empirical research and the publication of academic papers. In contrast, technology is often seen as an art that makes progress through trial and error, and it intentionally aims to hide its achievements. (de Solla Price 1965, pp. 561–562; Wise 1985, p. 235; about the nature of technology see also e.g. Rosenberg 1994, p. 15)

There are two reasons for the papyrophobic nature of technology. In the first place, there is an inextricable link between technology and industrial activities, namely that there is always an economic motive in developing technology. As a result, it is counter-productive to a business to publicize the advances it makes in its industry because doing so could potentially benefit a rival in the never-ending competition to maximize economic efficiency. The second reason is that developing technology largely involves practising the art of engineering. Whereas science entails conducting a series of tests and recording the results, technology is about know-how, learning by doing, developing technical skills and solving problems, processes that involve generating very little, if any, data.[5]

[3] HUT, Department of Forest Products Technology minutes 1941–1948.

[4] The first Professor of Paper Machine technology in Finland was Uolevi Konttinen at the University of Oulu 1969.

[5] Eugene S. Ferguson effectively describes this reality in his study of the nature of engineering (Ferguson 1977, p. 833; Ferguson 1993, pp. 1–40, 154).

In many cases the technology is not published or patented. A developer of technology must in every separate case predict if patenting technology is profitable, or alternatively whether the technology should just be launched and sold to customers. In many cases patent processes cause more harm than good for the company that is developing technology.[6]

The history of the art of engineering in the paper industry has been widely discussed. The literature describes how, after the Second World War, the rate at which technology became obsolete because of new developments quickened dramatically, and the art of engineering was gradually but surely almost forgotten. The question arose again in the 1960s, however, when the desirability of providing practical training and experience to formally-educated engineers became a hot topic at pulp and paper and engineers' conferences (Mardon et al. 1969).

After the Second World War, the general idea of many successful Research and development (R&D) programs in the Finnish industrial culture was to connect science and technology even though this aim was not explicitly declared. There were, however, a few exceptions to this rule, one of which was espoused by Petri Bryk, the CEO (1913–1977) of Outokumpu Ltd., a metallurgy and mining firm. He was renowned for his aphorisms, one of which captured this notion: 'to know how, you must know why'.[7] In practical terms this meant that Bryk required his personnel to appreciate the connection between science and technology.[8]

3.2 The Nature of a Paper Mill

Hughes (1987) used the term "Large Technological System" to describe the development of a complex technical structure, in which all the parts are developed and operated independently but they are connected to form a massive cluster. These dynamics are all present in a paper mill. The "Large Technological System" of a paper mill consists of several raw material flows, each of them grounded in the science of chemical engineering, and the paper machine's main structure with all its supplementary machinery. Because paper machines are such complex apparatuses to develop, maintain and operate, usually their components are built to the highest standards of engineering science.

These factors have made paper machines and mills very difficult subjects for historians to analyse. In particular, the aforementioned lack of paper trail on the technological development side and the degree to which mills represent such huge,

[6]About the patenting procedures see Kaataja 2010; Jukka Kilpeläinen, DI, MBA, Helsinki, comments on 10 March 2017.

[7]Tapio Tuominen, CTO Outokumpu Ltd (ret.), oral statement 26 October 2015.

[8]Bryk was one of a generation of Finnish engineers, educated during the 1930s in the optimistic atmosphere of the HUT. After the Second World War a group from the same generation formed the informal luncheon club of "Pilkkula", and it laid out a roadmap for modernizing Finnish engineering. The other members of this group were Jaakko Rahola, Heikki Miekk-oja and Erkki Laurila, all of whom would later become leaders in shaping modern Finnish technology policy.

complex technological systems that are exceedingly difficult to understand have caused historians to shy away from writing about these subjects. It has compelled most historians to focus on writing about these subjects only on their administrative or organizational levels (e.g., Tuuri 1999).

These factors explain why much of this chapter, specifically those parts that are devoted to the history of technology, is based on so few sources.[9] Only in rare cases did the engineers who practised in this field create documents that discussed their aims or intentions, and it is even rarer that these papers have survived. Hence, oral interviews formed an important part of the research upon which this chapter is based.

Many more sources were available for the section of this chapter that deals with the question of the research and development work. For these sections, research was conducted in the archives of the HUT,[10] the Finnish Pulp and Paper Research Institute (KCL, Oy Keskuslaboratorio—Centrallaboratorium Ab), Kangas paper mill of G. A. Serlachius Oy (GAS)[11] and United Paper Mills Ltd (UPM).[12] The archives of HUT reveal the role that education played in this process, especially during the 1940s and 1950s. In the archives of the Kangas paper mill the memoranda of Niilo Ryti, at that time the Technical Director of the mill, are preserved, and they provide an exceptional opportunity to describe the inner-workings of a Finnish paper mill from 1950 to 1960. The archives of UPM contain the papers of Niilo Hakkarainen, CEO of the company (1970–1991), and Ingmar Häggblom, its Technical Director. Examining them helps trace the roots of R&D strategic thinking during the 1970s and 1980s.[13] Unfortunately, the existing research literature says very little about R&D in the realm of paper making machinery.[14]

The problematic nature of research and development work in the industry is presented by Archimbault and Lariviere (2011), who investigated the scientific publications of Canadian companies over 25 years' time. Their work highlights the lack of a paper trail that plagues historians of industrial engineering, whereby outsiders are seldom able to see what insiders were doing in developing new processes.

[9]These include the archives of the board of United Paper Mills for the 1960s and 1970s, and the archives of Kangas paper mill during the 1945–1960 period.

[10]In the archives of Helsinki University of Technology (HUT), Aalto University, Espoo, Finland.

[11]In the Central Archives of Finnish Business Records, ELKA, in Mikkeli, Finland.

[12]UPM Central Archives, Valkeakoski, Finland.

[13]On Hakkarainen's role e.g., Ojala 2001. p. 83.

[14]It is quite revealing that, even the history of Beloit, in its time the leading manufacturer of large printing paper machines, is not publicly accessible. The history of Beloit is however to be found in Robert Hodge and Larry Ely, *Beloit Corporation. Your partner in papermaking*. 1979. See http://paperindustryweb.com/partner/partmenu.htm. Accessed 11 October 2017. In the context of Finland, the topic is discussed to a limited extent in a few sources, including Nykänen (2005) and Komulainen (2014). The latter is a history of the Finnish Paper Engineers' Association, and in it Komulainen discusses the emergence of an R&D policy for paper manufacturing especially after the Second World War. Nykänen (2005) discusses the history of paper making machinery from the view of Finnish machine shops.

3.3 The Early Stages of Paper Manufacturing and Wood Chemistry

The first Fourdriniers in Finland, bought for J. C. Frenckell's Tampere mill in 1842 and the Tervakoski paper mill in 1853, were operated by skilled craftsmen and used cotton fibres to produce quite expensive, high quality printing and writing paper. The first formally educated civil engineers came to Finland during the 1840s to work in the machine shops and textile mills in Tampere and Forssa. Specialized education in paper manufacturing began in the 1870s when mechanical engineering was separated as a special stream of education at the Helsinki Polytechnic Institute.

A major change in the method of production, and the technical and scientific support system behind it, occurred a short while later. The breakthrough came when the industry switched to using ground wood fibres and chemical pulp as the major raw materials for its rapidly expanding production. As a result, the cost of making these investments was significant when establishing new paper mills after the 1870s and thus represented a great risk for the industry's entrepreneurs and financiers.

3.3.1 Tar Manufacturers and the Scientific Approach to Wood Chemistry

Research into wood chemistry in Finland is older than the modern, industrial-scale paper industry. It was initially associated with pine and birch tar production. In fact, pine tar rose to be the first branch of mass production in Finland during the eighteenth century, and it was based on the tar pit process that yielded products of various densities from the pyrolysis process. In addition, birch bark tar was produced, mainly for use as a lubricant in maritime vessels. Research into pine and birch bark tar products came to be the oldest government-steered technology project in the country beside metallurgy and mining.

The pine tar industry's low profitability was well known by the government and the Economic Society in Finland. The latter had been founded in 1797 and was a formal think tank at the turn of the century, and in the 1830s it launched a research project to improve the economics of tar production. A pitch factory was built as a pilot project in Oulu (Blomenthal 1887, pp. 11–23). This enterprise was arguably the first chemical wood processing industry in Finland, but it was destroyed by the British Royal Navy during the latter's campaign in the Baltic during the Crimean War.

A second wave of tar production emerged during the 1850s after the founding of the Evo Forestry School in Southern Finland. Alexander af Forselles was the Director of the School for over one decade (1858–1869), and he launched a modern turpentine factory there. The purpose of the plant was to produce lamp oil, but the hopes surrounding it were crushed by the new standardized crude oil-based kerosene that entered the market after the American Civil War.

3.3.2 Formal Technical Education

The Helsinki Technical School was founded on 9 June 1847 based on the German model of "technische realschule" (i.e., a practical school for primary and secondary students that was geared toward educating future engineers). The school had three clear objectives in its operations. Before the elementary school system was organized in Finland, the technical school aimed to deliver basic skills of reading, writing, drawing and mathematics to the young apprentices working in the industrial shops. It also aimed to provide a "modern way" for apprentices to reach the status of craftsman instead of the professional guild system that still governed the life of small industries. The third goal was to raise Finland's level of competence in the chemical wood industry.

Its importance in the rapidly developing field of technical education is revealed by how it was organized. The technical school's curriculum provided for training in elementary subjects and two disciplines: technical chemistry and mechanical engineering. The lectures for both these disciplines were delivered in special laboratories, furnished and available for use from the beginning of the school's operation in 1849. Other technical disciplines were added to the curriculum after 1860 based on the German model of "technische hochshule" (i.e., a university-level institute of technology).

Anders Olivier Saelan was its first director (1848–1874), and he had a Master's degree in chemistry from the university in Helsinki. In addition, the chemists working in the field of wood chemistry played a leading role in recruiting new teachers to the school. For example, A. F. Soldan, H. A. Wahlforss and E. E. Qvist started as young researchers in the field of wood chemistry. Qvist and his student, H. A. Wahlforss, published a study in 1865 about *reten*, recin acids in the tar. This was arguably the first technical academic publication in Finland. H. A. Wahlforss completed his post-graduate studies, and published his Ph.D. work, *Bidrag till kännedomen af retén* (i.e., the investigation of recin acids), in 1868. He completed his studies in the laboratory of F. K. Beilstein in St. Petersburg. He is considered the father of the chemistry involved in the traditional Finnish production of turpentine, and he educated the first research-oriented generation of chemists for Finnish industry in the 1880s.

Gustaf Komppa, who studied under Wahlforss, continued down the same research path, and was able to complete in 1904 his studies into synthesizing camphor, a chemical component used in making early plastic. Komppa continued his studies at the laboratory of Johannes Wiliscenus, and he listened to the lectures of Wilhelm Oswald. Gustaf Komppa took over responsibility for teaching technical chemistry after H. A. Wahlforss's health forced him to step aside. Komppa was able to complete the construction of the new chemistry laboratory at the Helsinki Polytechnic Institute in 1898. As a result, by the turn of the century Finnish research in the field of wood chemistry had made considerable progress.

3.3.3 *The Origins of Research into Making Paper*

The pulp and paper industry really emerged in the Finnish economy during the 1870s, when the development of a process for using wood cellulose made it possible to exploit the country's forest resources as the industry's main raw material. The first pulp and paper factories were built on the shores of lakes and rivers in southern Finland, where it was possible to obtain the raw material needed from the vast forests reaching up to Central Finland. Mills built in Tampere (1866), Mänttä (1868), Kuusankoski (1872), Kymi (1873) and Voikkaa (1873) are all examples of using strategic inland locations.

From the very beginning the industry had a controversial approach to the research behind its production. The machinery was imported from Germany and Britain, and the Finnish machine operators had sufficient practical knowledge to know how to run this equipment. After 1886 the industry financed the creation of a technical school in Tampere. At this point, those involved in the operation of Finland's papermaking equipment were not in need of formal knowledge because they had enough practical know-how. This was the rationale of the commonly feared and notorious Director of the Mänttä factory, G. W. Serlachius, nicknamed "The Devil of the Forest", who stated emphatically that there was no need for scientific education or research in the paper industry (Nykänen 2007b, p. 50). His opinion probably reflected his own background, whereby as a penurious young man he had no chance to obtain a formal education. On the other hand, Serlachius was later able to obtain a background in chemistry by earning a pharmacy degree.

3.4 In Independent Finland

The official testing of paper products in Finland started in 1908 at the State Material Testing Laboratory, which was attached to the Helsinki Polytechnic Institute. After Finland gained its independence in 1917, its leaders showed greater interest in developing the country's scientific capacity, including constructing the institutional infrastructure to conduct technological research. As part of this process, in 1922 the paper research division of the State Material Testing Laboratory was founded, but the laboratory was merely involved in efforts such as inspecting the physical features of paper products needed for state purposes. There were no resources for deeper scientific analysis of products. (Michelsen 1993, pp. 46–47; Komulainen 2014, p. 81) Incidentally, the State Material Testing Laboratory became the State Research Centre (VTT) in 1942.

Research into wood chemistry followed a totally different path. In 1916, when the effect of the First World War's economic restrictions started to fetter the operation of the pulp and paper industry, the paper companies founded a central laboratory (KCL) to search for substitutes for the raw materials that were difficult to obtain during the conflict. After the war KCL became strictly a material testing laboratory for the paper industry. KCL's standing remained precarious for a prolonged period, but ultimately its activities would fundamentally affect the field of wood chemistry. The idea behind the KCL was that all the partners in its operation would provide part of its basic funding and share in the results of its research. Some of the research projects were financed only by individual partners, and the results stayed in the possession of the respective financial backers.

After the First World War KCL was not the only co-operative organization the Finnish pulp and paper companies established. There were two cartel-like organizations, the Finnish Cellulose Union (Finncell) and the Finnish Pulp and Paper Association (Finnpap), which facilitated the cooperation of the various producers in exporting their products. In many cases the Finnish pulp and paper industry functioned as one family for several decades. For a long time, the independent research work conducted by the forest companies was focused on material testing. A typical actor was the chemical laboratory that was established by G. A. Serlachius, and it concentrated on studying the properties of cellulose and features of wood alcohol.[15]

Scientific publications in Finland have been traditionally administered by the country's scientific societies, and Finnish paper engineers founded an association of their own just before the First World War. Called the Finnish Paper Engineers' Association (PEA), it launched in 1917 a periodical magazine, *Suomen Paperilehti*,[16] and already during the 1920s it had published scientific articles in four languages; German was the traditional language of technology in Finland until the 1950s. Because the new markets for Finland's pulp and paper industry were now in the English-speaking world, the other languages in which the publications appeared were English, Swedish and Finnish (Komulainen 2014, pp. 86–87).

PEA's international connections were initially established with both its German sister associations and, most importantly, Nordic organizations. The British Pulp and Papermakers' Association (BPPA) created formal relationships with the association in 1937 by nominating the chair of the PEA's board, Helmer Roschier, as an honorary member.[17] Roschier had been the chief chemical engineer of Kymi-corporation since 1919. He was the second student to complete his Doctor of Science degree at the Helsinki University of Technology in 1918, and he was a well-known member of the turpentine researchers' school in Finland. Kymi's main commercial contacts were with Britain, and this was likely the reason why the connection to BPPA was established. Helmer Roschier started as the Professor of wood chemistry at HUT in 1939.

---

[15]See http://www.papermakerswiki.com/innovations/mets%C3%A4klusterin-tutkimuslaitokset-ja-aktiviteetit/ga-serlachius-oy. Accessed 6 March 2017.

[16]Later *Suomen Paperi-ja Puutavaralahti*, *Paperi ja Puu* (*The Finnish Paper and Timber Journal*).

[17]Komulainen states that the relationship was only nominal (Komulainen 2014, p. 87).

3.4.1 Formal University Education in Paper Technology

Paper engineering was introduced to the curriculum in technical education after the Helsinki Polytechnic School was transformed into the Polytechnic Institute in 1879. Paper manufacturing was included as part of the general lectures in mechanical engineering during the next decade, and they were delivered by German Rudolph Kolster. The pedagogical method of the era was that the lecturer simply read mechanical engineering encyclopedias, and he alone had responsibility over all the courses in mechanical engineering for two decades. Kolster's successor Max Seiling declared a few years later that the task was an impossible responsibility to fulfil (Nykänen 2007a, p. 40).

The first attempt to create a separate discipline for studies in paper engineering was made at the turn of the twentieth century. The limited resources of the Polytechnic Institute prevented this endeavour from succeeding, however, although in 1904 the financing of the Institute was almost doubled to allow for the expansion of engineering education. Special instruction in machinery used in the paper industry started in 1908, and it was delivered under the rubric of general mechanical engineering.[18]

The professorship in paper technology was established in 1921, although it took another decade until A. J. Brax was hired as the first professor of paper technology in 1931. Paper technology was considered a strategic discipline, established to guarantee that technical research was conducted in fields of national importance. The other ones were shipbuilding technology, agricultural technology and wood chemistry (Nykänen 2007a, pp. 183–185).

Probably the first master's thesis in paper technology was written by Kaarlo Amperla in 1927. The topic was "General planning of a printing paper factory" (in Finnish "Sanomalehtipaperitehtaan suunnittelu"). Although the topic was rather general in nature, the paper was a turning point in the Finnish paper machine industry's history. Amperla made his career in Kone ja Silta Oy (in Swedish AB Maskin & Bro), and he is considered to be the father of Finnish paper machine engineering.

Before the Second World War in Finland twelve domestic paper machines were built. The major manufacturer was Wiborgs Mekaniska Verkstad, which built eight machines, and the Karhula Works produced two in the 1930s.[19] The overall size of the paper machine manufacturing sector was considerably larger, however, because long delivery times for obtaining foreign machinery created orders for the Finnish machine shops, which delivered parts to the existing machines that had been imported. For example, Kone ja Silta Oy in Helsinki entered into a contract in 1937 with the British machine shop Walmsleys Ltd. at Bury, Lancashire, which was the main manufacturer of the machines delivered for UPM. At the beginning of 1938 Kone ja Silta Oy, (later Wärtsilä Co.), established a paper machine division, the first of its kind in the country. Under Amperla, it first produced two paper machines for the Kauttua mill before the Second World War (Nykänen 2005, pp. 57, 60).

[18]HUT, Program of Polytechnic Institute 1879–1908.

[19]Two machines built by Karhula to Pankakoski mill on 1910s were primarily board machines.

The Finnish cluster of technical research took a giant step forward in the late 1920s when the HUT launched its Technical Laboratories program. The general idea was to create facilities to conduct the technical research the young independent state needed to support its industrial activities. The most important and first ones were the electric and power laboratories, and they were operational between 1927 and 1931. A huge laboratory building was erected on the Helsinki University of Technology's campus near Hietalahti Harbour.

Professor A. J. Brax got one floor of the new facility for his pulp and paper technology laboratories. There was room for two large pilot-plants, one for a cellulose-drying machine and another for a small paper machine. However, there was enough money for only the former piece of equipment because it was considered to be the more important of the two. After its delivery the university ran out of money due to the general economic depression. The researchers would wait for the pilot paper machine for several decades.

This delay did not have a major effect on the Finnish machine shops, however. By the mid-1930s, they were ready to start manufacturing modern, wide paper machines. At the time the Finnish machine shops lacked the expertise to start the effort from the ground up, and so all the major machine shops began their operation by depending upon foreign know-how. For example, Wärtsilä Co. planned to start with printing paper machines designed by Walmsley, and Tampella (Tampereen Pellava- ja Rauta-Teollisuus Osakeyhtiö) co-operated with widely respected American machine shop Black-Clawson to obtain the drawings for cardboard-making machines. The Finnish machine shops made an agreement over the division of products they would make, and the accord was named TAMAVAKA based on the initials of the corporations (i.e. Tampella, Maskin & Bro, Varkaus, Karhula, TAMAVAKA) (Nykänen 2005, pp. 58–61). The outbreak of the Winter War between Finland and the Soviet Union on 30 November 1939 interrupted these plans.

3.4.2 War Time Crisis and Tar Oil Production

The Winter War cut Finland off from its international connections. For the Finnish economy, the war was catastrophic, and only the quick end of the fighting on 13 March 1940 rescued the nation's economy from collapse. The industry had little chance to react to the situation, and the scientific work totally stopped because all the country's human and material resources were committed to the war effort. For the machine industry, the effects of the wartime economy lasted until 1952.

After the Winter War ended Finland keenly tried to re-connect to North American countries. For example, the Finnish Paper Engineers' Association applied for membership in the Technical Section of Canadian Pulp and Paper Association and the Technical Association of the Pulp and Paper Industry (TAPPI). The outbreak of the so-called Continuation War in the summer of 1941 ended these initiatives. Finland entered into a formal alliance with Germany against the Soviet Union, and this caused Britain to declare war against Finland in 1942. Nevertheless, the US and Finland had no open hostilities against each other during the war, which facilitated re-establishing relationships between Finland and the West after the war.

During the Second World War Finland was cut off from international trade, notwithstanding the goods Finland traded with Germany during the Continuation War (1941–1944). The most difficult challenge for the national economy was its lack of lubrication oils and greases, and Sweden shared this same problem. As a result, in 1941 Finland and Sweden launched a co-operative project to produce substitutes for lubrication oils from pine tar. The effort was based on the long-standing turpentine chemistry and wood processing industries. The know-how and scientific basis for the project were delivered from KCL in Helsinki, although a significant contribution also came from the Swedish side of the project (Nykänen 1999, p. 264).

For strategic reasons, the tar oil production in Finland was divided among several small factories. In 1944 Finland boasted 32 modern tar mills and four oil factories that produced tar oil. One of the major facilities was the Forsiitti-Dynamiitti Oy in Hanko, where young engineer Waldemar Jensen was trying to solve production problems (Nykänen 1999, pp. 268–269). After economic relations between Finland and Germany collapsed and all imports of petroleum based products ended in the summer of 1944, Finland was totally dependent on domestic tar oil production. The situation lasted until 1948, when the import of crude oil products started again from Britain and the US. The shortage of petroleum products during the war had clearly demonstrated the vulnerability of the Finnish national economy if any crisis closed international trade. It took nearly ten years before Finland learned to handle this problem effectively.

3.5 A New Beginning After the War

Before the Second World War the master's theses of the students at the HUT had addressed industrial issues only on an abstract level, and for good reason. It related to an intellectual battle that was being waged in Europe between general and technical universities, and it underscored the gaping chasm that separated traditional from modern approaches to higher education. The former focused on delivering a conservative, classical education in which theory was front and centre and practical matters were not considered worthy of study; the latter reversed these priorities.

The challenge for the students and professors at the university was the fact that the conservative, classical paradigm reigned supreme prior to the mid-1940s. This meant that their research work had to concentrate on theoretical issues and shy away from solving practical problems; otherwise, they would be subject to intense criticism. This difficulty was first widely discussed by Paavo Pero, who became a leading figure in shaping the education of the applied sciences in Finland (Nykänen 2007a, pp. 111–115).

The situation changed profoundly after the war. Finnish industry was pressed to its limits by the war reparation demands from the Soviet Union and the needs of the domestic economy. Students of technology came to occupy a central place in addressing both these demands. During the five years of war they had had little chance to complete their studies, but immediately after the conflict ended the number of M.Sc. Engineering degrees granted in Finland rose to an all-time high. In fact, the number of new engineers in the country in the mid-1940s was almost double the number that there had been before the war.[20] Many of these persons had been identified to work in the industry straight from the halls of the technical university.

The result saw many of the students' research topics focus on solving specific, practical industrial problems. Already in 1946 B. Immonen presented a final work concerning the post-processing of products at the Mänttä paper mill in Central Finland. The next year L. V. V. Sundström presented a work about the wire section of a rapid printing paper machine. The first generation of students, who started their studies after the war, finished their educations in 1951, and four of them had worked on topics concerning the production of paper. Clearly, a connection between the industry and the HUT had been established on a much higher level.[21]

One of the newly promoted engineers was Niilo Erik Ryti,[22] who earned his degree from the HUT as a machine engineer on 7 December 1944. His major was industrial technology, and almost immediately he started work as Production Engineer at the Kangas paper mill in Jyväskylä, Central Finland, and which was owned by G. A. Serlachius Oy Ltd.[23]

After the end of the Second World War major changes occurred in the way master's theses were finished at HUT. With Finland in dire need of technical capacity, the formal master's theses were often based on serving the industry's most urgent needs. The university's professors were also recruited to the Board of War Reparations Industry, called SOTEVA. A long-lasting discussion over the division between the theory and practice concerning technical education and also the crucial question of co-operation between HUT and industry was solved once and for all because of the crisis over the need to maintain Finnish independence, which was threatened by the onerous war reparation demands from the Soviet Union. The country had been put into a practically impossible situation by the Soviet Union's demands for war reparations (e.g., they included giving the Soviets 4 entire paper mills and 20

[20]Eeva Pitkälä. Statistics over the degrees and students at HUT. 2010. HUT archives.

[21]HUT, the minutes of the wood processing department. Anders Lund, Torkning av papper på en Yankee-cylinder, 16 April 1951; Matti Vihinen, Voimapaperin kreppaus, 10 September 1951; Jori Pesonen, Sanomalehtipaperikoneen tuukiväli laboratoriotutkimusten valossa, 19 November 1951; Arvo Reipas, Äänekosken paperi- ja kartonkitehtaan laajennus-suunnitelma, 21 May 1951; Nils Lindberg, Över arkformeringsproblemet i allmänhet och speciellt skagningens inverkan på arkformering.

[22]Niilo Ryti was a son of Risto Ryti, Finland's Second World War-time president (1940–1944). His mother was Gerda Serlachius, whose grandfather was G. A. Serlachius, founder of the Mänttä Paper Mill. After his graduation in 1914 Risto Ryti founded a law firm with Eric Serlachius, brother of Gerda, and this led to his interest to the wood-processing industry. Both of his sons became professors at the Helsinki University of Technology; Henrik Ryti was later a professor of thermodynamics and machinery.

[23]HUT, a name record, Niilo Erik Ryti.

additional paper machines), and adopting this new approach to technical education was essential if Finland were to survive.

3.5.1 Two Paths for Research and Development

It was clear that, during the 1950s, the science of papermaking in Finland was starting to undergo a slow but steady and dramatic change. The traditional paper consumers did not provide much of an impetus to the mills to develop new products, and so the long-established craftsman-like thinking and procedures remained entrenched. Two crucial developments fundamentally altered this situation, however, namely the competition between the international printing houses and the new approach to photojournalism. Demand for printing paper expanded when magazines like *Time* and *Life* spread all over the world, and the publishers started to seek better quality paper on which to print their products. Also, the increasing speed of papermaking machines and printing presses necessitated producing paper products that were more uniform in quality. Furthermore, growing demand for printing paper led to competition between the machine manufacturers, who had to address the need to produce paper that possessed greater printing capacity.

Need for better quality in printing paper required pilot tests with paper machinery capable of being used in trials that involved innovative constructions and special runs. The experiments were traditionally run with production-scale machinery, but this was getting to be too expensive after the speed and size of the machines had grown so quickly. The problem was solved by building laboratory-scale pilot paper machines for paper machine manufacturers' research and development programs.

Their laboratories really began emerging during the 1950s. In Britain there had been pilot machines already at the beginning of the century. T. J. Marshall built in 1906 a miniature paper machine that conducted the entire process. Also, Manchester University had a small paper machine for educational and research purposes by the end of the 1930s (Komulainen 2014, p. 111).

The atmosphere in the research climate changed drastically during the 1950s. American Beloit was the first to build a special research paper machine for developing new and better processes. In Finland, the idea of testing new types of machines was based on the close contact between the machine shops and the paper mills. It was quite normal to use the mill's narrow paper machines for testing the new design, but usually the speed of the machines in these tests was restricted. The Tervakoski paper mill in southern Finland was the first one to construct a narrow pilot plant for testing the processes in the 1950s, and it used parts of several old machines for the project (Nykänen 2005, p. 149).

Finnish machine shops began conducting laboratory tests during the 1960s, which is quite late in comparison to similar developments in other countries. Alhlström, the state machine shop, and Valmet and Tampella corporations constructed pilot plants that were designed to develop papers with different printing capacities, and during the 1970s research and piloting had become a standard procedure in development work

(Nykänen 2005, pp. 199–207). The HUT received its long-awaited pilot machine in the 1970s when it was finally built as a co-operative project with the university, the KCL and VTT.

The late emergence of pilot plants in machine shops did not necessarily mean that the development work was totally absent from the paper industry in the 1950s and 1960s but these developments are difficult to discern from archival sources. Nevertheless, one can trace the history of two typical projects in the archives of the Kangas paper mill. The first, a co-operative project that occurred in 1955, was the result of Valmet offering a new paper machine to the Kangas paper mill. After reviewing the product specifications sought by the customers, Valmet produced preliminary drawings for a new machine.

Engineer Ryti noted the high quality of engineering of the planned structure, but from the plans he also pinpointed problems in how the paper was formatted and dried (i.e., how the fibres were arranged). The close relationship between the customer and machine shop produced fruitful result that affected numerous paper machine projects in the future.[24]

It is difficult to discern how a new paper product is developed because the demand for it is usually outside the scope of the paper mill, and thus the archival sources are difficult to locate. But one example has been uncovered, specifically the development of a paper that was suitable to be used as insulation in electric transformers for the Dutch company, Willem Smit Transformatorenfabrik N. V., in the late 1950s. This paper had very exact specifications, because it if were too thick it would prevent the transformer from working properly. The project started in 1955, when Willem Smit's engineers first laid out the technical requirements for paper that could act as an insulator in building a new transformer. Four years were allotted for the research and development work, and production of it was scheduled to begin in 1960. Because the Kangas paper mill had already been producing transformer paper for Willem Smit, the company turned to its known supplier. After several months of testing, the first samples of the new kind of paper were sent to Holland, where the machine shop could start the tests on a large scale with the new product.[25] Though the result was a brand-new type of insulating paper for transformers, and the research and development process took several years and consumed many resources in two technical laboratories, it did not represent a specific innovation. Instead, the end product, a new type of transformer paper, merely represented the refinement of an existing product.

---

[24]ELKA, GAS, memorandums of Engineer Niilo Ryti 1950–1960, paperikonetarjoukset (in Finnish), 22 June 1955.

[25]ELKA, GAS, memorandums of Engineer Niilo Ryti 1950–1960, a memorandum of Engineer Frowein (representing Willem Smit Transformatorenfabrik N.V., Nijmegen, Holland), regarding his visit at Kangas paper mill on 21 November 1956.

3.5.2 Rebuilding the Connections

After the Second World War the Finnish war reparations administration, SOTEVA, started to work on solving the problems involved in providing the Soviet Union with the goods it had demanded. Because the list of demands consisted of several paper machines that could not be produced in Finland, SOTEVA turned to western manufacturers that, prior to 1948, had delivered several large items to the Soviet Union as Finnish orders. One of the manufacturers was the well-known American machine shop Bagley & Sewall. The western machine shops were soon caught in the political games being waged between the super powers and also pressed by the need to rebuild European industry. When the Cold War began in earnest in 1948 the western producers refused to trade machinery with the east. The only possibility for Finland to survive in this situation was to create a domestic paper machine-making industry.

Kaarlo Amperla from Wärtsilä had first proposed this possibility in the autumn of 1945. The next year Finnish industry prepared to develop and produce Finnish paper machines. The Karhula machine shop and the Wärtsilä-corporation seized the moment, while the state machine shop Valmet also started to investigate both the possibility of producing paper machines and improving the efficiency of this production. Still in this new situation the TAMAKAVA agreement from the 1930s over the division of production that steered the plans of Finnish machine shops was still strictly followed. Tampella in the first place did not initially conceive of producing the entire machines, and instead aimed to make only the auxiliary machinery for Bagley & Sewall machines (Nykänen 2004, p. 88).

Beginning in the late 1940s, countries in Western Europe used the financial and material aid delivered by the US under the Marshall Plan to rebuild their infrastructure. Finland officially stayed out of the Marshall Plan, and since it was not a member of the NATO, many considered Finland as part of the Eastern bloc. But the reality was very different. Relationships between Finland and the West were actively built on many levels, and some of them were clandestine or nearly so. In fact, many organizations in Finland received unofficial but meaningful economic support from the US. This aid was often hidden behind the aegis of normal trade whereby surreptitious aid was delivered to the Finns (Nykänen 2008). Although Finland was wooed by both sides in the Cold War, it ended up on the western side of the Iron Curtain near the beginning of the Cold War.

Contacts among universities were an essential part of the relationship that developed between Finland and the West, and this was integral to the exchange of ideas and information. A major first step in this process occurred in the spring of 1948, when the Massachusetts Institute of Technology (MIT) invited six to eight Finnish graduate engineers to pursue a course at this institution in an effort to further their education.[26]

---

[26]HUT, OK 23 March 1948, § 16. The grants were given to Olli Pöyry, Juha Alhojärvi, Pekka Rekola, Osmo Korvenkontio, Bjarne Huldén, Nils Björklund, Leo Toivonen, Stig Landgren, Mauri Tanttu. About Risto Hukki Nykänen 2009, p. 97ff.

An important player in these events that involved reconnecting Finland to western economies was the PEA. During the 1930s PEA had been a leader in fostering relationships with the British profession of paper engineers, and had concluded a formal relationship with the Technical Section of the Paper Makers' Association of Great Britain and Ireland in 1937. The relationship fell into abeyance during the war, but in 1948 the organizations found each other again, and the representatives L. G. Cottrall and J. Chapman of the Technical Section took part in PEA's annual meeting that spring in Helsinki. Also, Bertil Nybergh from the KCL travelled to London to visit the Technical Sections meeting, at which he presented a paper.[27]

In 1949 the connections between Finland and the US were formalized through the execution in the US of Public Law 81–265, which created a remarkable situation. Finland had been the only country that had repaid in full the loans it had received from the US during the First World War, and it had gained the respect of the American government as a result. Even while Finland had been fighting the Winter War, it had continued to deliver its payments to the US. Curiously, the US Treasury had collected these funds in a special account and had not simply deposited them into its general revenues. The aforementioned law that the US had enacted in August 1949 allowed the money the Finns had repaid the US after the First World War to be used as travel grants, thus creating the financial resources to support a massive student exchange between Finnish and American universities. The program in Finland was named Amerikan Suomen Lainan Apurahaohjelma (ASLA). In addition, these funds were used to purchase scientific instruments and the latest literature for the Finnish universities. Then in 1951, Finland joined the Fulbright program. Through both these programs, between 1949 and 1957 Finland sent 419 scholarship holders, 65 graduate engineers, 130 visitors and 5,454 interns to the US to further their technological educations (Teknillinen apu 1958; Hietala 2002, pp. 538–539).

Clearly, the demands to pay war reparations and of the Cold War forced Finland to modernize and greatly expand its machine-making industry. The major problem it faced in doing so was its lack of research and technical development capacity. Before the 1940s very few Finnish machine shops dared to compete with major foreign producers in building large industrial machinery. Nevertheless, during the Continuation War Finnish engineers were forced to solve problems that had previously seemed impossible, including those in the field of aviation. The State Aircraft Factory (in Finnish Valtion Lentokonetehdas, VL) educated some 500 competent constructors (i.e., engineers who create new technology) during the war for the engineering industries in Finland. After the war VL became part of Valmet. This was the major boost that made it possible to establish the Finnish paper machine-making industry on a large scale after the war.

On 19 November 1949, a very important meeting was arranged by the Paper Engineers Association (PEA) in Helsinki. For the first time in the history of the paper industry the PEA called a meeting of both the paper makers and paper machine manufacturers in Finland. The main subject that was discussed dealt with the ability of domestic machine shops to fulfil the needs of the pulp and paper industry both

[27]ELKA, PEA, Francis Bolan to Torolf Lassenius on 24 March 1948; 21 April 1948.

in the present and future.[28] This subject anticipated the time when Finland had paid its war reparations to the Soviets, and it was free to start marketing its industrial products to the western world again.

After Finland had paid its war reparations in 1952, the industry enjoyed a boom from the Korean War and the rising international demand for pulp and paper. But after the conflict ended, the Finnish paper industry again seemed to be falling behind its international competitors. It was given a huge boost in 1957, however, when the Finnish Mark was devalued by 30%, and a new era of investment began in the Finnish paper industry. The other reason that Finland was in dire need of modernizing its pulp and paper research and development program was the need to improve the quality of its products. During the Second World War, its international competitors—especially those in the US—had developed new, high-tech products for military purposes and conducted significant research into improving production. As a result, research and quality control had become an elementary part of production in those countries. This gave them a major advantage over paper mills in northern Europe, where production still relied heavily on "the art" of papermaking.

Modernization in the Finnish paper industry began at the Kangas paper mill in Jyväskylä in the early 1950s. Niilo Ryti started a project to modernize the mill's laboratory facilities immediately after his return from the US in 1949. With assistance from the KCL in Helsinki he furnished a material testing laboratory during the early 1950s.[29] The purpose of the investment was clear: the Kangas mill was focusing its production on high-quality writing and specialty papers. One of its main products was the cardboard for punch cards used in computer technology. By the end of the decade, IBM and Remington Rand were buying a great deal of their supplies of punch cards from the Kangas mill. It also manufactured a number of specialty products such as waxed paper.

Kangas mill was a pioneer in R&D work, but in general the attitude in the industry was becoming more geared toward this type of activity. In the late 1940s long-distance travel was quite expensive, and only a few Finns had the funds to cross the Atlantic to study the modern R&D laboratories in the US.[30] The status of the paper engineers was quietly discussed among the PEA in the late 1950s, when the effect of the currency devaluation in 1957 was having a major impact and Finland was at the peak of its huge investment boom in its paper industry.

The international connections enjoyed by the Finnish the paper industry created the possibility for it to make several inquiries about the differences between the Nordic producers and those in the West. Finnish salesmen and leading engineers could travel much more readily to Britain and North America to see the "modern ways" of production. Also, foreign officials began to visit the Finnish industry, specifically forest cities such as Mänttä, Jyväskylä, Tampere and Valkeakoski.

The exchange of information related to pulp and paper technology moved in both directions. This was clearly noted when Charles F. Payne from the Eastman Kodak

[28]ELKA, PEA, a circular, Torolf Lassenius, 6 November 1948.

[29]ELKA, GAS, memorandums of Engineer Niilo Ryti 1950–1960.

[30]E.g. ELKA, GAS, travelling documents, 1943.

Company visited Niilo Ryti at the Kangas mill in August 1957. Payne wondered how the Finnish and Swedish paper mills could operate successfully even though they employed so few engineers and researchers. Payne noted that a Swedish paper mill typically employed two engineers: a technical senior expert and his younger assistant. In contrast, in Rochester, New York, the Kodak paper mill had a research and development division with 35 scientists who were all formally educated.[31]

The need for further education of younger staff was clear, but in practice the recently graduated engineers and scientists were not able to take the time to travel abroad to enhance their professional development. In 1962 the secretary of the PEA, C. J. Ganszauge, made a study of the role of the paper engineers in the state bursary systems targeted to further studies overseas. The result was depressing. Of the Fulbright bursaries, only four of 84 had been received by paper engineers since 1949. E. Aaltio, Lea Valtasaari, Y. Hentola and R. von Konow had deepened their scientific knowledge after their researchers' degree (i.e., a licentiate, or first degree for a researcher between a masters and doctorate in Finnish universities). Of the more important ASLA bursaries that supported graduate students furthering their educations by gaining practical and work-related experience, only four of the total 416 had been received by paper engineers. These were L. Hummelstedt, T. Ulmanen, A. Esilä and K. Sormanto, who had continued their studies in the US.[32]

Although Finland was attempting to make the pulp and paper industry the backbone of its national economy after the war, the situation was still in dire need of improving. It still lacked researchers, both in terms of process engineering and machine construction. One of the major problems in terms of both formal and practical education was that the country's rapidly expanding industrial sector was aggressively recruiting engineers with M.Sc. degrees, and so they were in high demand. This meant that these young professionals were able to obtain well-paid jobs in industry without undertaking further education. (Komulainen 2014, p. 171) Nevertheless, the builders of the large technological systems, who were looking to the future of the paper industry, were in dire need of engineers whose education had gone beyond the M.Sc. stage.

3.5.3 Over the Atlantic Ocean

One of the few paper engineers who had personal connections overseas and real possibilities to travel to international conferences was Niilo Ryti, who was appointed the director of the Schauman's Pietarsaari pulp mill in 1962. Ryti took part in the Oxford Symposium in Britain in that year. After returning home, Ryti wrote a letter to the PEA stating that something should be done to improve research resources in

---

[31]ELKA, GAS, memorandums of Engineer Niilo Ryti 1950–1960, a memorandum of Engineer Charles F. Payne, Asst. Superintendent, Paper Mills Division, Eastman Kodak Co, Rochester NY regarding his visit at Kangas paper mill on 27–28 August 1957.

[32]ELKA, PEA, C. J. Ganszauge to Niilo Ryti on 25 August 1962.

Finland; otherwise, he argued, the result would be disastrous. (Komulainen 2014, p. 172) Another Finnish paper engineer dealing with the same problem was Tapio Ulmanen, who was working for the KCL.

An ideal contact was the Institute of Paper Chemistry in Appleton, Wisconsin. It had been established in 1929 as a partnership between Lawrence College and the local paper industry. It was designed as an institute in which students could gain specific practical and formal training on the road to becoming paper chemists, and it was organized and administered by the college but financed by the paper industry. The Institute had its own Board of Trustees and its own budget. Lawrence College awarded Master of Science and Doctor of Philosophy degrees to graduates of the Institute.[33]

The Finns knew they needed a liaison between themselves and the Americans during the Cold War, and specifically a go-between who could help establish a relationship between the Finnish paper industry and its engineers and the Institute in Appleton. Niilo Ryti asked Professor Waldemar Jensen, CEO of KCL, to try to be that contact with the Institute, because Jensen personally knew Dr. Roy Whitney, the Institute's vice president of academic affairs. Whitney responded quickly but guardedly. He could not promise anything but was open to the possibility of Finnish students applying for spaces at the Institute. The problem for the Finns was the cost of attending the school. One term cost about 2,500 USD, and for a Finnish student or young M.Sc. that represented an astronomical sum.[34] The task for the PEA's administration was thus to find the financing for the interested students. To realize this goal, three older funds in the PEA's custody, namely Uno Albrecht's, Frenckell & Son's and Georg Holm's, were brought together to form a new travelling grant for young scholars.[35] In 1963, the first Finnish student attended the Institute, with financial backing from the PEA.[36] The aim of the PEA's program for further education was clear, because the Institute in Appleton dealt exclusively with wood chemistry, as did most of the Finnish fellows.

3.5.4 Engineering for the Industry

The art of engineering in the Finnish paper industry followed a different path than it did for wood chemistry. Jaakko Murto succeeded Helmer Roschier as the Professor of wood chemistry at the HUT in 1959. He found a partner in Jaakko Pöyry, who had graduated from the HUT in 1948. Pöyry was first recruited straight from the university for Wärtsilä Company as a construction engineer in the war reparations

[33] Lawrence University Archives (LUA), Seeley G. Mudd Library, Appleton, Wisconsin, Institute of Paper Chemistry Records, LU-RG10-001.

[34] ELKA, PEA, Niilo Ryti to C. J. Ganszauge on 10 April 1963.

[35] ELKA, PEA, C. J Ganszauge to Niilo Ryti on 16 April 1963.

[36] Appleton WI, PRI Archives PRI Board of Trustees minutes from 1963–10–03. Dix correspondence on 12 November 2015.

industry. In 1958 Jaakko Murto founded an engineering company with Jaakko Pöyry, and it specialized in paper industry engineering. From Wärtsilä Pöyry also took with him Matti Kankaanpää, who specialized in the design of paper machines; specifically, his work involved incorporating scientific principles into technical drawings. Kankaanpää was soon sent to work at Beloit in the US, however, and stayed there until 1963. During that period, he was involved in designing and constructing huge paper machines for UPM's mill in Kaipola and the Veitsiluoto Company. Murto left the company in 1961, and it was renamed Jaakko Pöyry Oy. The company became the world's leading engineering company in the forest industry during the 1960s, and Matti Kankaanpää became its vice president in 1963 (Oinonen 2002).

Also during the 1960s the HUT, Åbo Akademi, and the new Technical Department of the University of Oulu were constructing new links to the world's wood processing industry. The university now had a dedicated professorship for paper machine engineering, and the first person appointed to the post was the former chief of Valmet's Rautpohja machine shop in Jyväskylä, Uolevi Konttinen, in 1969 (Nykänen 2005, p. 79).

These developments were enormously important to Finland's paper-machinery making industry. By the latter part of the decade, Finnish machine shops were able to break into the highly competitive North American markets and by the 1980s they were the undisputed and dominant leaders in the world market (Nykänen 2005, pp. 180ff). Having mastered the R&D part of the equation, the Finns increasingly focused on refining their machinery.

3.5.5 Building the Research Organization

A real turning point in the research and development culture of the Finnish paper industry occurred during the early 1960s. The decade saw the different research organizations and industries form a quasi-official network that fostered rapid advancements in many branches of the industry. The key figure in the network was Niilo Ryti, nominated to be the Professor of paper technology at HUT at the beginning of 1963. His appointment meant that he would be at the nexus of a network of all the organizations working with the pulp and paper industry. Ryti's right hand man was Pertti Aaltonen, who worked for the KCL in the 1950s and was nominated to be Ryti's assistant in the spring of 1964. Aaltonen's task was to convert Ryti's dreams into reality. His working time was divided evenly between the HUT and Jaakko Pöyry Oyj, so he formed a true connection between the industry and the university.[37]

One of Ryti's first tasks as a professor was to establish a link between the HUT, the KCL and the PEA. They were all needed to launch the scholarship program to educate young research oriented engineers in the paper industry. Ryti's role in the program was to select the applicants for the Paper Engineers' Association. The

[37] HUT, a name record, Pertti Aaltonen; Hannu Paulapuro, emer. prof. Helsinki, interview, 1 April 2015.

number of students of paper technology at the HUT was annually around 45 after the rapid expansion in the number of freshmen at the beginning of the 1960s. He hinted to some of the newly graduated students to apply for the stipend to travel to the Institute in the US to study.[38]

The first scholarship holder to study at the Institute of Paper Chemistry in Appleton was Tapani Kaila in 1963. He completed his Ph.D. degree at the Institute, and this would prove to be common for Finnish students who attended it. Thereafter, the PEA sent one student yearly. The next ones were Matti Stén and Kari Ebeling (Komulainen 2014, p. 302).

Then a change in the culture of Finnish paper engineers occurred during the 1970s. Those belonging to the "old school" had recently begun retiring, and the new class of more research-oriented engineers gradually took over their responsibilities. One great benchmark was passed when the last of the great directors, Juuso Waldén, retired in 1969, and Niilo Hakkarainen took his place as CEO of UPM. As the firm's new leader, he led it through a period of massive innovation. The result saw industry now searching for data and research prior to making a decision about its operation; a new era of empirical-based decision-making had begun.

R&D in the factories that produced machinery for the pulp and paper industry was developing at the same pace. Until the 1950s the machine shops were completely dependent on feedback from the paper industry in developing new methods and equipment. Paul Ohlström from the Karhula works wrote in 1954:

> The launching of new inventions and constructions on the market absolutely requires the support and cooperation of industry. After we have made, installed and started up these innovations, we are forced to adjust some small details and therefore the cooperation of industry is extremely important. A new construction will, in any case, have been so carefully designed and tested that, when it is introduced in an industrial process, only a few small details remain to be adjusted. (Ohlström 1954, p. 200)

The 1960s thus saw the evolution of industrial research on two fronts—the R&D carried out by the machine shops themselves and the R&D that was dependent on the technical universities and the handful of university laboratories that were developing papermaking technologies to generate new knowledge. In addition, the machine shops established their own R&D laboratories in the late 1960s. Research facilities and laboratories were also built by Tampella at Inkeroinen, by Valmet at Rautpohja and Jyväskylä, and by Ahlström at Karhula (Nykänen 2005, pp. 196–204).

A huge problem for the research laboratories in the paper and the machine industries was a lack of venture capital. UPM had invested heavily in basic production technology to raise the quality of its products. Many parts of the "secondary factory" activities, however, like handling materials and waste-water treatment, were left behind. The investments also aimed to increase employment levels and benefit the national economy.[39] The government tried to back up the progress and the Bank of

[38]Komulainen (2014) states that the low number of applicants was due to the lack of interest in participating in the program, (p. 173) but clearly some pre-selection of applications had already been done at the university before they were sent to the PEA. (Antti Arjas. Ret. Gen. Director, KCL. Espoo, interview, 14 April 2015).

[39]UPM, memoirs of Niilo Hakkarainen on 14 June 1971.

Finland decided to devalue the Finnish Mark by 23.8% in 1967 in an effort to boost export industries, but by the end of the decade UPM was mired in deep debt because of its heavy investments.[40]

3.5.6 *Time for Innovation*

Near the end of the 1960s, it became increasingly common for Finnish R&D projects to be supported by public funding. Such possibilities had arisen for economic and regional development reasons, and the same phenomenon happened in all industrialized countries.[41] A prime economic funder was SITRA, the Finnish National Fund for Research and Development, which had been founded in 1967; it soon became a major source of R&D funding in Finland. In addition, the 1970s also saw the establishment of TEKES, the Finnish Funding Agency for Innovation.[42]

The emergence of the national R&D policy eventually fuelled a larger discussion over the need for research and development organizations and the innovation processes. In 1970 a major conference over the problems involved in facilitating innovation was arranged. The Finnovation 70—seminar[43] was organized by the Asko Foundation, and the main topic was how best to coordinate all the work that was going on. Since the mid-1940s Finland had seen significant R&D, but these efforts had largely been conducted independently of each other, and the goals now were to continue to foster a culture of innovation as well as develop a systematic approach to R&D. One of the organizers of the conference was Professor Erkki Laurila, an academic who had been steering Finnish technology policy since the early 1960s.[44]

The new approach to the innovation processes soon affected the paper industry. As a new policy, Niilo Hakkarainen, the newly appointed CEO of UPM, was well versed about the Finnovation meetings and he launched a program to modernize the information organization around the company's board. UPM's mills thus received special research, development and planning units, and a research division was created within the technical department of its central administration in 1971. UPM had high hopes regarding its decision to take a modern approach to research and development. The company appointed Antti Arjas as director of its research organization, and

[40]UPM, memoirs of Niilo Hakkarainen on 21 January 1970.

[41]Of the early discussion over innovation processes, see Esilä 1968.

[42]A general view for the Finnish technology policy, see Murto et al. 2006.

[43]Before the Finnovation 70—seminar was Innovation 68—seminar arranged by a group of young engineers and architects in Helsinki. This is probably the first meeting discussing over the innovation policy in Finland, and it was funded by newly established SITRA (Jaakko Ihamuotila, oral communication on 28 February 2017). The history of Finnovation seminars is under research by the author and Sampsa Kaataja.

[44]UPM, memoirs of Niilo Hakkarainen.

Ingmar Häggblom, who already had lengthy experience leading its research activities, stayed on to continue working in this field with the company.[45]

A similar process occurred with the firm G. A. Serlachius. Its new laboratory facilities in Mänttä[46] worked with samples from all the firm's mills, and it became a major site for the preparation of master's theses under the auspices of HUT's Department of Paper Technology. The Mänttä laboratory also had close connections to other research organizations, like the State Research Centre, VTT and KCL in Espoo (where it was relocated from Helsinki in 1972).[47] Niilo Ryti from HUT was behind these co-operative efforts. Although the mill in Mänttä already boasted a material testing laboratory, it established a separate R&D department in 1971. It gathered all the separate laboratories under the same umbrella organization, and in 1974 it was given a modern facility in Mänttä.[48]

The need for research in the pulp and paper industry came largely from outside the industry itself. The first reason was the growing population in the developing world and its increasing appetite for paper. In addition, there was an increasing awareness of the global environmental issues during this period. Furthermore, increasing competition in the printing paper market made it clear that challenges were coming.[49] In terms of environmental issues, discussions over wastewater conditions had already started in the 1930s in Finland but it was not until the late 1950s that the first meaningful public consultations about the problem began. One manifestation of this development was a conference over water treatment that was conducted in Mänttä at G. A. Serlachius' paper mill.[50]

There was also an interest in conducting research into producing food as a by-product of the sulfite pulping process. During the war experiments had been conducted into using torula yeast as a fodder, and KCL developed in the mid-1960s a method to grow protein using *Paecilomyces varioti* bacteria. The process was named Pekilo and patent rights to it were reserved by the Tampella. In conjunction with UPM, it established a co-operative pekilo factory in 1972 in Jämsänkoski. Pekilo protein, which the yeast had produced by converting harmful substances to valuable ones, was used for animal fodder, but there was another benefit to this activity as well. In the process of recovering the protein researchers discovered that the conversion

[45] Antti Arjas. Ret. Gen. Director, KCL. Espoo, interview, 14 April 2015; UMP, Häggblom archives, Oy Mec Rastor Ab to Ingmar Häggblom on 7 February 1974.

[46] G.A. Serlachius had already enlarged material testing laboratories in the 1950s at Mänttä Paper Mill.

[47] See http://www.papermakerswiki.com/innovations/mets%C3%A4klusterin-tutkimuslaitokset-ja-aktiviteetit/ga-serlachius-oy. Accessed 6 March 2017.

[48] See http://www.papermakerswiki.com/innovations/mets%C3%A4klusterin-tutkimuslaitokset-ja-aktiviteetit/ga-serlachius-oy. Accessed 6 March 2017.

[49] Of the environmental issues impact see e.g. UPM, Hakkarainen archives, E. Gray King, research Director, Georgia Pacific to Osmo Aho on 13 November 1972; see also Kaje 1968.

[50] See http://www.papermakerswiki.com/innovations/mets%C3%A4klusterin-tutkimuslaitokset-ja-aktiviteetit/ga-serlachius-oy. Accessed 6 March 2017.

process had also decreased the harmful substances in the effluents from pulp mills.[51] Pekilo production was discontinued at the beginning of the 1990s, however, when the production of sulphite pulp ceased because of environmental concerns.[52]

The sulphite process was being intensely debated already in 1975, when the ministry for agriculture and forestry collected information from the sulphite industry's daily operations. UPM declared then that the production of sulphite pulp was already diminishing because of the environmental issues it created. Nevertheless, the industry tried to slow down the transition to the sulphate pulping process by justifying its sulphite operations on the grounds that they generated strategic products such as alcohol and proteins.[53]

A general change in the role of the paper industry was on its way. The first steps in the globalization of the world economy were taken in the early 1980s. In a presentation in October 1981, Niilo Hakkarainen emphasized the value of the general understanding of technology as part of human society's evolution. The previous decade had seen a wave of research laboratories created by UPM. Progress had been rapid, and the discussion over the role of the R&D operations was well under way.[54]

At the same time, technologies involved in industrial processes experienced a profound change. The emergence of automation and control technologies slowly began creating the possibility of improving the quality of production in Finnish paper mills. They had first been introduced at the beginning of the 1950s, when Erkki Laurila had taken the initiative to begin producing control systems for paper machines. This effort did not succeed, however, and it was only in the 1960s when the modern inspection and control devices were attached to high-speed paper machines. A clear indicator of the rising importance of the automation technology was the discussion over the new technology's advantages and restrictions, a debate that really began in the early 1980s when Antti Arjas unleashed a public controversy (Arjas 1982). At that time major innovations in paper machine technology were already well underway, and computers were now powerful enough to manage the complex tasks of process control. As a result, automation and control technologies were still considered in the 1960s and early 1970s to be supplementary technologies in running a paper mill, but became leading factors by the 1980s (Nykänen 2005, pp. 90–94).

After Niilo Ryti had departed the scene, Hannu Paulapuro became the person who connected the different organizations. He took over responsibility for the R&D work conducted at HUT and education in the new global situation. This passing of the torch mirrored the turning point in Finland's research and development environment. Paulapuro was promoted from the international tradition that Ryti had initiated two

[51]UPM, Hägglund archives, e.g. Osmo Aho to Reijo Salminen, Georgia Pacific, Bellingham, Washington, USA on 25 April 1973.

[52]See https://www.papermakerswiki.com/innovations/sivutuotteet/pekilo-proteiini. Accessed 13 February 2017.

[53]UPM, Hägglund archives, UPM to ministry for agriculture and forestry on 7 April 1975; see also PM 19 March 1976; 13 April 1976 of the sulphite production. PM Sulfiittiteollisuus maa-ja metsätalousministeriölle 6 February 1975; *Talouselämä* 7 November 1975.

[54]UPM, Hakkarainen archives, Niilo Hakkarainen 28 October 1981. The topic of the meeting where the speech was given is not known.

decades earlier. After earning his master's degree, he worked first for Pöyry and then in the US for four years. After returning to Finland he gained a position at KCL, but at the same time started lecturing in paper technology at HUT. He was promoted to full professor of paper technology in 1990, and over 20 years he supervised about 350 master's theses and 50 doctoral theses.[55]

During Paulapuro's time as a professor of paper technology, the research topics changed due to the new demands that were placed on the industry. He published papers, based on his students' research, mostly on process diagnostics, mechanical pulping, wet pressing and fibre and paper physics. This reflected the status of paper machine engineering at the time. Slowly, the age-old differences between the two traditions of R&D faded and they merged into one stream by the 1990s. The gulf between the art of engineering and science of paper chemistry no longer formed a barrier to progress.[56] Hannu Paulapuro continued the approach that integrated the pulp and paper mills, research laboratories, and academic research and education. He practically formed a personal union between KCL and HUT due to his positions in both of the organizations. Also, the State Research Centre, VTT, was central to the cooperative research.[57]

The international economy was changing, though, and creating numerous challenges for the pulp and paper industry. Specifically, this meant a decreasing demand for printing papers of all sorts (particularly newsprint), and a concomitant rising demand for cardboard, and personal hygiene and packaging papers. At the same time Finnish paper companies were undergoing significant internal changes, a process that was driven by their focus on maximizing profits. These firms were no longer guided by technical experts because the latter had lost their decision-making power to the companies' financial managers. The growth of the companies and the economy was due to the merger of the factories and a strong belief in globalization. As a result, R&D changed as well, whereby the aim was no longer the search for new products or new ways of producing them but rather low-level research and improving production without doing much to improve the mills' operations. A paper mill was now aiming to be a global leader by making only one or a few top of the line products. In essence, the wheel had come full circle in a century. The aim of R&D was again to exploit the existing technology, often found in one paper mill, and development was largely concerned with the technology transfer from other units of a global company or from the scientific community.[58]

---

[55] See http://www.paperhall.org/hannuviljamipaulapuro/. Accessed 6 March 2017; HUT, a name register, Hannu Paulapuro.

[56] Jukka Kilpeläinen, DI, MBA, Helsinki, comments on 10 March 2017: 'I did not notice a difference in between the engineering and paper chemistry'.

[57] Hannu Paulapuro, emer. prof. Helsinki, interview, 1 April 2015.

[58] https://www.puunjalostusinsinoorit.fi/biometsateollisuus/innovaatiot/12-metsäklusterin-tutkimuslaitokset-ja-aktiviteetit/yhtyneet-paperitehtaat-valkeakosken-tutkimuskeskus/. Accessed 27 December 2017. Jouni Huuskonen 10 February 2009. Yhtyneet Paperitehtaat, Valkeakosken tutkimuskeskus. Accessed 20 February 2017; Jukka Kilpeläinen, DI, MBA, Helsinki, comments, 10 March 2017.

The national organization of R&D changed rapidly in Finland, and the reason was simple. Competition was now occurring in this realm among the firms because they had internalized R&D to prevent their results from leaking out. In addition, the merging of several companies reduced the number of actors who owned KCL. At one time it had been owned by 24 companies but by the turn of the millennium this number had shrunk to four and then three after Myllykoski Paper was purchased by UPM. At the same time a wide debate over commercialization of R&D research dominated discussions in the Finnish innovation community. Also, the co-operation with international actors was unsuccessfully tried for a short time.[59]

3.6 Conclusions

The movement to teach Finnish society to value educating legions of engineers was not enough by itself to change the country's research culture. To realize that goal, several interested parties from industry and broader society were needed. Industry, research laboratories, universities and engineering companies came together to form a network to facilitate research and development, and a scientific society, the Paper Engineers' Association, acted as a catalyst in the process.

These entities emerged over a long period, and they were founded for various reasons. For example, KCL was created to look after substitutes for exported materials and because of the need for industrial standardization and material testing. Similarly, the universities had tried to respond to the need for engineering education. The machine shops, such as Valmet, Tampella, Ahlström and Wärtsilä, all of which were on the vanguard of this campaign, developed their activities based on the needs of the pulp and paper mills as well as their other customers. Jaakko Pöyry Oy was an independent actor in engineering and planning.

The crucial factor that made it possible for Finland to develop a cluster of organizations and interests around wood chemistry and paper manufacturing technology was the informal co-operation among the different actors. In a small profession in a small country, most of the actors knew each other. This co-operative, unofficial network lasted until the 1990s and was concentrated around key persons, of whom Niilo Ryti and Erkki Laurila were clear examples. One major reason why the networking succeeded was the war economy and war reparation industry, both of which compelled Finland to strive to create a lean and efficient means of domestic industrial production using minimal resources. The importance of Niilo Ryti as a pioneer of the modern R&D culture in Finland is obvious. He exercised his influence through a series of actions due to his ability to create networks in the community of paper making engineers, both in academia and industry.

[59]At the end of the first decade of the new millennium VTT bought the R&D division of KCL. The division was later returned to KCL Oy. See https://www.puunjalostusinsinoorit.fi/biometsäteollisuus/innovaatiot/12-metsaklusterin-tutkimuslaitokset-ja-aktiviteetit/keskuslaboratorio-kcl, Jan-Erik Levlin, 31 March 2009. For more details Levlin 2010, pp. 24–25; Jukka Kilpeläinen, DI, MBA, Helsinki, comments on 10 March 2017.

This type of teamwork was unique to Finland and had a long tradition there, and it is probably without parallel models in other countries. Only the basic division between science and technology divided the papermaking profession, and that division was significant until the 1990s, when paper machine technology peaked. Near the end of the century, improvements in production seemed unattainable by taking steps that involved either technology or science; that aim would only be achieved by taking a holistic approach to the problem.

The nature of the relationship between the industry and the universities has not been a clear one. The concept of university-based research was well known in Finnish industry already by the beginning of the twentieth century. In fact, a basic model for this type of activity was enshrined in the original strategy for the Helsinki Polytechnic Institute in 1892, and there are good examples of dynamic and long-term co-operation between the university and industry at that time.[60] A new approach to the connections among actors affected all branches of industry from the 1950s. For example, engineering physics under the leadership of Professor Pekka Jauho and his protégé, Ph.D. graduate Olli Lounasmaa, enjoyed strong and important relationships to the generation of modern industries that used electronics in their new lines of products. Corporations like Nokia, Neste and Outokumpu took a fearless attitude towards undertaking innovation in the 1960s. The atmosphere in the more conservative pulp and paper industry was more complicated. Quite often it maintained a traditional mentality that referred to theoretical education and research as being unnecessary (Nykänen 2007b, pp. 230–235).

One of the limiting factors in the development of the relationships between the universities and the industry was the lack of financing for the basic research infrastructure in the universities. The new facilities at the HUT's Otaniemi campus were seen as being antiquated already by the 1960s (Nykänen 2007b, p. 230). As a result, Finland's pulp and paper industry often decided to conduct the research it needed in foreign, well-equipped laboratories.

The evolution of research and development in the Finnish pulp and paper industry was not a straight-forward and linear process. These activities must be understood as operations that affected components of large technological systems, which have been described by Thomas Hughes. The process of R&D cannot be seen or explained as an entirety, because all the independent projects have a unique nature due to their different purpose and context.

Consumers typically note the presence of paper only when the product runs out. The aim of R&D in the pulp and paper sector has often been to improve the efficiency—and thus the profitability—of the production process. Typical R&D programs have been aimed, for instance, at minimizing the consumption of energy or raising the efficiency of the drying section of a pulp or paper machine. The results of these efforts are not visible to consumers, but if a large paper mill can lower its use of water the result might be felt at the level of the national economy. A good example

---

[60] Karl Axel Ahlfors was one of the major developers of Finnish machine industry during the first half of the century. A great deal of know-how over fluid mechanics was established by Ahlfors (Kaataja 2015).

of the important R&D work that is not well known is the use of raw materials during the 1980s and 1990s at the Helsinki University of Technology.[61] One major reason that the pulp and paper industry moved so slowly in terms of R&D is that it is a very capital-intensive industry. The investments needed for new machinery are enormous and take decades to re-pay. This leads to a strong path dependence (Nykänen 2007b, p. 230).

One of the divides in the field of papermaking separates the traditions of chemical engineering from machine engineering. The fissure follows the general idea of de Solla Price, who has identified a difference between science and technology, as stated in the introduction. These two disciplines, both elementary for the paper industry, act in different ways. The research and development work in the pulp industry, including the by-products of processes, is a typical discipline of science. It is papyrocentric, based on published scientific studies. On the other hand, the mechanical part of pulp and paper mills and the manufacturing process of paper is papyrophobic. The results and even the R&D work is intended to be hidden as much as possible.

The challenges inherent in chronicling the history of technical research and development in the paper industry can be seen in the available archival sources, or more precisely, the lack thereof. In contrast, the history of technical chemistry is relatively easy to chronicle because the sources are generally available and there is a fair amount of literature published in this field. In contrast, writing about the history of R&D work in the papermaking machinery industry and the production process itself is very difficult. Because of the papyrophobic nature of the work, the companies rarely document their activities, and the ideas and experiences have remained largely confined within the enterprises that conducted the work.

Heikki Toivanen (2005) states that the basic components of paper machines have not changed dramatically since the Second World War. He explains most of the development in the speed and width of the paper machines came with the rapid development of automatic data processing that was surely one of the major components in progress (p. 79). This approach ignores large areas of technical development if it is considered to be the only or major explanation for what occurred in this field. During the 1950s and 1960s new former constructions like Beloit-Walmsley's Inverform and Bel Bond formers or Black Clawsons Vertiforma, changed the main structure of the wet end of the paper machine. Also in the 1950s research on new material technology, like new forms of cast iron and ceramic structures, was published and taken to be an elementary part of machine technology. In addition, the theoretical understanding of a structure's physics leaped forward. The latter really made it possible to widen the rolls and increase their speed, thereby giving new possibilities to increase the efficiency of the modern paper machine.

[61] Hannu Paulapuro, emer. prof. Helsinki, interview, 1 April 2015.

References

Archambault E, Lariviere V (2011) Scientific publications and patenting by companies: a study of the whole population of Canadian firms over 25 years. Science and Public Policy 38(4):269–278
Arjas A (1982) Kenen käsissä prosessikehitys. Paperi ja puu 11:643
Blomenthal EL (1887) Kertomus valtion avulla vuonna 1885 tehdystä matkasta pohjoisen Wenäjän metsäseuduille tiedon saamista varten sikäläisen yhteiskunnan metsäkeinollisista teollisuushaaroista. Hufvudstadsbladett, Helsinki
de Solla Price D (1965) Is technology historically independent of science? A study in statistical historiography. Technol Culture 6:553–568
Esilä A (1968) Innovaatiot ja pääoma. Teknillinen Aikakauslehti 23–24
Ferguson ES (1977) The mind's eye: nonverbal thought in technology. Science 197(4306):827–836
Ferguson ES (1993) Engineering and the mind's eye. The MIT Press, Cambridge, MA
Hietala M (2002) Kansainväliset yhteydet. In: Tommila P (ed) Suomen tieteen historia, vol 4. Tieteen ja tutkimuksen yleinen historia 1880-luvulta lähtien. WS Bookwell Oy, Porvoo, pp 525–561
Hodge R, Larry E (1979) Beloit corporation. Your partner in papermaking. http://paperindustryweb.com/partner/partmenu.htm. Accessed 11 October 2017
Hughes TP (1987) The evolution on large technological systems. The social construction of technological systems. New directions in the sociology and the history of technology. The MIT Press, Cambridge, MA
Kaataja S (2010) Tieteen rinnalla tekniikkaa. Suomalaiset korkeakoulututkijat kaupallisten sovellusten kehittäjinä 1900-luvulla. Suomen Tiedeseura, Juvenes Print, Helsinki
Kaataja S (2015) Tekniikkaa yliopistosta teollisuuteen. Karl Axel Ahlfors vesiturbiinien kehittäjänä 1900-luvun alkupuolella. Tekniikan Waiheita 3–4:7–19
Kaje M (1968) Teollisuus—ympäristö. Teknillinen Aikakauslehti 19–20
Komulainen A (2014) Paperimaailman valloittajat. Paperi-insinöörit ry 1914–2014. Siltala, Helsinki
Layton E (1971) Mirror-image twins: the communities of science and technology in 19th-century America'. Technol Culture 12(4):562–580
Levlin J-E (2010) KCL: n vuosikymmenet Otaniemessä 1962–2009. Helsinki
Marton J, Cripps WC, Matthews GT (1969) How to maintain the education standard of practising engineers. A paper presented at the 55th annual meeting of the technical section, Canadian Pulp and Paper Association, Montreal, Jan 28–31 1969
Michelsen K-E (1993) Valtio, teknologia, tutkimus. VTT ja kansallisen tutkimusjärjestelmän kehitys, Painatuskeskus Oy, Espoo
Murto E, Niemelä M, Laamanen T (2006) Finnish technology policy from the 1960s to the present day. Libris Oy, Ministry of Trade and Industry
Nykänen P (1999) Bensiinihiilivetyjen valtiaat. Voitelu- ja moottoripolttoaineiden tutkimus Suomessa vuoteen 1948. Gummerus, Saarijärvi
Nykänen P (2005) Around the roll. A century of Finnish paper machine and paper making technology, Gummerus, Jyväskylä
Nykänen P (2007a) Kortteli sataman laidalla. Teknillisen korkeakoulun historia, vol 1. WSOY, Porvoo
Nykänen P (2007b) Otaniemen yhdyskunta. Teknillisen korkeakoulun historia, vol 2. WSOY, Porvoo
Nykänen P (2008) Toiveet ja todellisuus. Teekkaritoiminnan Edistämisyhdistys ry: n 60 vuotta. Gummerus, Jyväskylä
Nykänen P (2009) Jos kultaa kaivannet. Outokumpu Oy: n Säätiö vuoritekniikan, metallurgian ja geologian opetuksen ja tutkimuksen edistämistä varten 1937–2010. Gummerus, Jyväskylä
Ohlström P (1954) Eräitä uusia kotimaisia konstruktioita selluloosa- ja paperikonealalla. Paperi- ja puu 4

Oinonen H (2002) Matti Kankaanpää, the man behind Valmet's expansion. Paperi ja Puu 4:222–227
Ojala J (2001) Haastajasta voittajaksi: Yhtyneet Paperitehtaat 1900-luvulla. Kilpailu, keskittyminen ja kasvu pitkällä aikavälillä. Painopörssi, Helsinki, Metsäteollisuuden strategiset kehityspolut
Paper Makers' Wiki. https://www.papermakerswiki.com/innovations/mittaus-ja-s%C3%A4%C3%A4t%C3%B6tekniikka/painesihdin-aiheuttama-neli%C3%B6massavaihtelu/yhtyneet-paperitehta. Jouni Huuskonen 2009. Yhtyneet Paperitehtaat, Valkeakosken tutkimuskeskus. Accessed 20 Feb 2017; http://www.papermakerswiki.com/innovations/mets%C3%A4klusterin-tutkimuslaitokset-ja-aktiviteetit/ga-serlachius-oy. Accessed 6 Mar 2017; http://www.paperhall.org/hannuviljamipaulapuro/. Accessed 6 Mar 2017
Rosenberg N (1994) Exploring the black box: technology, economics and history. Cambridge University Press, Cambridge
Selluteollisuus kaavailee uusia väkiviinatehtaita (1975) Talouselämä 36–43
Tuuri A (1999) Eteenpäin katsomisen taito. Tekniikan edistämissäätiö 1949–1999. Otavan Kirjapaino Oy, Keuruu
Teknillinen apu (1958) Yhdistyneiden kansakuntien pohjoismainen tiedotustoimisto, Kööpenhamina
Toivanen H (2005) Sorvareista palvelijoiksi. Maailman paperi- ja sellukonetoimialan strateginen peli ja logiikka 1970-luvulta 2000-luvulle, Tampereen teknillinen yliopisto, Tampere
Wise G (1985) Sci Technol 1: 229–246. 2nd series

Chapter 4
The Greening of the Pulp and Paper Industry: Sweden in Comparative Perspective

Ann-Kristin Bergquist and Kristina Söderholm

4.1 Introduction

Since the 1960s, concern about environmental degradation has greatly altered the commercial conditions for most industries in the Western world, but the global pulp and paper industry (PPI) has encountered more serious challenges than most other industries. The reason is that pulp and paper production, especially bleached pulp manufacturing, is extremely damaging to the environment. Thus, pulp mills cause, among other things, emissions of odorous gases and sulphur compounds into the air and of organic and inorganic chemicals including chlorinated compounds and other substances into waterways. Thanks to technological developments, however, the environmental situation has been greatly improved during the last five decades with reduced emissions; in many cases the reduction has been more than 90%. This chapter concerns the environmentally driven transformation of the PPI and examines it in a country which has pioneered parts of the greening process in this industry, namely Sweden. This nation is one of the world's leading pulp and paper countries, as it ranks as the ninth largest producer of paper and the fourth largest producer of pulp (Lamberg et al. 2012; Järvinen et al. 2012; Bergquist and Keskitalo 2016).

For any polluting industry, technology is at the very core of the challenge to reduce its environmental impact (Jaffe et al. 2005) and indeed this has been the case for the PPI (Bajpai 2011). Technological strategies and timing for investments have differed between countries and regions depending on the different national jurisdictions, organizational solutions, demand characteristics and geographical circumstances. The

A.-K. Bergquist (✉)
Department of Geography and Economic History, Umeå University, Umeå, Sweden
e-mail: ann-kristin.bergquist@umu.se

K. Söderholm
Department of Business Administration, Technology and Social Sciences, Luleå University of Technology, Luleå, Sweden
e-mail: Kristina.Soderholm@ltu.se

T. Särkkä et al. (eds.), *Technological Transformation in the Global Pulp and Paper Industry 1800–2018*, World Forests 23, https://doi.org/10.1007/978-3-319-94962-8_4

literature on corporate environmentalism, which has stressed companies' modes of responding to environmental issues, has generally identified a movement along an evolutionary, adaptive learning process that formed specific attitudes or means of response during certain periods (Hoffman and Bansal 2012). An important conclusion from this research when it comes to technological strategies is that before the 1990s, firms typically employed end-of-pipe approaches to achieve regulatory compliance, which practically means external treatment of effluents after they have left the plant (Lee and Rhee 2005; Frondel et al. 2007; Hoffman 1997). The alternative technological approach, internal process changes (e.g., in-plant measures that prevent or reduce effluents before they leave the plant, sometimes also referred to as 'cleaner production technologies'), has subsequently been acknowledged and adopted as a more advantageous and effective strategy (Sinclair 1990).

As will be discussed in this chapter, Swedish PPI has not followed the textbook example of the learning process from end-of-pipe technology to "cleaner" production processes. The Swedish PPI, in contrast to both its North American and Finnish counterparts, embarked already in the 1970s on developing internal process alterations rather than end-of-pipe solutions, as the main technological strategy for pollution control (Bergquist and Söderholm 2015; Harrison 2002). This early strategy became formative for the Swedish PPI's green technological development for decades to come. Kemp and Soete (1990) have argued that, compared to 'normal' technological efforts in industry, pollution control efforts are generally more focused on incremental improvements in 'cleaning' technology, following relatively well-established technological strategies of 'progress'. This argument suggests that pollution control technologies are particularly 'path dependent' and breaking the 'path' demands a lot of investments and even a new way of thinking. The notion of path dependence can be seen from several perspectives; from the simple assertion that 'history matters' to the problem of imperfect information and the appreciation that institutional arrangements may have limits, and the phenomenon of increasing returns (e.g. Liebowitz and Margolis 1995; Pierson 2000). When it comes to technology, path dependence can provide very powerful incentives for limiting actions to incremental technological solutions, which in turn strengthen established lines of action (Geels 2004; Kemp and Soete 1990; David 1994). Technological approaches in one country, as well as divergent technological strategies between different pulp and paper producing countries, might therefore be considered in the logic of technological path dependence.

For several reasons the greening of the PPI has been a core challenge for the sector. Environmental regulations have greatly altered the commercial conditions for the PPI over the past five decades (as for most industries), and since the 1980s, market pressure has also interacted with environmental regulation as a driver towards greener production processes and products. In addition, since the 1970s, the drive to use energy more efficiently has been increasingly linked with environmental issues, not least through the issue of climate change. Overall, the greening of the PPI has been a complex process. It has evolved along several different paths in combination with other business challenges, with the latter stemming from changed input prices on, for instance, energy, wood and labour and also changes in market competition, governmental policies, and new technologies outside the immediate realm of pollution control.

4.2 Environmental Impacts of Pulp and Paper Production: A Short Background

The environmental impact of pulp and paper mills has been, and still is, serious from many aspects (Table 4.1). It causes or produces human toxicity, ecotoxicity, photochemical oxidations, acidification, nutrification, solid wastes and climate change (Bajpai 2011). Still, the situation is much better today than it was in the past; wastewater from a modern pulp mill contains only a small fraction of the contaminants that it did in the 1960s. One of the most serious problems is the very large quantity of wastewater resulting from pulp manufacturing, and it is contaminated with a number of organic and inorganic chemicals including lignin, cellulosic compounds, phenols, mercaptans, sulphides and chlorinated compounds (Thompson et al. 2001). The PPI is also a large user of both energy and water, and toxic chemicals. For example, in the United States (US) the paper industry is the third largest energy user among manufacturing industries, accounting for 11% of domestic energy consumption in 2010 (World Watch Institute 2015). The most significant environmental problems occur in the stage of the pulping process (in relation to paper production), which is why the most severe problems in the past were found in pulp producing countries, such as Sweden, Finland, Canada and the US although pulp importing countries, such as the United Kingdom and Germany also have struggled with severe problems (Bajpai 2015; Bergquist and Söderholm 2015; Mutz 2009).

Pulp can be roughly divided into the processes (and products) of mechanical and chemical pulp. Each process varies in terms of immediate environmental impact depending upon the amount and type of energy and chemicals used. Pulp mills are, however, traditionally associated specifically with the discharge of high levels of wastewater contaminated with organic matter. In the chemical pulp process, wood chips are first cooked in chemicals and then the solution is washed to produce "clean pulp", which consists of only cellulose fibres. Dirty "wash water", containing cooking chemicals and dissolved lignin, are thereafter sent to a recovery boiler[1] from which the cooking chemicals are recovered for reuse and the lignin is burned for power generation. Wastewater containing the remaining organic matter and chemicals is then discharged (Gunningham et al. 2003, pp. 10–11).

The most serious environmental problem related to pulp manufacturing in the past was caused by the bleaching process. It takes place in stages, generally alternating between acid and alkaline stages. The use of elemental chlorine (Cl_2) as the bleaching agent became dominant in the 1950s (Norberg-Boom and Rossi 1998) and was

[1]The invention of the recovery boiler by G. H. Tomlinson in the early 1930s was a milestone in the advancement of the Kraft pulp process.

Table 4.1 Major pollution problems caused by pulp and paper production

Suspended solids (SS)	Have its origins in bark, pieces of fibres and filling, and coating agents. Consumes oxygen when decaying and can be carriers of poisonous substances
Organic matter in general (BOD and COD)	Uses oxygen from water. May cause oxygen deficiency in waterways, which leads to the death of, i.e., fish and cause severe damage to the ecosystem
Chlorinated organic compounds (AOX)	From mills using elemental chlorine in their bleaching sequence. Wastewater containing organic matter formed by elemental chlorine reacting with wood products to form absorbable organic halide (AOX). AOX has become an accepted measure of chlorinated organic material, and is used to monitor and regulate bleached Kraft pulp mill effluents. AOX is used as a surrogate parameter of dioxins in wastewater (and stack gas) from pulp mills
Water consumption	Pulp mills are major water users. Consumption of fresh water can seriously harm habitats near mills from reduced water levels (necessary for fish) and changed water temperature (also a critical environmental factor for fish)
Sulphur dioxide (SO_2) and reduced sulfur compounds	Leads to acid rain and causes soil degeneration
Nitrogen oxides (NOx)	Gases composed of nitrogen and oxygen formed during combustion. In moist air, the substances are converted into nitrogen oxides and then nitric acid, which creates acid rain. Originates from recovery boilers in the Kraft pulp process
Sulphur compounds (TRS gases)	Smell

Source Ince et al. (2011), Skogsindustrierna (1995), Nilsson (2007)

subsequently replaced after the 1980s with new bleaching agents and methods such as the elemental chlorine free (ECF) and total chlorine free (TCF) bleaching.

4.3 Environmental Regulation as a Driver for Technology Development

The Swedish PPI has been central to the Swedish economy ever since the introduction of mechanical pulp in the 1850s and chemical pulp in the 1870s. Already at the time of the First World War Sweden was the world's second largest pulp producer after the US and the world's largest pulp exporter (Fahlström 1948; Rydberg 1990; Järvinen et al. 2012). As the PPI represented both one of the largest industrial sectors in Sweden and one of the biggest polluters, it has received much attention in Swedish policy to control industrial pollution, including joint state-industry efforts in R&D to control it (Bergquist and Söderholm 2011).

Over time environmental policies have grown more complex with a long-term shift in focus from local conflicts over water pollution and odour problems in the late nineteenth century to transnational and global problems such as climate change in the late twentieth century. This development has taken place in stages (McNeill 2000). Hence, in the 1960s, Sweden along with many other western societies saw the initiation of serious governmental action to control industrial pollution in parallel to a growing environmental awareness in society (Lönnroth 2010). In contrast to the US, environmental awareness in Sweden grew mainly amongst experts and not the public in the 1960s and 1970s. It is often assumed that heavy polluting industries did not focus seriously on mitigating their environmental impacts before this environmental awakening. There is plenty of historical research, however, which illustrates how companies and even industrial sectors undertook action to control pollution due to local conflicts and governmental intervention before the 1960s and sometimes they did so even proactively to avoid a bad outcome for business (for an updated literature overview, see Bergquist 2017). In fact, the Swedish PPI undertook action to control pollution long before the 1960s due to increasing local criticism and the industry's own awareness of the negative environmental effects from production. This in turn spurred joint efforts to develop new and cleaner technologies already in the early 1900s (Söderholm and Bergquist 2012). For example, by 1908 all Swedish sulphate pulp producers had united to form the so-called *Sulphate Pulp Committee* to develop technology to lessen the odour problem stemming from the sulphate pulp process.[2] Central reasons behind these initiatives were rising concerns about industrial pollution that had been expressed in the Swedish Parliament in the early 1900s, along with court cases concerning the contaminating activities of single pulp mills. These court cases built on early health- and water protective regulation, which to some degree included regulation of industrial pollution. For instance, the Public Health Act of 1874 constituted the first governmental "all-embracing" attempt to address sanitary issues. In the 1910s, the Swedish government presented a proposal to implement stricter legislation against air and water pollution. However, the proposal was rejected in the early 1920s due to an economic recession (Söderholm 2009; Lundgren 1974).

Pollution problems, however, increased in tandem with a growing Swedish economy in the 1930s, whereupon the Swedish government initiated a process of tightening the regulation of water pollution. Hence the Water Act was reformed in 1942, whereby it introduced a concession system according to which enterprises such as chemical pulp and sugar mills and textile factories had to apply to the Water Court for permission to operate. Still, air pollution was not addressed properly until 1963, and then through the formation of the State Air Pollution Control Board (Lundgren 1999). Finally, in 1969, the Environmental Protection Act (EPAct), the first uniform Swedish framework for regulating air and water pollution, noise and other disturbing activities from industrial plants, was passed by the Swedish Parliament. Years before the final bill was enacted, the construction of a modern system for environmental

[2]For a more detailed overview of regulation and organization of the Swedish PPI during this period, see Söderholm and Bergquist (2012).

protection had begun in Sweden. Thus, in 1967 the government created the National Environmental Protection Agency (EPA), a unified body for almost the entire area covered by the Environmental Protection Act (Lundqvist 1971). Soon other countries followed this lead, such as the US, which established its own EPA in 1970. Still, in the neighbouring pulp producing country of Finland, environmental policy developed at a slower pace (Söderholm et al. 2017). The National Board of Waters was created in 1970, but it was not until 1983 that a unified body to handle a major part of the environmental issues founded there (i.e., the Ministry of the Environment) (Joas 1997).

From the 1970s until the 1990s, the Swedish EPAct was the main tool to control pollution and other environmental problems related to Swedish industry, and it fundamentally changed the conditions for the operations of the Swedish PPI. Thus, in the 1970s and the 1980s it forced the sector to increase considerably its green R&D activities and to undertake deep emission cuts (Bergquist and Söderholm 2011). The regulatory approach was based on case-by-case assessments whereby permits had to be reassessed and renewed every 10 years or after production increased. It relied on so-called performance standards rather than technology standards and these were negotiated with each plant owner, sometimes over extended periods of time. The standards were typically implemented in combination with extended compliance periods, meaning that the companies were giving necessary time to develop and test technology. In these ways, the regulatory approach provided scope for environmental innovation and permitted the affected companies to coordinate pollution abatement measures with productive investments (Bergquist et al. 2013). In 1999, the EPAct along with 15 other acts were amalgamated into the Swedish Environmental Code and the responsibility for issuing permits was thereby transferred from the Franchise Board of Environmental Protection (the organ responsible for issuing permits since 1969) to the Environmental Courts (Michanek and Zetterberg 2007).

The Swedish PPI has also been governed by European Union (EU) environmental legislation ever since Sweden became an EU-member in 1995. One example is the Integrated Pollution Prevention Control Directive (IPPC Directive) of 1996 and 2008. The IPPC Directive is a key instrument in the EU's environmental legislation, and its purpose is to achieve 'integrated pollution prevention' and control of the pollution occurring at large industrial installations (Schoenberger 2009).[3] Further, in 2011, the Industrial Emission Directive (IED)[4] came into force, meaning all Member States before 7 January 2013, had to incorporate the IED into national leg-

[3]The conditions of required permits have to be based on Best Available Techniques (BAT), and in 1997, the Sevilla Process was established to develop BAT. Since then, 33 BAT Reference Documents (BREFs) have been drafted, adopted and published containing ambitious consumption and emission levels which cannot be found anywhere else (Schoenberger 2009).

[4]The IED is the successor of the IPPC Directive and, in essence, is about minimizing pollution from various industrial sources throughout the EU. Operators of industrial installations are covered by Annex I of the IED and are required to obtain an integrated permit from the relevant EU country authorities. Permit conditions including emission limit values (ELVs) must be based on the Best Available Techniques (BAT) as defined in the IPPC Directive.

islation. In Sweden the directive was included in the Industrial Emission Regulation (*Industriutsläppsförordning* 2013, p. 250) (SEPA 2016).

Since the 1990s in Sweden, policies targeting the PPI sector have embraced a blend of mandatory governmental and voluntary stipulations concerning social and environmental requirements. Thus, in line with international trends, environmental management systems (EMAS) have been implemented in virtually all Swedish pulp and paper mills. And most mills have chosen to certify their environmental management systems under either the global standard ISO 14001 and/or the European EMAS scheme. Thus, in 2013, 97% of the pulp and 98% of the paper produced in Sweden was manufactured under certified environmental management systems (Swedish Forest Industry Federation 2015). Essentially the EMAS scheme is an opportunity for producers to formally demonstrate from year to year their improved environmental performance to customers and partners, and is thus a form of 'eco-labelling'. The opportunity for the Swedish PPI to engage in 'eco-labeling' first appeared in the 1980s (see below). Regulatory measures to stimulate improvements in energy efficiency and lower carbon dioxide emissions have developed as an additional control box parallel to other emissions (see e.g. Thollander and Ottosson 2008; Henriksson et al. 2012). Also, in 2005 the voluntary Swedish Program for Energy Saving (PFE) came into force (Henriksson et al. 2012).

4.4 The "Spring-Cleaning" of the Swedish PPI

4.4.1 Internal Process Changes and Structural Rationalizations

The magnitude of pollution control implied by the EPAct in 1969 made the Swedish pulp and paper producers conclude that it was only through reconstructed and new plants embodying the most novel techniques that the discharges could be substantially reduced and requirements met (Wohlfart 1971b, p. 320). In this context, the Swedish pulp producers strategically aimed for production expansion based on the adoption of the sulphate process in the late 1960s, since the calcium-based sulphite mills had many environmental disadvantages in terms of their discharges of BOD, lignin, gases and dust, and chemical recovery. The discharges of the sulphite mills required radical external purification works while the sulphate process had the potential to recover chemicals and at the same time generate electricity (Wohlfart 1971b). Thus, the enforcement of the EPAct made it economically impossible for some mills to continue operating, especially small, inefficient mills, many of which were forced to shut down. And it was in this context that Swedish sulphite pulp mills were almost totally phased out. The Domsjö sulphite mill owned by the Modo Group is, however, one of few sulphite mills that survived in Sweden after the 1970s (Söderholm and Bergquist 2013).

Contributing to the reduction of discharges from the PPI was a process of structural rationalization, and Sweden was not the only country to experience this phenomenon. The OECD stated in 1973 that replacing small, uneconomical and environmentally obsolete units with larger more energy- and resource-efficient mills would create cost-effective means of reducing discharges per production ratio opportunity without deviating from the business inclination to seek productivity gains through capacity expansion. The approaches, however, differed between countries. In most cases, oxygen depletion of waterways required the rapid adoption of stringent Biochemical Oxygen Demand (BOD) limits, which typically locked industry into end-of-pipe measures (Rajotte 2003; OECD 1973). In Sweden, however, the pulp and paper producers at the time of the enforcement of the EPAct had a basic approach developed, one that already defined decades of environmental development activities and which they aimed at maintaining (i.e., to cope with the pollutants inside the mills and not when they "leave the plant") (Wohlfart 1971a, pp. 432–434).

The Swedish PPI's approach to dealing with pollution through changing internal processes can be traced back to the 1940s and 1950s and the pioneering R&D activities then taking place within collaborative platforms established jointly by the industry sector (see Sect. 4.4.3 below). While the first collaborative initiatives of the sector to deal with pollution problems had taken place already at the turn of the twentieth century, the R&D activities thereafter increased in tandem with the development of environmental policy up to the 1960s. During this novel period, attention was foremost directed towards improved efficiency and reduced fiber emissions. The practice of discharging fibers was highly inefficient as it resulted in 15% waste. Increased efficiency was therefore closely linked to the ability to reduce discharges of organic materials (Söderholm and Bergquist 2012). Also Swedish authorities and other industrial sectors, such as the metal smelting industry, early on chose to focus on in-plant measures and primary effluents instead of external measures (Lindmark and Bergquist 2008). For example, in describing this strategy, the environmental manager of Stora Enso in the 1990s explained that 'internal process changes have a completely different potential for the business. An end-of-pipe is a cost. An integrated solution can provide so much more. It can provide opportunities to increase production, reduce operating costs, and simply increase productivity. The key to all the improvements we have seen over the years is that the industry has been able to expand. The basic rule is that environmental improvements are a consequence of the corporate will to invest and its propensity to do so in the future. You take care of environmental costs as an integrated part.'[5]

4.4.2 *Emission Reductions and Costs*

During the 1970s and 1980s, costs for environmental investments were considerable for the Swedish PPI and accounted for 9–14% of the total investments (see Table 4.2).

[5]Interview with Per G Broman, Falun, 9 May 2006.

Table 4.2 Environmental investments in the Swedish pulp and paper industry, 1968–1988. Real prices (2000) million SEK (deflated using investment price index for the pulp and paper industry)

	68–70	71–75	76–79	80–84	85–88	Total
Environmental investments	930.2	4254.7	2372.9	2235.1	3388.9	13101.5
Environmental investments (share of total investments)	6	12.15	13.67	9.9	13.5	na

Source Swedish Forest Industries' Water and Air Pollution Research Foundation (SSVL) 'SSVL 74–85' (1989), see *Miljöskyddskostnader inom Svensk Skogsindustri 1985 t o m 1991. Appendix 2 and 20.* The estimates are based on questionnaire data collected by SSVL during the periods 1968–1970, 1971–1975, 1976–1979, 1980–1984, and 1985–1988. For a detailed accounting, see Söderholm and Bergquist (2012)

Over 60% of these investment costs involved altering internal processes that aimed to decrease water use and improve chemical and fibre recycling while about 14% of the costs concerned external wastewater purification measures. The cost share for air-purification measures amounted to only 15% (SSVL 1991, p. 93).

Costs were the highest among pulp producers (in comparison to paper producers), a sector in which the share of environmental investments in 1985–1988 accounted for as much as 17.5% of total investments (SSVL 1989, Appendix 10). This share was even higher among producers of bleached pulp at the beginning of the 1990s due to the discovery in the mid-1980s of the formation of dioxin in the process of pulp bleaching. Reported expenditures of US mills during the 1970s were on average 24% (i.e., higher than for the Swedish mills). However, during the 1980s capital expenditures on pollution control in the US PPI declined to 8.1% (Smith 1997, pp. 109ff).

The environmental investments undertaken by Swedish pulp and paper producers generated significant emission cuts. One example occurred in terms of Chemical Oxygen Demand (COD), which over the period 1970–1995 decreased from approximately 2.3 to 0.4 million tons annually. Moreover, from the mid-1980s to 2010 emissions of absorbable organic halides (AOX) declined by 97%.[6] AOX is associated with dioxin and is a measure of total halogens (chlorine, bromine, iodine). Emissions of sulphur, AOX and COD are shown in Figs. 4.1, 4.2 and 4.3, while pulp production is shown in Fig. 4.4.

[6]http://www.skogsindustrierna.se/skogsindustrin/branschstatistik/hallbarhet/. Accessed 19 January 2018.

4.4.3 Firm Collaboration

A central part of the technologies that generated the significant emission cuts illustrated in Figs. 4.1, 4.2, 4.3 and 4.4 were developed within collaborative R&D arenas of the Swedish PPI. And at the time of the enforcement of the EPAct (1969) these measures were supported by environmental authorities and involved various actors of importance to the innovation process. Two new collaborative environmental R&D

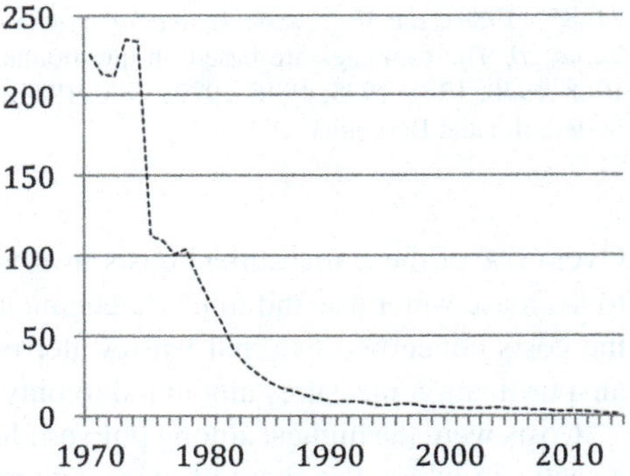

Fig. 4.1 Emissions of sulphur 1970–2014 (tons). *Source* Skogsindustriernas miljödatabas

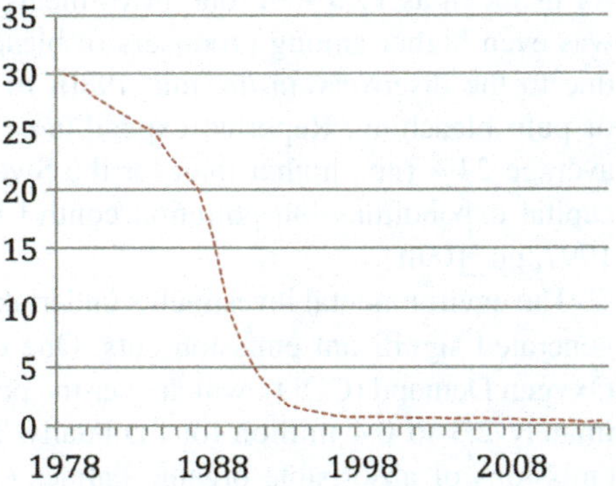

Fig. 4.2 Emissions of AOX 1978–2014 (tons). *Source* Skogsindustriernas miljödatabas

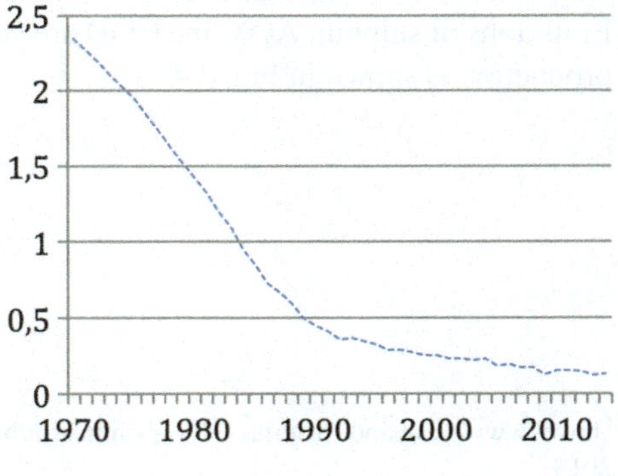

Fig. 4.3 Emissions of COD 1970–2014 (million tons). *Source* Skogsindustriernas miljödatabas

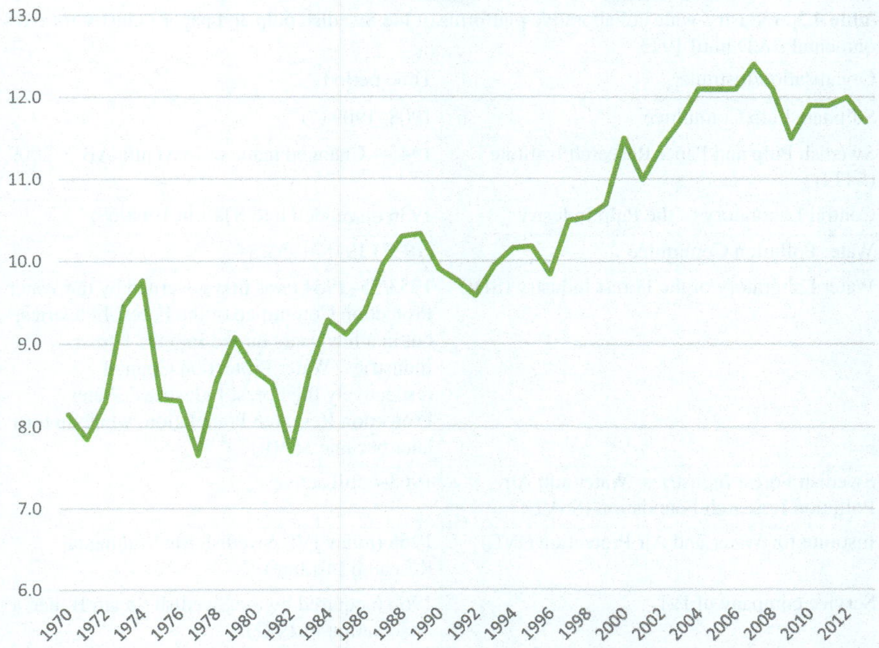

Fig. 4.4 Swedish pulp production 1970–2013 (million tons). *Source* Skogsindustriernas miljödatabas

platforms were established by the Swedish PPI in the 1960s. These were the state- and industry-funded Institute for Water and Air Protection (IVL), created in 1966, and the Forest Industries' Water and Air Pollution Research Foundation (Stiftelsen Skogsindustriernas Vatten- och Luftvårdsforskning, hereafter SSVL), established in 1969. The motive for creating the two new platforms was the recognized need for effective collaborative efforts in environmental R&D to manage costs and risks related to emerging stricter environmental requirements. SSVL consisted not only of a broad set of representatives from private companies, research institutions, and industry interest groups, but also of consultants, equipment suppliers, and research institutions outside the immediate sector (Bergquist and Söderholm 2011).

The furthering of green knowledge was accomplished in collaboration also with state agencies, particularly within IVL. The Institute was jointly founded in 1966 by the Swedish government and the Swedish pulp and paper industry in collaboration with other industrial sectors. Its primary assignment was to conduct research on the relationship between industrial production and environmental problems and to identify effective solutions. Closely affiliated with the Institute was the Industry Water and Air Protection Agency (Industrins Vatten och Luftvård AB), a service company that was also established in 1966. Together with SSVL, IVL supported knowledge diffusion and technology development of decisive importance to the green reconstruction of the Swedish PPI from the 1960s through the 1970s and 1980s

Table 4.3 Industry-wide collaborative platforms of the Swedish pulp and paper industry for environmental R&D until 1973

Organisation/Institute	Time period
Sulphate Pulp Committee	1908–1909 (?)
Swedish Pulp and Paper Research Institute (STFI)	1945—Changed name to Innventia AB in 2009
Central Laboratory of the Pulp Industry	1936 (ascended into STFI in 1968/69)
Water Pollution Committee	(1937) 1945–1953/54
Water Laboratory of the Forest Industry (SIV)	1953/55–1964 (was first governed by the Water Protection Committee of the Forest Industries, but in a few years turned into the Forest Industries' Water Protection Council respectively the Forest Industries' Water Protection Research Foundation, which in turn later became SSVL)
Swedish Forest Industries' Water and Air Pollution Research Foundation (SSVL)	1963—still active
Institute for Water and Air Protection (IVL)	1966 (today IVL Swedish Environmental Research Institute)
Service company of IVL	1966 (acquired by the Swedish Steam Boiler Association in 1982)
Energy Committee	1973—still active

Sources Söderholm and Bergquist (2012, 2016)

(Bergquist and Söderholm 2011). Table 4.3 provides an overview of the collaborative platforms.

It should be mentioned that the knowledge generated within IVL and SSVL was applied by the Franchise Board of Environmental Protection (FBEP) while enforcing the EPAct through the issuing of individual permits. The network connected to the environmental R&D activities, including both industry and environmental authorities, hosted technical competence and contributed to a shared understanding of reasonable courses of action. There were, however, conflicting opinions on how far industry could and should go in terms of environmental compliance.

4.5 Towards ECF and TCF Bleached Pulp

The replacement of elemental chlorine as a bleaching agent with the alternative ECF and TCF methods is an internationally well-known, environment-related technological shift of the PPI. Technically the ECF-process means a complete replacement of chlorine (Cl_2) with chlorine dioxide gas (ClO_2). By replacing Cl_2 with ClO_2 the levels of absorbable organic halogens (AOX) are greatly reduced. The TCF-process in turn completely eliminates the use of both Cl_2 and ClO_2 and instead typically

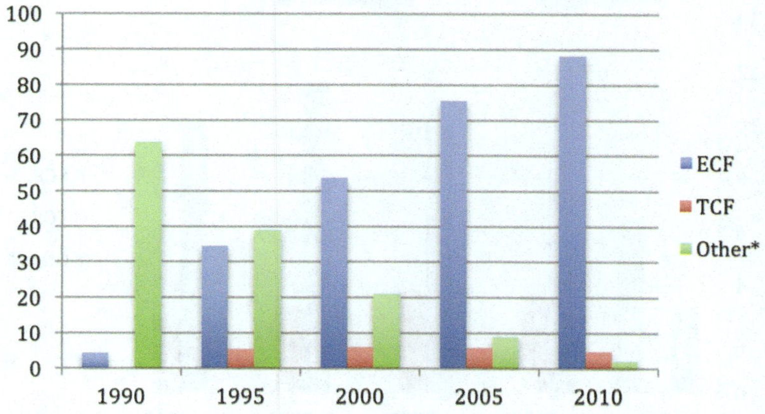

Fig. 4.5 Bleached chemical pulp production in the world 1990–2010 (million tons). *Source* Alliance for Environmental Technology (2012). *Pulp bleached with some molecular chlorine

uses hydrogen peroxide (H_2O_2) and/or ozone (O_3) as substitute bleaching agents. The development of ECF and TCF was further aided by improvements in the oxygen delignification technology, which is used in a pre-bleaching stage.

The shift towards alternative bleaching methods was dramatic. Soon after the US EPA in 1985 detected low concentrations of the extremely toxic and highly chlorinated hydrocarbon group dioxins in fish caught downstream from a few pulp mills, Swedish environmental authorities immediately followed up with their own investigations. In 1986, they concluded that crabs collected near the effluent site of a Swedish pulp mill contained dioxins, too. In the years to come the issue of bleached pulp was raised on both public and governmental agendas throughout the world and regulatory processes were initiated in pulp producing countries. Hence, the risks of dioxin were intensively discussed during the European environmental debate and Sweden was, already in 1988, the first country in the world to set up a regulatory standard for controlling dioxins discharged from pulp mills (Bergquist and Söderholm 2015). Until recently the dioxin alarm has caused a major shift in bleaching technologies worldwide. In 1990, 93% of the bleached pulp produced in the world was bleached by elemental chlorine and in 2010, this figure had dropped to only 2.7%. Figure 4.5 illustrates this development.

It is widely recognized that the technologies diffused more rapidly in the Nordic countries compared to North America, and it is mainly explained by the rise of green consumerism in key markets for Nordic producers while the demand for chlorine free paper was absent in North America (Marcus 1999; Smith and Rajotte 2001; Harrison 2002; Norberg-Bohm and Rossi 1998; Reinstaller 2005; Smith 1997; Popp et al. 2011). Figure 4.6 illustrates the diffusion of ECF and TCF in the Nordic countries since 1990.

Yet, it gives a highly simplified picture to bundle the Nordic countries in this respect. It is rarely highlighted that the great majority of the pulp producers a few

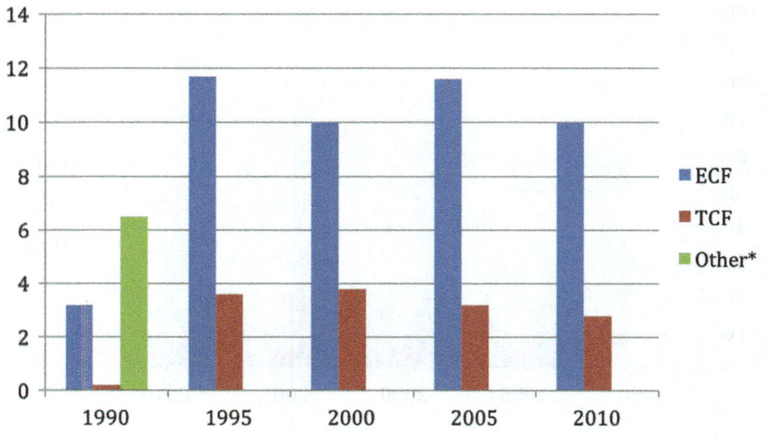

Fig. 4.6 Bleached chemical pulp production in the Nordic countries 1990–2010 (million tons). *Source* Alliance for Environmental Technology (AET) (2012). *Pulp bleached with some molecular chlorine

years into the 1990s that had switched to TCF were of Swedish origin and only two were Finnish (Reinstaller 2005, p. 1373). Moreover, Swedish producers, unlike Finnish ones, pioneered in the development of several core technologies for the ECF and TCF processes in the 1970s and 1980s, such as oxygen delignification, advanced batch cooking and ozone bleaching (Kramer 2000). The pioneering development of the Swedish producers in this matter should be viewed in the collaborative context of the long-term environmental R&D activities discussed above, and where important technological steps were achieved through networks of actors rather than by single firms. Thus, already from the mid-1970s, projects were initiated within both SSVL and the Swedish Pulp and Paper Research Institute (STFI) that focused largely on improving the pulp bleaching process.

The projects typically aimed at decreasing the use of chemicals while increasing the level of delignification in the boiler and in the subsequent oxygen delignification. The main motives in these instances were both to decrease costs and cut emissions from the bleaching process. Thus, already in the early 1970s a pilot plant for oxygen delignification was constructed at a Swedish mill in relation to a SSVL-project (Jerkeman 2007), and two decades later the technology was installed at all Swedish pulp mills (SSVL 1991). At the time, the Nordic Council of Ministers considered the oxygen delignification technology the single most effective process-oriented device to reduce chlorinated organic compounds (Nordic Council of Ministers 1989, pp. 21–23). Also during the 1980s, methods of boiling the pulp to very low levels of lignin, which in turn reduced the need for bleaching chemicals even further (Modified Continuous Cooking/Super Batch Cooking), were developed jointly (and later widely diffused) in the Swedish PPI (Jirvall and Noodapera 1995).

Industrial pollution was regulated only in a limited way in Finland before the late 1980s, and the preferred method for controlling it was by end-of-pipe technology and particularly the activated sludge method. In 1989, it was applied by a large number of Finnish sulphate/kraft pulp mills, whereas Swedish mills instead were better outfitted with pollution prevention methods such as oxygen delignification, modified cooking and high chlorine dioxide substitution (Auer 1996). Hence, before green consumerism grew in key markets for Nordic producers in the 1990s, regulation was the main trigger for the technological shift towards ECF and TCF among Swedish pulp mills. The Swedish FBEP took an initial and clear standpoint on the dioxin issue already in 1986, when it delivered final conditions for a sulphate pulp mill that declared that the wastewater from the bleaching process caused significant local damage to water-based organisms. In 1988, the Swedish Parliament in turn adopted a proposal which stipulated that industrial emissions of chlorinated organic compounds must be reduced to 1.5 kg AOX/ton of pulp (the normal emission level was about 4 kg per ton of pulp at this time). At this stage the Swedish Forest Industries Federation identified necessary investments to meet the requirements that were now being enforced to be 4 billion SEK, and for some mills it represented up to 50% of the annual investments until 1992 (Bergquist and Söderholm 2015).

The final important technological steps towards TCF pulp were taken by the Swedish PPI in the very last years of the 1980s due to rather forceful regulatory action. In 1988 the Swedish government imposed on an individual mill the most stringent standards ever, namely a maximum of 0.5 kg AOX per ton. The mill, Aspa, was a small producer of market pulp, however, with a sensitive location on a lake that supplied drinking water for several communities. Already in the spring of 1989 the mill had installed a method which produced these low emissions, however, after further development work and only a year later the so called Lignox process[7] allowed for the production of TCF pulp. Hence, the small Swedish firm ASPA was the first mill in the world to produce TCF pulp. Still, it is important to note that while the development work physically took place at ASPA, the knowledge underpinning the transition stemmed from the long-term, industry-wide R&D collaboration described above (Bergquist and Söderholm 2015). In 1991, ASPA would, in collaboration with Greenpeace, produce an exact copy of "Das Plagiat", the weekly *Der Spiegel* printed on paper made of such pulp. It was the first time a magazine was published on chlorine-free paper and this accomplishment has been recognized as having had a great impact on the German public and, by extension, also the greater northern European market (Waluszewski and Håkansson 2004). For the first time it was possible to market paper products with eco-labelling, an opportunity which the Swedish producers quickly embraced. At this time the Swedish Nature Conservation Association had launched two environmental classes ("low chlorine" and "chlorine-free") for bleached paper qualities which got major impact and helped the Swedish pulp producers in their first-time marketing of green products by this new 'eco-labelling' standard (i.e., chlorine-free pulp and paper products) (Bergquist and Söderholm 2015).

[7]The Lignox process: oxygen-bleached pulp is treated with hydrogen peroxide at high temperatures after the removal of heavy metals with a complexing agent. Subsequent final bleaching takes place with peroxide and chlorine dioxide.

In 1994 the majority of the Nordic mills that had switched to TCF were of Swedish origin (remaining Swedish mills had all switched to ECF). In contrast, only two mills in Finland had opted for TCF by this time. Also in the US, the companies showed little interest in shifting to TCF. Only two mills had adopted TCF in 1994 (Reinstaller 2005, p. 1373). Thus, also in the US the policy process moved more slowly and when the Swedish PPI aimed to produce ECF and TCF pulp in the early 1990s the US EPA still worked on defining the standard settings for controlling dioxin emissions. For the US PPI, the dioxin issue meant something completely new as the mills could not compel only by installing end-of-pipe treatments, but had to invest significant funds in adopting new bleaching technology. The US industry had never had a strategy to deal with pollution with internal process changes (Gunningham et al. 2003, p. 16; see also Smith 1997). Thus, resistance to making the change was strong within the sector because it would come both at a very high cost and without any advantages on the American market (Reinstaller 2005, 1380). This helps explain why Greenpeace failed in its attempt in 1992 to convince the American *Time* magazine to switch to publishing on TCF paper. Further, Georgia Pacific, one of the leading pulp and paper producers in the US, announced in 1992 to consumers that it would not market TCF pulp, having decided to invest in chlorine dioxide substitution instead. Customers who wanted TCF pulps would have to seek other suppliers. This was possible for the company to do and say as as the demand for TCF pulp was very low on the US market (Smith 1997, p. 131). And together with a slower regulatory process, US firms could wait to alter the technology. In this context, in 1989 the US Office of Technology Assessment (OTA) reported on the compliance of Swedish pulp mills to more stringent standards of chlorinated organics, whereas it was unreasonable to hold US firms to such standards due to both economic risks and scientific uncertainties (Powell 1997, p. 12).

In sum, although the shift towards ECF and TCF apparently happened swiftly in the Swedish case, the development-path towards environmentally friendlier bleaching methods was the outcome of decades-long industrial efforts to increase efficiency and reduce the use of chemicals by changing internal processes, in combination with long-term and continuously strengthened environmental legislation.

4.6 Energy Transition

In parallel to the "greening" of industry with respect to increased pollution control, two other issues have had a major impact on the greening of the PPI since the early 1970s. The first is the increased energy prices in the wake of the oil crises of the 1970s and the second is the fear of a shortage of wood. Neither issue was initially related to environmental concerns, but became increasingly so as they drove serious concerns about energy use and triggered energy savings as well as recycling.

Indeed, pulp and paper production is energy-intensive. The PPI is the largest energy user in Sweden and accounts for 52% of the total industrial energy use (SEA 2017, p. 21). In the EU the PPI accounts for 14% of the total industrial energy use (Jönsson 2011). Although energy usage was always important to the PPI due to its cost, the 1970s caused the need for energy savings and finding substitutes for oil to become preeminent concerns, and both the Swedish PPI and the Swedish government turned their attention to phasing out oil from pulp and paper production. As a consequence, the energy mix of the Swedish PPI underwent radical changes over the 1970s and 1980s, and a large-scale substitution of oil took place. In 1973, for instance, oil accounted for 43% of the total external energy use and in 1984 the share of oil had decreased to 16%. Moreover, in 2011 the corresponding number was 5% (Skogsindustrierna 2012).

Oil reduction was mainly achieved through increased use of internal biofuels (external biofuels remained fairly constant); between 1973 and 1984 the share of energy generated from internal biofuels increased from 55 to 72%, and in 2011 the share of biofuels was 79% (Fig. 4.7). The biofuels mainly consisted of by-products from the pulp manufacturing process, where the biggest share is generated from black liquor[8] and the rest from bark and wood residues (Federation of Swedish Forest Industry 2012). The overall reduction in the use of oil was also made possible through other energy efficiency improvements and increased internal production of electricity through back-pressure turbine power generation (Lindmark et al. 2011; Bergquist and Söderholm 2016). Of central importance for this development, which included a great deal of incremental technology development, was the already established collaborative strategy of firms within the industry and between the PPI and state authorities. Thus, a large number of energy projects were subsequently conducted through inter-firm and state-industry collaborative R&D platforms established after the 1940s, such as STFI and SSVL (Table 4.3). Still, to manage the new energy challenge the Swedish PPI had already appointed in 1973 a standing Energy committee consisting of 12 members from among management and technical personnel within the sector.

Between 1973 and 1977, no fewer than 51 new energy projects had been initiated or at least proposed within the sector. These involved 37 energy conservation and 14 energy generation projects, most conducted/proposed in collaboration with organisations closely associated with the sector, such as STFI, the Swedish Cellulose and Paper Mill Association (SCPF), and the Steam Generator Association (in Swedish Ångpanneföreningen), but also universities and research institutes outside the sector (e.g., the Thermal Engineering Research Institute (in Swedish Värmeforsk) (Marklund 1994, p. 143). Underpinning the energy transition of the Swedish PPI was also a proactive governmental strategy to emphasize knowledge management and collaboration with industry along with the substitution of oil with internal biofuels. The Swedish government assigned significant subsidies for prototypes and demonstration plants by its energy policy decision in 1975, and such activities increased greatly over time. In the 1970s, new processes and technologies on a factory-wide scale could be subsidised by as much as 50% (Bergquist and Söderholm 2016).

[8]In the pulping process, cooking chemicals, known as white liquor are used to break out and dissolve the lignin whereas the white liquor becomes black liquor. There are basically two main process streams coming out of a pulp mill: cellulose fibres and black liquor.

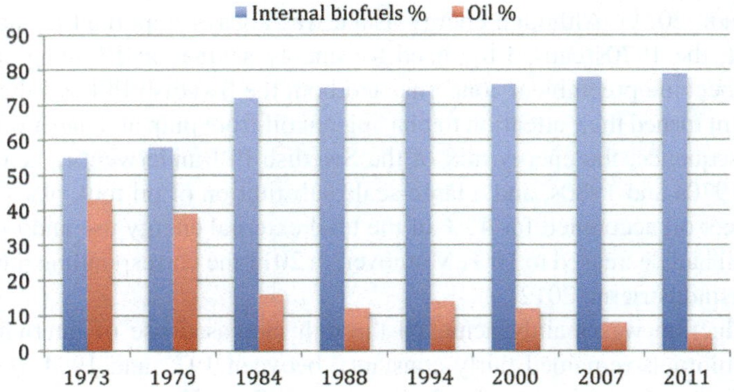

Fig. 4.7 Consumption of internal biofuels and oil in the Swedish pulp and paper industry 1973–2011 (% of total energy use). *Source* Skogsindustrierna (2012)

The energy transition of the Swedish PPI was further driven by the fact that focus was directed towards unutilized potential energy sources, whereby a previous waste problem with bark and chips now could be transformed into energy savings and improved energy efficiency. Moreover, the PPI feared shortages of wood and sought means of increasing its energy efficiency, and these forces led it to increase its use of recycled paper in pulp production; producing pulp from recycled paper requires only one-fifth of the energy needed produce it from virgin wood. Thus, from not have being utilised at all in Swedish pulp production prior to 1975, recycled paper came to be an important raw material. Of central importance for this development were improved methods for removing ink and other contaminants, such as plastic, from the recycled paper, and the introduction in 1975 of a compulsory collection system for old newsprint from households (Bergquist and Söderholm 2016).

The improved energy efficiency over the 1970s and 1980s coincided with an on-going structural change in industry, namely a trend towards fewer and larger production units and an improved ability to take advantage of economies of scale (Järvinen et al. 2012). The concentration of production units allowed, among other things, a higher degree of integrated pulp and paper production whereby the energy intensive step of drying the pulp (at the pulp mill, only to dissolve it again at the paper mill) was avoided (Bergquist and Söderholm 2016).

Today the energy efficiency of the Swedish (and Finnish) PPI is higher than that of other major pulp producing countries such as Brazil, the US and Canada (Fracoro et al. 2012). However, the sector's total energy use has stayed relatively constant due to considerable production expansion, not the least in mechanical pulp, which has contributed to increased use of electricity beginning in the early 1980s (Bergquist and Söderholm 2016). As a result, between 1973 and 2011 the total energy consumption of the Swedish PPI increased from 55 to 57 TWh (Skogsindustrierna 2012).

4.7 Environmentally Driven Transformation—The Past and the Future

Over the second half of the twentieth century, environmentally related issues have played a central role in the overall transformation of the Swedish PPI, essentially as new or altered technologies were required to comply with tightening environmental regulations. In the Swedish context, the greening of polluting manufacturing industries occurred within a collaborative regulatory framework. Knowledge was flowing quite openly between Swedish PPI companies and between environmental authorities and the regulated industry (Bergquist and Söderholm 2011). Thus, the significant environmental improvements accomplished within the Swedish PPI over the period can only be fully comprehended by acknowledging the role of collaborative environmental R&D activities among firms within the sector and between the sector and the state. Furthermore, the development after the 1980s illustrates how the emergence of green consumerism came to play a role as a driver towards environmental improvements.

The PPI has developed from being considered one of the greatest polluters of the twentieth century to an industry that is essential for the transition towards a more sustainable (bio-based) economy. In the EU, the PPI today already constitutes the biggest single industrial producer and user of renewable energy; 56% of the industry's primary annual energy consumption is bio-based (CEPI 2013, p. 50). And there are big hopes, especially within the Nordic countries, that the PPI will take the lead in clean-tech innovation and green growth in terms of new products and the second generation of "green" fuels. The trend is global and most pulp and paper companies are working in this direction. The sector, particularly in Sweden, is at the same time undergoing substantial change due to stricter environmental regulations, unstable oil price, energy policies, global competitiveness and structural changes. In addition, the increasing use of electronic rather than paper-based communications are pushing the PPI towards self-renewal to improve its profitability (Backlund and Nordström 2014), whereby it appears as if parts of the pulp industry have experienced a shift towards becoming more "biorefineries" than pulp producers (Hamaguchi et al. 2012). One promising technology in this context is black liquor gasification (BLG[9]) for the production of fossil free transportation fuels. Here the Swedish company Chemrec has been a key player. Thus, in 2009 and after 20 years of R&D on BLG, Chemrec constructed a pilot plant with financial support from the European Community's Framework Programmes and the Swedish Energy Agency. However, after the BLG demonstration technology did not develop as expected and there was a lack of investment interest in the project, the pilot plant was shut down and remains so. One suggested reason for the lack of interest was the uncertain market conditions for biodiesel and methanol (SEA 2016). Still estimates suggest that BLG could supply as much as 7% of Sweden's total—not merely its industrial—demand for electricity, or as much as 30% of the nation's demand for transportation fuels. Outside the

[9]BLG is in its essence a process in which a clean synthesis gas is produced from black liquor by converting its biomass content into gaseous energy carriers (Bajpai 2014, p. 3).

Nordic countries, BLG technology is also developing in the US and Asian pulp producing countries (Bajpai 2014, pp. 4–5). Another example of on-gong renewals is the Finnish Metsä Group's investment in the 'bioproduct' mill in Äänekoski, Central Finland, which represents the largest forestry investment ever made in Europe, with a value of EUR 1.2 billion. It is the first next-generation product mill in the world, were the production of energy, along with the pulp production, play a crucial role (Metsä Group 2017).

Another trend in the PPI's development towards environmental improvements, partly from new products within the area of bioenergy, is a continued expansion of the traditional pulp and paper production driven by an increased global demand for cardboard and sanitary papers. For example, the Swedish pulp and paper company SCA is currently rebuilding its Östrand mill (due to start up in 2018) in Sweden and has publicized that a large-scale investment programme will lead to 'the largest production line for bleached softwood sulphate pulp in the world'. Above all the firm has stressed that the mill will become 'world class in terms of product quality, environment and competitiveness' by becoming 'the leader in terms of resource management' and 'generate surplus energy' to be sold in the form of 'green electricity and district heating'. Also, 'both TCF and ECF pulp' will be produced with raw materials from 'sustainable forestry'.[10] Overall the Swedish example illustrates, however, that the transition towards cleaner and more energy efficient pulp and paper production is the result of long-term and incremental processes, whereby the technology development has been shaped by both evolving markets, long-term collaborative R&D and shifting concerns in society about the natural environment and human well-being.

References

AET (Alliance for Environmental Technology) (2012) Trends in world bleached chemical pulp production 1990–2012. http://www.aet.org/science_of_ecf/eco_risk/2013_pulp.html. Accessed 18 Jan 2018

Auer MR (1996) Krafting an agreement: negotiations to reduce pollution from the nordic pulp industry, 1985–1989. Yale University Press, Yale University

Backlund B, Nordström, M (2014) Nya produkter från skogsråvara. En översikt av läget 2014. Innventia Rapport nr. 577 november 2014

Bajpai P (2011) Environmentally friendly production of pulp and paper. John Wiley & Sons, Inc, Hoboken

Bajpai P (2014) Black liquor gasification. Elsevier London, Waltham

Bajpai P (2015) Management of pulp and paper mill waste. Springer, Cham

Bergquist A-K (2017) Business and sustainablity. New business history perspectives. Harvard business school working paper, pp 18–043

Bergquist A-K, Keskitalo C (2016) Regulation versus deregulation. Policy divergence between Swedish forestry and the Swedish pulp and paper industry after the 1990s. For Policy Econ 73:10–17

[10]https://www.sca.com/sv/massa/aktuellt-inom-pulp/expansion-av-ostrand/. Accessed 19 January 2018.

Bergquist A-K, Söderholm K (2011) Green innovation systems in Swedish industry 1960–1989. Bus Hist Rev 85(4):677–698

Bergquist A-K, Söderholm K (2015) Transition to greener pulp: regulation, industry responses and path dependency. Bus Hist 57(6):862–884

Bergquist A-K, Söderholm K (2016) Sustainable energy transition: the case of the Swedish pulp and paper industry 1973–1990. Energy Efficiency 9(5):1179–1192

Bergquist A-K, Söderholm K, Kinneryd H, Lindmark M, Söderholm P (2013) Command-and-control revisited. Environmental Compliance and Technical Change in Swedish Industry 1970-1990", Ecological Economics, vol 85, pp 6–19

CEPI (The Confederation of European Paper Industries I) (2013) CEPI sustainability report 2013. http://www.cepi.org/system/files/public/documents/publications/sustainability/2013/e-mail_sustainability%20report%202013_SHORT_pages_LOW.pdf. Accessed 18 Jan 2018

David PA (1994) Why are institutions the 'carriers of history'?: path dependence and the evolution of conventions, organizations and institutions. Struct Change Econ Dyn 5(2):205–220

Fahlström JM (1948) Några drag ur den svenska pappers- och cellulosa industriens ekonomiska utveckling under 1900-talet. Cellulosa och papper. Svenska Pappers- och Cellulosaingeniörs-föreningen, Stockholm, pp 69–80

Fracoro G, Vakkilainen E, Hamaguchi M, Nelson S, de Souza M (2012) Energy efficiency in the brazilian pulp and paper industry. Energies 5:3550–3572

Frondel M, Horbach J, Rennings K (2007) End-of-pipe or cleaner production? An empirical comparison of environmental innovation decisions across OECD countries. Bus Strategy Environ 16(8):571–584

Geels FW (2004) From sectoral systems of innovation to socio-technical systems: insights about dynamics and change from sociology and institutional theory. Res Policy 33:897–920

Gunningham N, Kagan RA, Thorthon D (2003) Different shades of green. Business, regulation, and environment. Stanford University Press, Stanford, CA

Hamaguchi M, Cardoso M, Vakkilainen E (2012) Alternative technologies for biofuels production in kraft pulp mills—potential and prospects. Energies 5(7):2288–2309

Harrison K (2002) Ideas and environmental standard setting: a comparative study of regulation of P&P industry. Governance 15:65–96

Henriksson E, Söderholm P, Wårell L (2012) Industrial electricity demand and energy efficiency policy: the role of price changes in private R&D in the swedish paper & pulp industry. Energy Policy 47:437–446

Hoffman AJ (1997) From heresy to dogma. An institutional history of corporate environmentalsim. The New Lexington Press, San Francisco

Hoffman AJ, Bansal P (2012) Retroperspective, perspective and prospective: introduction. In: Bansal P and Hoffman AJ (eds) The oxford handbook of business and the natural environment. Oxford University Press, Oxford, New York, pp 140–157

Industriutsläppsförordningen (2013): SFS 2013:250

Ince BK, Cetecioglu Z, Ince O (2011) Pollution prevention in the pulp and paper industries. In: Broniewicz E (ed) Environmental management in practice. https://www.intechopen.com/books/environmental-management-in-practice/pollution-prevention-in-the-pulp-and-paper-industries. Accessed 17 Dec 2017

Jaffe AB, Newell RG, Stavins RN (2005) A tale of two market failures: technology and environmental policy. Ecol Econ 54:164–174

Jerkeman P (2007) Bokslut. Några lönsamma forskningsresultat, Skogsindustrins Forskningsstiftelse, Stockholm

Jirvall N, Noodapera M (1995) Miljöinfo från Skogsindustrierna. Media express, Stockholm, Skogsindustrierna

Joas M (1997) Finland: from local to global politics. In: Andersen MS, Liefferink D (eds) European environmental policy: the pioneers. University Press Manchester, Manchester pp 119–160

Järvinen J, Ojala J, Melander A, Lamberg JA (2012) The evolution of pulp and paper industries in finland, sweden, and norway 1800–2005. In: Lamberg J-A, Ojala J, Peltoniemi M, Särkkä T (eds)

The evolution of global paper industry 1800–2050. A comparative analysis. Springer, Dordrecht, pp 19–48
Jönsson J (2011) Analyzing different technology pathways for the pulp and paper industry in a european energy systems perspective. Chalmers University of Technology, Gothenburg
Kramer JD (2000) Pulping/bleaching technology view shows North America Lagging. Pulp Paper 74:51–59
Kemp R, Soete L (1990) Inside the "green box": on the economics of technological change and the environment. In: Freeman C, Soete L (eds) New explorations in the economics of technological change. Pinter, London, pp 248–258
Lamberg J-A, Ojala J, Peltoniemi M, Särkkä T (eds) (2012) The evolution of global paper industry 1800–2050: a comparative analysis. Springer, Dordrecht
Lee S-Y, Rhee S-K (2005) From end-of-pipe technology towards pollution preventive approach: the evolution of corporate environmentalism in Korea. J Clean Prod 13(4):386–395
Liebowitz SJ, Margolis SE (1995) Path dependence, Lock-In, and history. J Law Econ Organ 11(1):205–226
Lindmark M, Bergquist A-K (2008) Expansion for pollution reduction? Environmental adaptation of a Swedish and a Canadian metal smelter 1960–2005. Bus Hist 50(4):530–546
Lindmark M, Bergquist A-K, Andersson L-F (2011) Energy transition, carbon dioxide reduction and output growth in the Swedish pulp and paper industry: 1973–2006. Energy Policy 39(9):5449–5456
Lönnroth M (2010) The organisation of environmental policy in Sweden. a historical perspective. Swedish environmental protection agency, Report 6404. https://www.naturvardsverket.se/Documents/publikationer6400/978-91-620-6404-4.pdf?pid=3712. Accessed 18 Jan 2018
Lundgren LJ (1974) Vattenförorening: debatten i Sverige 1890–1921. Bibliotheca historica Lundensis, Lund
Lundgren LJ (1999) Från miljöproblem till miljövård. ett genombrott med fördröjning. Deadalus, Yearbook of the Technical Museum, Stockholm
Lundqvist LJ (1971) Miljövårdsförvaltning och politisk struktur. Uppsala: Verdandi; Stockholm: Prisma Bokförlaget Prisma/Föreningen Verdandi
Marcus A (1999) TCF is alive and well and living in scandinavia. Papermaker 21:20–23
Marklund G (1994) Institutions and appropriation in Swedish technology policy. Graphic systems. Stockholm
Metsä Group (2017). http://bioproductmill.com/about-the-bioproduct-mill. Accessed 15 Sep 2017
McNeill JR (2000) Something is new under the sun. W.W Norton, New York
Michanek G, Zetterberg C (2007) Den Svenska Miljörätten, 2nd edn. Iustus Förlag, Uppsala
Mutz M (2009) Managing resources: water and wood in the german pulp and paper industry 1870s–1930s. Jahrbuch für wirtshaftsgeschiticthe/economic history yearbook 59(2):45–68
Nilsson N (2007) Cleaner production: technologies and tools for resource efficient production. Baltic University Press, Uppsala
Norberg-Bohm V, Rossi M (1998) The power of incrementalism: environmental regulation and technological change in P&P bleaching in the US. Technol Anal Strateg Manag 10:225–245
Nordic Council of Ministers (1989) Reduction of cloro-organic discharge in the Nordic pulp industry. Jaakko Pöyry for the Nordic Council of Ministers, Copenhagen
OECD (1973) Pollution by the P&P industry—present situation and trends. OECD, Paris
Pierson P (2000) Increasing returns, path dependence, and the study of politics. Am Polit Sci Rev 94(2):251–267
Popp D, Hafner T, Johnstone N (2011) Environmental policy vs. public pressure: innovation and diffusion of alternative bleaching technologies in the pulp industry. Res Policy 40:1253–1268
Powell MR (1997) Control of dioxins (and other Organochlorines) from the P&P industry under the clean water act and lead in soil at superfund mining sites: two case studies in the EPA's use of science, discussion paper 97-08. Resources for the Future, Washington DC
Rajotte A (2003) Paper production technology and environmental performance in Sweden and finland: policy, science and market share. Soc Natur Resour 16:719–728

Reinstaller A (2005) Policy entrepreneurship in the co-evolution of institutions, preferences and technology: comparing the diffusion of totally chlorine free pulp bleaching technologies in the US and Sweden. Res Policy 34:1366–1384

Rydberg S (1990) Papper i perspektiv—Massa- och pappersindustri i Sverige under hundra år. The Swedish Forest Industries, Stockholm

Schoenberger H (2009) Integrated pollution prevention and control in large industrial installations on the basis of best available techniques—the sevilla process. J Clean Prod 17:1526–1529

Smith M (1997) The US paper industry and sustainable production. The MIT Press, Cambridge, MA, An Argument for Reconstruction

Smith A, Rajotte A (2001) When markets meet sociopolitics: the introduction of chlorine-free pulp bleaching in the Swedish P&P industry. In: Coombs R, Green K, Richards A, Walsh V (eds) Technology and markets. demand, users and innovation. Edward Elgar, Cheltenham, pp 136–155

Sinclair WF (1990) Controlling pollution in the Canadian pulp and paper manufactureres: a federal perspective. Minister of Supply and Services, Ottawa

Skogsindustrierna (1995) MiljöInfo från skogsindustrierna. Skogsindustrierna, Stockholm

Skogsindustrierna (2012) Energiförbrukning i massa och pappersindustrin. http://docplayer.se/20505932-Energiforbrukning-i-massa-och-pappersindustrin-2011.html. Accessed 18 Jan 2017

Swedish Environmental Protection Agency (2016) Vägledning om industriutsläppsbestämmelser. Rapport 6202. http://www.naturvardsverket.se/Documents/publikationer6400/978-91-620-6702-1.pdf?pid=17109. Accessed 17 Jan 2018

SSVL (Swedish Forest Industries' Water and Air Pollution Research Foundation) (1991) Miljö 90. Final report, Swedish Forest Industries Federation, Stockholm

SSVL (Swedish Forest Industries' Water and Air Pollution Research Foundation) (1989) SSVL 74–85—miljöskyddskostnader inom svensk skogsindustri 1985 t o m 1991. Swedish Forest Industries Federation, Stockholm

Swedish Energy Agency (2016) Energimarknadsrapport för biodrivmedel och fasta bränslen. http://www.energimyndigheten.se/contentassets/4ea2e3aa160647e68df2bc3f18bd361b/energimarknadsrapport-biobranslen_160602.pdf. Accessed 18 Jan 2018

Söderholm K (2009) Environmental awakening in the Swedish pulp and paper industry: pollution resistance and firm responses in the early 20th century. Bus Strategy Environ 18(1):32–42

Söderholm K, Bergquist A-K (2012) Firm collaboration and environmental adaptation: the case of the Swedish pulp and paper industry 1900–1990. Scand Econ Hist Rev 60(2):183–211

Söderholm K, Bergquist A-K (2013) Growing green and competitive: a case study of a Swedish Pulp Mill. Sustain 5(5):1789–1805

Söderholm K, Bergquist A-K, Söderholm P (2017) The transition to chlorine free pulp revisited: Nordic heterogeneity in environmental regulation and R&D collaboratio. J Clean Prod 165:1328–1339

Thompson G, Swain J, Kay M, Forster CF (2001) The treatment of pulp and paper mill effluent: a review. Biores Technol 77:275–286

Thollander P, Ottosson M (2008) An energy efficient Swedish pulp and paper industry—exploring barriers to and driving forces for cost-effective energy efficiency investments. Energ Effi 1(1):21–34

Waluszewski A, Håkansson H (2004) Das plagiat. In: Widmalm S (ed) Artefakter: industrin, vetenskapen och de tekniska nätverken. Gidlunds förlag, Möklinta, pp 219–244

Wohlfart G (1971a) Svensk skogsindustri i omvandling: utvecklingen sedan 1950, Part 1. Strukturutredningen, Grafikon AB, Stockholm, The Cooperation Committee of the Forest Industries (Skogsindustriernas samarbetsutskott)

Wohlfart G (1971b) Svensk skogsindustri i omvandling: utvecklingen sedan 1950. Part 2. The cooperation committee of the forest industries (Skogsindustriernas Samarbetsutskott). Strukturutredningen, Grafikon AB, Stockholm

World Watch Institute (2015) Paper production levels off, environmental footprint still high. http://www.worldwatch.org/paper-production-levels-environmental-footprint-still-high. Accessed 21 Jan 2018

Part II
Regulations and Institutions

Chapter 5
Varieties of State Aid and Technological Development: Government Support to the Pulp and Paper Industry, the 1970s to the 1990s

Jari Ojala, Niklas Jensen-Eriksen and Juha-Antti Lamberg

5.1 Introduction

In this chapter, we look at how governments have supported the growth of the pulp and paper industry in the Organisation for Economic Co-operation and Development (OECD) countries. Scranton and Fridenson (2013) have recently argued that business historians should recognize that the 'state is always "in"'. Government consumption, support, protection and regulation have indeed played an important role in the evolution of industries over time. Yet, scholars have argued that industrial policy, despite its significance, is a relatively neglected field of historical and economic research (Grabas and Nützenadel 2014; review of literature e.g. in Pack and Saggi 2006).

The period from the 1970s to the 1990s witnessed a gradual change in a number of countries from coordinated market economies with a strong role played by the state and regulation towards more liberal market economies and de-regulation. Today European Union legislation, for example, is aimed at hindering "unfair advantage" through governmental support for industries—even though the European Commission has rather flexible rules on which kind state aid is allowed and which is not.[1] The World Trade Organization (WTO) has similarly tried to limit state support for

[1] On EU competition policy see, for example, http://ec.europa.eu/competition/consumers/government_aid_en.html. Accessed 7 August 2016.

J. Ojala (✉) · J.-A. Lamberg
Department of History and Ethnology, University of Jyväskylä, Jyväskylä, Finland
e-mail: jari.ojala@jyu.fi

J.-A. Lamberg
e-mail: juha-antti.lamberg@jyu.fi

N. Jensen-Eriksen
Department of Philosophy, History and Art Studies,
University of Helsinki, Helsinki, Finland
e-mail: niklas.jensen-eriksen@helsinki.fi

T. Särkkä et al. (eds.), *Technological Transformation in the Global Pulp and Paper Industry 1800–2018*, World Forests 23, https://doi.org/10.1007/978-3-319-94962-8_5

industries. However, the governments of many emerging economies have been eager to support their companies, and since the financial crisis the popularity of industrial policy has grown in OECD countries as well. Grabas and Nützenadel (2014) have even described this development as representing 'a true renaissance of industrial policy, not only in Europe, but also in other parts of the world' (p. 2). Whether industrial policies in different countries have been successful or not is a moot point, especially as the 'relevant counterfactuals are not available' (Pack and Saggi 2006, p. 268). Thus, it is vital to expand our understanding on what kind of tools governments have used and can use in their efforts to promote the growth of their national companies, industries, and economies.

Industrial policy is not an easy topic to study because the governments often have a motive to shield it from outsiders. The state aid can also take many forms and has varied over time and space (Bianchi and Labory 2006; Buigues and Sekkat 2009). Therefore, observers have either focused on individual countries, or have looked at how certain tools, usually financial aid, were employed (e.g. Aydin 2007; Ford and Suyker 1990). We have adopted a more holistic approach. We aim to capture the entire industry's policy system in different countries, and look at the significant instruments used by governments to support one specific industry over a certain period of time. As far as we know, no one has tried this before. Though we look at the various forms of the state aid holistically, we will especially pay attention to ones that hand on impact to technology development.

We use OECD information on various mechanisms of state aid. These data have a number of problems, which we will highlight later, but they will, at the very least, give us an overview of the role played by governments in the development of the pulp and paper industry. Our holistic approach is warranted as previous studies are not unanimous as to why and how certain industries are supported and why certain nations engage in industrial policies more and others less. Public choice scholars (e.g. Frey 1984), for example, assume that the magnitude of governmental support correlates with the bargaining power of a specific industry. Other literature on industrial policy, however, suggests that governments are not just eager to "pick winners" but also often to "help losers" (Foreman-Peck 2014), in particular if it is important to do so for employment or electoral reasons. Accordingly, we are motivated to explore the de facto governmental activity in the pulp and paper industry to understand the mechanisms either resulting in governmental support or lack of it.

Regulation and governmental support for industries are among the major issues used in classifying countries according to different varieties of capitalism and business systems (Whitley 1999; Hall and Soskice 2001; Iversen and Thue 2008). As Whitley (1999) has argued the role played by the state varies in different business systems in time and in space. According to him the states that were established during or after industrialization has occurred tend to be more involved in industrial development in general than the states that were established earlier. This generalization, though, might not hold in a case of one single industry—as can be seen in the case of pulp and paper in this chapter.

We argue that government support and overall institutional arrangements played an important role in influencing investments and technological development in this particular industry. Investments in the pulp and paper industry are very large and thus socially important and widely publicly debated. This has been witnessed most recently in the cases of Uruguay (since the early 2000s) and Finland (2010s). Financing the investments, infrastructure (e.g. energy and transport), regional policies, and environmental considerations, among other things, are issues that have gained attention also from the governmental side (Söderholm and Bergquist 2012; Wagner et al. 2002). On the one hand, governments are eager to promote investments in their respective countries in the face of global competition, but on the other hand investments are regulated in order to prevent possible environmental or societal challenges (Recently, e.g. Bergquist and Söderholm 2015; Jensen-Eriksen and Ojala 2015).

In the following sections, we will first describe the data used and offer some important critical considerations regarding our sources. Thereafter, we will briefly map the importance of pulp and paper industries in OECD countries from the 1950s until the 1990s. Next we will analyse the government support for the pulp and paper industry in OECD countries during the 1970s and 1980s, and follow it with our conclusions.

5.2 Data Considerations

In this chapter we will make a comparative macro analysis of state aid to the pulp and paper industry in western countries mainly during the 1970s and 1980s by using the data compiled at the time by the OECD. The Organisation for European Economic Cooperation (OEEC) was established after the Second World War in 1948 to run US-finance reconstruction program (Marshall Plan) in Europe. The name was changed to OECD in 1961 after US and Canada had joined the organisation. It was particularly interested in government support for and regulation of industries during this period, and it is a research-based organization that compiles data and conducts research to promote a better understanding of economies and industries to policy makers and industrialists (e.g. Godin 2004).

The pulp and paper industry was among the lines of businesses followed by the OECD and OEEC. A special pulp and paper industry committee was established to OEEC already at the turn of the 1940s and 1950s to compile production statistics from European countries; this committee continued also after OECD was established a decade later. Therefore, this committee compiled production figures from different countries from the late 1940s onwards, made reports on production and productivity, and evaluated the current and prospect trends of paper consumption. Moreover, some comparative material was even compiled from the pre-war period.

A specific ad hoc committee of the OECD, established in the early 1970s, compiled information from member countries about government support for the pulp and paper sector. This committee continued its work until the late 1990s. This data and other OECD documentation will be used here as the key source for the study. The fact that the OECD put the forest industries on its agenda early on witnesses the need of the producing countries and industries to discuss this sector and its development.[2]

The impact of governmental measures on the pulp and paper industry was topical in the OECD at the turn of the 1970s. Around that time, for example, in Finland and Sweden investments in the pulp and paper industries were regulated through a system that had been created in the late 1960s and were, to a certain extent, related to environmental considerations (details in Jensen-Eriksen and Ojala 2015; Bergquist and Söderholm 2011, 2015). Within the OECD, however, the main focus in analysing state support did not concern investments or technology development as such but rather forest ownership (i.e., company, private citizen or public), cartelization, and production. The special "ad hoc committee" analysed the topic further. The member countries were asked to answer a questionnaire focusing on both general and specific areas of concern that the governments might have in order to regulate—or rather promote—the pulp and paper industry in each respective country.[3]

The OECD reports can be seen as summaries of the available information and outlines of the evaluation exercise of government policy affecting the pulp and paper industry. Though surveys do have a number of inherent deficiencies in the source data, they are, nevertheless, valuable sources for studying the regulation of the pulp and paper industry from the early 1970s up to the late 1990s. In particular, they offer a shortcut to drawing international comparisons. As the surveys concentrated on all state aid and regulation in the field, investments and technology development, for example, do not show up specifically in the reports. Nevertheless, in this chapter this governmental support is seen as a motivation for investments, which in turn, were related to technology development. Moreover, government support in each respective country might be underestimated in the reports sent to the OECD because it could have been perceived as an organization whose aim was to hinder government regulation.

The OECD data also has other obvious shortcomings. Foremost, the OECD data includes only the (Western and developed) countries that were members of the association. Thus, we cannot say with the OECD data whether, for example, Asian or South American countries had governmental measures that attracted the pulp and paper industry before the turn of the millennium. Also the OECD countries considered that reports were to certain extend unclear and incomparable, which led ultimately to the end to this kind of data gathering. Namely, during the late 1980s European member countries in particular got frustrated with the work of the ad hoc committee and were cautious about the quality of the data they compiled; Japanese participants even questioned the whole rationale of the data gathering rather early on.[4] In contrast, North American participants were more willing to continue this

---

[2]The collection of documents of OECD (OEEC) Special Committee For Pulp And Paper (1950–1970); Industry Committee, Pulp and Paper Section (1971–1998) can be found, for example, from the archives of the Finnish Forest Industry Association at the Finnish Central Business Archives (ELKA), Mikkeli, Finland.

[3]ELKA. OECD Industry committee, Pulp and Paper Section, 22 June 1972.

[4]ELKA. OECD Industry Committee, Pulp and Paper Section, letters between Castrén and Neal 18 and 24 April 1973; report by chairman Neal, 29 March 1973.

committee, as they saw it as a means for discussing more openly with European member states, as European integration was its way to make European countries more unite and one negotiation partner. The high hopes of the early 1970s changed to frustration as the type of data, which the committee hoped to compile was seen to be too ambitious, not representative enough and not really comparative. The reports from the data were revised time and time again as the member countries were not satisfied with the results. Moreover, there were even some clashes within the group, and the chairmen and consultants used were frequently subject to intense criticism. The focus of the pulp and paper committee was changed several times as the "holistic" approach did not work out. Thus, first the committee concentrated on some "hot topics" at the time (e.g., energy and waste paper), and lastly during the 1980s some more general policy issues were discussed in seminars. The results both from the "hot topics" and from seminars were rather general and obvious, and so the criticism of the committee intensified. Nevertheless, the OECD continued to compile important statistical information that was not criticized by the member country representatives.[5]

By the mid-1990s the rationale for having this type of specific committee became even more challenging, as there were no resources for compiling accurate data from either former Eastern Bloc countries or emerging Asian economies. The North American representatives, however, were still more than willing to continue the work, but the Europeans more or less put a halt to it.[6] As a source critical consideration one must bear in mind that most of the data compiled by the OECD came either directly from member countries' officials or private companies. There was no compulsion to provide the information, and thus its accuracy might be questioned.[7]

In this chapter we use also some other sources besides the ones provided by OECD. The Food and Agricultural Organization (FAO) of the United Nations (UN) also compiled data and analysis that is useful in comparative studies. The OECD was more interested in the role of governments in protecting their respective industries.[8] In contrast, the FAO's reports are more concerned with global development, environmental considerations and changes in inequality. Moreover, the OECD data are more specifically aimed at analysing the pulp and paper industries, whereas the FAO reports concern forestry (and forestry politics) in general (e.g. King 1974/1975).

[5]ELKA. OECD Industry Committee, Pulp And Paper Section, summary record of the 2nd session held at the O.E.C.D., 24–25 November 1975; Working Party memorandum (in Finnish), 11 February 1975; Industrial deregulation: synthesis paper, 24 July 1986.

[6]ELKA. OECD. Ad hoc working party on pulp and paper, memorandums by V. Ungern-Sternberg, 1 November 1988 and 28 October 1996. About the initial discussion to compile comparative data on pulp and paper industry on OECD level, see OEEC, Pulp and Paper Committee, 26 November 1959.

[7]ELKA. OEEC. Pulp and Paper Committee, 26 November 1959.

[8]ELKA. OECD. Industry committee, Pulp and Paper Section, 6 December 1971.

5.3 The Pulp and Paper Industry in the OECD Countries

The pulp and paper industry has received less government support and been subject to less regulation in OECD countries than many other industries, such as food, textiles, clothing, steel, and shipbuilding. An explanation for this situation might be the small rate of change in employment in pulp and paper industry when compared to other industries, and the pulp and paper industry's ability to raise private capital.[9] Moreover, pulp and paper industry did not experience such a decline than many other industries did during the 1970s and 1980s (Lamberg et al. 2018). Nevertheless, the pulp and paper industry was an important line of business in OECD countries during the post war era. The industry was also highly globalized. The chairman of industry committee for pulp and paper industry, F. A. Neil, though, might have exaggerated this point when he stated in an industry committee meeting in 1974 that the 'paper industry is perhaps one of the most international that exists', as all its big players were engaged in foreign trade. At that time, the annual value of the paper industry trade was 7,000 million US dollars,[10] whilst in the 1950s its total value in European OEEC countries was 3,500 million US dollars; it was produced in 2,000 paper mills using a labour force of 420,000. Moreover, the industry had grown rapidly since the Second World War. During the 1950s alone, the growth in paper consumption in OEEC countries was 70%,[11] and the growth continued during the 1960s–1990s. Among European countries the industry was especially important in Finland, Austria, Sweden, and Norway, and these nations produced already by the late 1950s two-thirds of all wood pulp produced in OEEC countries, and exported over half their paper and over 60% of their pulp production respectively. The other European OEEC countries were, on the contrary, dependent on imports: 59% of their pulp and 14% of their paper was imported in the late 1950s.[12] Although the production figures increased considerably from the 1960s to 1980s, the size of the labour force decreased due to the growth in productivity. In Germany, for example, the labour force in pulp and paper industry in 1960 numbered 77,000 whilst the figure in 1985 was 47,000 (Table 5.1). At the same time, though, the value of production rose roughly five fold.[13]

During the late 1980s, there were around 9,000 paper machines in the world, of which around 2,500 were located in the OECD countries. However, the machines in the OECD countries were larger and newer. Between 1971 and 1988, for example, in the OECD countries 1,300 new machines were built and 1,800 had gone through major rebuilds. Moreover, during the 1970s and 1980s no fewer than 3,000 small machines were shut down; these machines averaged a capacity of 7,500 tons a year,

[9]ELKA. OECD Ad hoc working party on pulp and paper application of new technologies in the paper industry, 21 January 1988.

[10]ELKA. OECD Ad hoc working party, speech by F. A. Neal, 30 August 1974.

[11]ELKA. Including Finland, though Finland was not a member state of OEEC at the time. OEEC Pulp and paper committee, 3 July 1959.

[12]ELKA. OEEC. Pulp and paper industry committee, July 3, 1958.

[13]ELKA. OECD. Ad hoc working party on pulp and paper application of new technologies in the paper industry, January 19/21 1988.

Table 5.1 Share of Pulp and Paper Industry employment of total manufacturing employment in the OECD countries 1967–1982 (% share)

Country	1967	1974	1982
Austria	n/a	3.7	3.2
Belgium	2.4	2.5	2.3
Denmark	2.9	2.5	2.3
Finland	11.1	10.0	9.2
France	2.5	2.6	2.4
Germany	2.6	2.4	2.3
Greece	2.3	1.9	2.3
Ireland	2.9	2.8	2.3
Italy	2.4	2.1	2.0
Netherlands	3.0	3.0	2.7
Norway	6.9	5.9	4.6
Portugal	3.9	2.3	2.6
Spain	2.4	2.5	n/a
Sweden	6.9	6.6	7.2
Turkey	2.2	2.3	2.2
UK	3.0	3.1	3.1
Canada	7.1	7.3	7.0
USA	3.5	3.5	3.4
Japan	3.2	2.9	2.7
Australia	2.8	2.4	2.2
New Zealand	3.9	3.6	3.9
Total av.	3.9	3.6	3.5

Source OECD, DSTI/IND/PP/87.2/(2nd revision). Based on UNSO—New York data
Note Employment of pulp, paper and paper products included; Table includes only OECD countries from which we have enough data available

and this figure rose in the period 1981–1985 to 20,000 tons a year. The industry also went through a major restructuring during this period. In European Economic Community (EEC) countries alone, between 1975 and 1985 roughly 45% of companies disappeared and 723 mills were shut down (43%). The closed ones were, however, small and medium sized firms; among the top 100 producers there were no significant changes before the late 1980s. Moreover, production was concentrated among the big firms. The top 100 firms in the OECD countries produced one third of paper in the world in the mid-1970s, and in 1985 this share was already 46%.[14]

As Table 5.2 illustrates, the share of pulp and paper production compared to a nation's total manufacturing activity declined in all OECD countries from the mid-1960s to the mid-1980s, with exception of Australia and New Zealand. Thus,

[14]ELKA. OECD. Ad hoc working party on pulp and paper application of new technologies in the paper industry, 19/21 January 1988.

Table 5.2 Share of pulp and paper products of total manufacturing in selected OECD countries 1964–1984 (% from output value)

Country	1964	1974	1984
Austria	n/a	4.4	3.8
Denmark	3.3	2.7	2.4
Finland	20.4	19.2	16.5
France	2.5	3.2	2.9
Germany	2.6	3.7	2.4
Greece	2.1	2.4	1.8
Ireland	2.4	2.9	1.8
Italy	n/a	3.0	2.4
Norway	9.9	8.5	5.7
Portugal	4.1	4.3	3.9
Spain	2.1	4.0	2.8
Sweden	9.8	11.2	9.8
Turkey	2.1	2.2	2.0
UK	n/a	3.4	2.8
Canada	8.2	8.4	7.4
USA	3.8	4.1	3.9
Japan	3.9	3.7	3.0
Australia	3.3	2.8	2.8
New Zealand	4.8	5.6	6.6
Total av.	5.3	5.2	4.5

Source OECD, DSTI/IND/PP/87.2/, 2nd revision, based on data by UNIDO
Note Share of output (at producers' prices) of pulp and paper industry in total manufacturing output. Table includes only OECD countries from which we have enough data available

although total production rose, the pulp and paper industry did not keep pace with overall industrial growth at the time. By the 1980s the pulp and paper industry was, in this respect, already a declining industry.

5.4 Government Support for the Pulp and Paper Industry

Not surprisingly, countries have promoted development and investments in the pulp and paper industry using direct industrial, technology and innovation policy measures (see e.g. Clapp 1995; Christensen and Caves 1997; Ghosal and Nair-Reichert 2009). Moreover, regional[15] and environmental policies, together with more general governmental measures on trade negotiations, taxation, labour policies, and infras-

[15]The regional policies, though important in some countries, were, according to an OECD report in 1988, not important for the general development of pulp and paper industries in OECD countries.

tructure (e.g. roads, energy) had also an impact on investments in this particular sector. The industry policy literature usually divides industry policy measures into horizontal and vertical categories (e.g. Buigues and Sekkat 2009, p. 5; Bianchi and Labory 2006). Horizontal ones consist of those measures that concern all industries in general whereas vertical ones are aimed at supporting a specific industry or company. In the following discussion we will use this same division and look at the government aid in different phases of production, from upstream raw materials to production and finally to downstream markets (Galbraith 1983; Ojala et al. 2006)

Previous studies suggest that in countries where the pulp and paper industry was a dominant line of business, attention was paid to creating a favourable regulatory environment, as the companies had bargaining power in shaping governmental policies or were embedded tightly in the political power system. The Nordic countries (i.e., Finland, Norway and Sweden) are primary examples of this kind of pattern (Hazley 2000; Järvinen et al. 2012; Kuisma 2008; Lamberg and Ojala 2005; Jensen-Eriksen 2007). Yet, even in the UK, where the paper industry was not dominant, the government was willing to protect it for a long time until the British membership in the European Free Trade Association (EFTA) made this impossible (Jensen-Eriksen 2008).

Internationalization and recent globalization of the production in paper and pulp industry companies have made the regulative environment more complex from the companies' perspective as they had to adjust to different institutional environments. Moreover, international institutions such as the EU, the UN, and World Bank also have their stakes in protection and regulation. Although the international trade in forest industry products is indeed an ancient business, the internationalization of production is still rather a late phenomenon (Björklund 1984; Ojala et al. 2006; Zhang 1997; Nagubadi and Zhang 2008). Institutions have played a crucial role in the evolution of the global pulp and paper industry, not only in investments but also in other activities as well. This can be seen, for example, in the case of the German paper industry in which cartelization presumably hindered the possibilities of individual firms to grow in size (Turunen 2012). Today, environmental regulation is a crucial factor that determines in part the location of new plants, for example in fast growing paper-producing areas in South America (Lima-Toivanen 2012). As a part of institutional development, property rights play a significant role, whether in terms of forest ownership or possibilities to industries to operate at large, i.e. to be sure that their investments are secured also in the future and are not threatened e.g. by socialization. Forest ownership is among the crucial questions in Nordic countries and differences in terms of property rights partly explains the different paths in North American and East Asian development (Järvinen et al. 2009, 2012; Kuhlberg 2012; Palo and Lehto 2012; Toivanen 2012).

In the following discussion we will analyse specifically the governmental support for the pulp and paper industry during the 1970s and 1980s by using the OECD documents and questionnaires as sources for the study. During the early 1970s the OECD

ELKA. OECD. Ad hoc working party on pulp and paper application of new technologies in the paper industry, 21 January 1988.

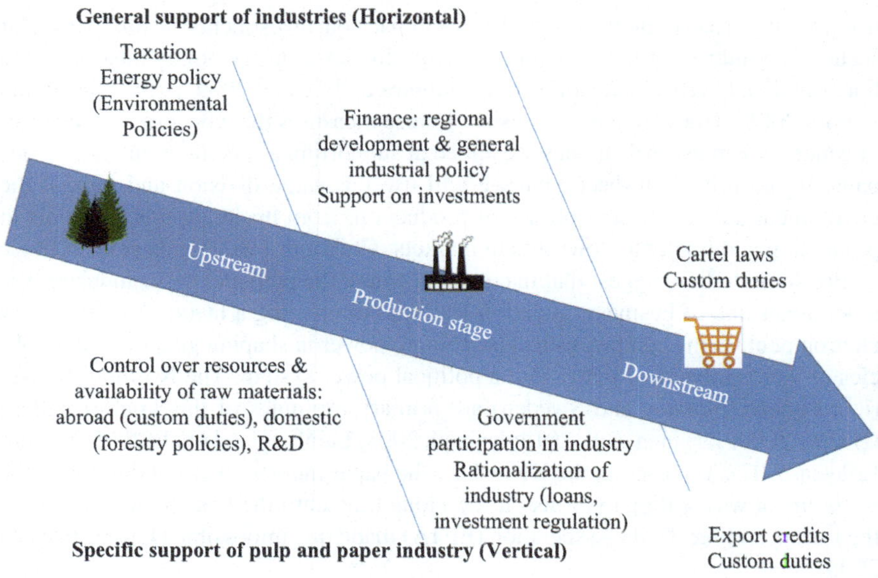

Fig. 5.1 Various ways of regulation and governmental support on pulp and paper industry in different phases of production. *Source* ELKA. OECD Industry committee, Pulp and Paper Section, ad hoc working party of the industry committee. *Note* Classification in accordance to OECD questionnaires from the early 1970s

surveys created certain categories for different types of government regulation. First, the questionnaires and reports were divided into upstream, production, and downstream stage support and regulation (Fig. 5.1). The first two included all measures to enhance production (from raw materials and energy to actual production in mills), whereas downstream strategies referred to markets, which was an especially important topic at the time because of concerns about collaboration within the industry and the existence of possible cartels. Although cartels were considered too varied and difficult to study, the ad hoc committee left the issue to yet another committee in the OECD, which specialized in studying cartels.[16]

Both up and downstream production were further divided into more general (horizontal) and more specific governmental measures (vertical), each having a number of sub-categories. General measures included, roughly, all sorts of industry policies not specifically aimed at promoting the pulp and paper industry but industries in general, whilst specific measures were more directly concentrated on pulp and paper industries.

The distinction drawn between general and specific measures shown in Fig. 5.1 did not reflect their economic importance for the pulp and paper industry. This was also underlined in the reports: the specific measures were not necessarily the most

[16]ELKA. OECD Industry committee, Pulp and Paper Section, Memorandum by Aarne Castrén, 20 May 1972.

important ones and their impact was not always the greatest, though they were easier to assess. Moreover, the OECD reports do not take a clear stand on whether these government measures were effective or not. Also for historians, though, the causal relations between the government measures and outputs—in this case their effect on investments and technology development—might be impossible to pinpoint as the specific and more general (industry) political measures might have affected the industry at the same time, but differently. The categorization schema by the OECD is, however, valuable in identifying national differences in industrial policies.

5.4.1 Government Measures to Support Upstream Production

In the OECD report from the early 1970s, the government measures applied to upstream production (broadly, raw materials and energy use) could affect industries in general and the pulp and paper industry in particular. This was the case, for example, with the Japanese special tax rates apply to the consumption of electric power (see Seung-Joon and Ruuskanen 2015). Similarly, in Finland during the 1970s and 1980s, energy policies played an important role in determining certain massive investments. United Paper Mills (UPM), for example, concentrated on energy-intensive thermomechanical pulp (TMP) in its paper production partly as a consequence of the nuclear power capacity built in the country at the time (Michelsen and Särkikoski 2005; Ojala and Lamberg 2006; Ruostetsaari 1989).

Besides general measures affecting upstream production, governments in different OECD countries applied more specific means of support to industries. This support—usually in the form of subsidies—considered different fibre furnish (i.e., wood, vegetable matter and recycled paper) used as raw materials for the pulp and paper industry. Again, the direct link between the regulation and governmental support and investments might be difficult to show, but at least some initial observations can be made. In order to save the availability of these fibres governments could either: (1) ensure their availability from abroad; (2) ensure their availability from domestic sources; or (3) stimulate research on raw materials and encourage the salvaging of fibrous material for recycling. Of these three measures the first one is not analysed in detail in the reports; they only refer to the lowering of customs duties on imported raw materials in Japan and duty free quotas in the EEC. The second one—especially the use of forest resources—is widely discussed in the reports. The third one, the importance of renewable materials, was one of the focus areas of pulp and paper industry committee during the turn of the 1970s and 1980s. Especially the committee was interested in technological development in this particular sector.

The second measure, the availability of a domestic supply of raw material fibres was analysed in greater detail in the OECD reports, and the reason was simple. This was among the most common measure the governments had, for example, to affect the ownership of forests through legislation, regulate the use of forest resources, and prevent over-exploitation and pollution. In order to enhance the availability of domestic raw materials some countries restricted their exports. According to the reports Austria was controlling exports of wood and chemical pulp, and certain Canadian provinces had limited the exports of round wood. However, the reports does not mention in this context the measures made in Finland and Sweden to control investments that in practice also reduced the exports of raw materials and semi-finished (pulp) products (Jensen-Eriksen and Ojala 2015; Bergquist and Söderholm 2015).

Moreover, the OECD reports list numerous government policies that apply to maintaining or developing both private and public forest resources and facilitating their efficient exploitation. According to the report the governments (either state, regional, or local) controlled quite significant shares of forest resources available in the domestic markets, namely in Canada 92%, Australia 76%, Germany 70%, Belgium 47%, France and UK 40%, Japan 32%, Turkey and Denmark 30%, USA and Spain 25%, The Netherlands and Sweden 20%, Austria 14%; but Finland and Italy only 10%, Norway 7%, and Switzerland and Portugal 5% respectively. Exactly how this control was exercised is not described in detail in the OECD reports, and most likely the strength of government control varied from direct forest ownership to rather general rules on the use of forest resources. Thus, these figures are not as comparative as the OECD reports suggests.

Also the reports indicate that other measures were taken by the governments to ensure the availability of a domestic supply of raw fibre. These included securing the use of the governmental-owned forests by supporting forest road-building (especially in the USA, Australia, Canada, the UK, and Finland); by supporting forestry (e.g. by funding for removing dead and diseased trees); and by encouraging the rationalization of exploitation procedures. Similarly, various measures were taken to support the use of privately owned forests as well, such as information services, improvements in tree cultivation and favourable taxation policies in Canada; subsidies to building forest roads in Japan; reforestation and tax incentives in the Netherlands; the purchase of capital goods in Portugal; education in Canada and Portugal; and technical assistance in the United States. Moreover, in most of the OECD countries governments gave aid for tree planting through tax rebates, loans or subsidies, technical assistance, and through information services. These measures secured the raw material base for the forest industry companies and thus made greater investments possible.

The third measure to ensure the availability of a domestic supply of raw fibre was, according to the OECD reports, to stimulate research on raw materials and encourage the salvaging of fibrous material for recycling. All countries gave aid to technological research into wood, mainly through specialized offices and services run by the state or handing out subsidies. These included forest technology services in Austria, Belgium and Portugal; research institutes in the Netherlands and Belgium; a forest products laboratory in the USA; wood technical centres in France and Belgium; a school of agriculture in Denmark; and research laboratories on wood use in Germany, Finland, and Australia. Waste paper as a potential raw material was studied in only a handful of countries (Austria, Canada, the Netherlands, Turkey, UK, and USA), although its importance was rising at the time (see especially Bouwens 2004, 2012). Non-wood crops were studied in Denmark as the country had a well-developed agricultural sector and lacked forest resources. The use of non-wood raw materials was also studied intensively in Austria, USA, and Turkey.

The government's indirect role in investments and technology can also be seen in the form of inputs to research and development, and this was mainly done through universities. During the late 1980s, though, most of the OECD countries had rather small shares of government financed R&D in the pulp and paper industry, with the exception of Canada and Finland, which both had over 6% of government spending in this area. Sweden had a share near 5%, and other OECD countries had much less government funding devoted to R&D in pulp and paper.[17]

5.4.2 *Measures to Support the Production Stage*

There were also various government measures that were applied to pulp and paper production in the early 1970s. They were designed to expand or rationalize the production capacity of the pulp and paper industry. The OECD grouped these into two main categories, depending on whether or not they were specifically concerned with the pulp and paper industry. Among the general measures to support the production stage, the most typical ones were concerned with finance; they came under either regional development policy or general industrial policy. These measures were directly related to investments by the companies. Regional policies played an important role in a number of countries, as the raw material and also the mills were quite often situated in the remote and less-developed areas of each respective country. The general measures in support of regional development included loans at reduced rates of interest in Australia and Germany; special arrangements for loan repayment in Germany and Italy; guarantees of loans in Belgium, Germany, Italy, the Netherlands, and Sweden; subsidies or grants in Belgium, Canada, Italy and Sweden; investment credits in Italy, Sweden, Japan; and more special measures such as reimbursement of water or land freight transport costs in Italy.

More general industrial policy measures affected the pulp and paper industry on the same basis as other industries. Whilst the above-mentioned regional policy measures might have had an effect on the geographical location of investments, the regulations applying to international trade in capital and capital goods had a decisive effect on a number of companies, especially those in small countries to make the initial investments in these locations. Namely, with regional policy measures governments could have on affect to geographical situation of investments; thus, to better match, for example, needs for employment in different parts of the country. The government's role was important in small countries from this perspective as it played a role in financing (directly or through various indirect measures) the investments, as collecting capital from domestic markets was challenging on the one hand, and getting loans from abroad was regulated on the other (Jensen-Eriksen

[17]ELKA. OECD. Ad hoc working party on pulp and paper application of new technologies in the paper industry, 21 January 1988.

and Ojala 2015). For example, Norway and Portugal supported investments in new machinery by exempting it from customs duty. Moreover, most of the countries had restrictions on inward FDI and foreign ownership at the time, and some governments also had regulations concerning outward foreign investments. The general governmental policies also included measures of financial aid to manufacturing industries other than those taken under regional development policy. These included rates of depreciation in Canada, Denmark, Portugal, Spain, Sweden, and Australia; credits for investments granted in Belgium, Spain, Belgium, Denmark, Finland, Turkey and UK; and subsidies granted in Belgium, the Netherlands, Spain, and the UK.

5.4.3 Vertical Measures to Support Pulp and Paper Industry

The specific (vertical) measures to support the pulp and paper industry are less numerous than the more general ones mentioned above, and the reason this is the case is obvious: the industry was not dominant in most of the OECD countries. The countries in which the pulp and paper industry played a more central role in the economy were most likely also to support and regulate this specific industry with more direct measures. Moreover, these measures also included ones that were directly related to investments and technology.

The specific government measures to promote and regulate the pulp and paper industry during the 1970s were related specifically to the government's direct participation in ownership of pulp and paper industry companies, or were aimed at rationalizing existing capacity or restructuring industry structure. Government ownership in the companies was, indeed, the ultimate measure to control and regulate the industry. However, the enterprises with government participation did not, except in Turkey and to a smaller degree in Finland and Spain, account for a very high percentage of the country's total pulp and paper production capacity: Canada 1.5%, Denmark 8%, Spain 24% (pulp) and 5% (paper); Norway 10%, Sweden 4% (pulp), 5% (paper); Finland approximately 25%; Turkey 100% in pulp and 82% paper. With the exception of Denmark (3%) and Norway (10%), the governments held very high percentages of the total share of capital of enterprises in which they were concerned. In Sweden, for example, ASSI that was founded in the 1930s was entirely government owned (Melander 2005). Even though the OECD was quite sensitive about government participation in company ownership, its reports were still quite positive on the subject. Its perspective was related to the fact that, in most cases, the authorities did not play an active role in determining company policy and so government ownership did not greatly affect the activity of the pulp and paper industry. Thus, according to the reports from the early 1970s, only in Sweden and Turkey could the activities of companies with government holdings be described as constituting government intervention.

In some OECD countries governments also aimed to restructure their respective pulp and paper industries using direct measures. In Germany, for example, the federal government had been guaranteeing loans since 1966 for investments designed to rationalize the pulp industry. Similarly, in Italy a 1966 law provided for financial aid for the same purpose, though with more limited capacity than the German one. Spain and Turkey reported to the OECD as the only countries that government helped establish new structures (larger units, integration of sector) in the paper industry; Spain in turn enforced the concentration of industry; and the Netherlands had started in the early 1970s a governmental-led study on the structure of the country's pulp and paper industry.

Moreover, most of the countries with pulp and paper industries had some sort of measures in the early 1970s to support downstream production, principally marketing and especially exports. In practice all countries had special tax arrangements that applied to exports and 16 out of 19 countries used export credit insurance (with the exception of Austria, the Netherlands, and Turkey). Moreover, Norway, Portugal, Finland, Australia, and to a certain extent Sweden had direct export credits. General measures to protect the pulp and paper industry at the downstream stage included custom duties, countervailing duties, and quasi taxes on imports. Furthermore, some countries applied measures designed to cover consumption requirements, including EEC duty-free tariff quota for newsprint.

Related to investments, a number of OECD countries also had direct measures aimed at regulating investments in pulp and paper industries, and most of these related to securing the raw material base. For example, any increase in capacity required prior authorization from competent authorities in Japan and in Spain. Moreover, the OECD reports were critical of the withdrawal of government investment grants, which, in the UK, were judged to have affected investment decisions. Also the decisions preventing pulp and paper production capacity from expanding beyond the annual cutting possibilities in Finland were listed by OECD (more on the issue in Jensen-Eriksen and Ojala 2015).

5.5 Conclusions

The OECD surveys on the pulp and paper industry creates a picture that is to some extent obvious: in countries in which the forest industry in general and pulp and paper industry in particular were especially important in terms of share of GDP, governments applied more favourable regulation to promote this industry. Obvious examples are Finland, Sweden and Canada.[18] As previous studies have shown the countries indeed tend to move resources to areas where they have comparative advantages in terms of factor endowments, although results from this type of support are not necessarily dynamic (Succar 1987; Pack and Saggi 2006). Moreover, countries also have a tendency to support "infant industries" as the production costs for new industries might be higher than already established industries in competing countries such that the state support is needed (Baldwin 1969; Pack and Saggi 2006).

[18]ELKA. OECD. Ad hoc working party on pulp and paper application of new technologies in the paper industry, 21 January 1988.

The picture is, however, not simple. Even within the OECD countries there were a number of governments that supported the pulp and paper industry even though it did not play a significant role in the economy nor was the industry at its "infant" stage with promising future prospects. In the countries with fewer "natural" factors of production (especially raw materials) in the pulp and paper industry, though, government aid fell mainly to the horizontal level. It was part of the more general industrial policies rather than directly aimed at the vertical level of this particular industry. Moreover, in the case of pulp and paper industries in OECD countries during the 1960s–1990s "infant industry" support hardly occurred, but it was most likely the case in emerging industries in this particular sector in Asia and South America.

Unfortunately, the OECD data enables us to show only the various means of state aid, not the real magnitude of this support in terms of volumes or values. Nevertheless, the OECD data also clearly shows that even "liberal market economies" had programs to promote state aid to this specific business sector. Moreover, even in countries in which the pulp and paper industry played rather insignificant role (like Denmark) there was government support for this industry, although in these countries the state aid was more often a part of more general industrial policies. Moreover, the data indicates that even from the 1970s onwards, a period that is usually considered as one of deregulation, there were no significant decline in state support for this industry. These outcomes, though, are highly tentative as more in-depth analyses ought to be carried out with this subject matter.

Our view in this chapter is biased, as we had accurate data from the OECD committee only from the 1970s and 1980s. Thus, the development of government policies and their effect on investments before and after that period are not analysed in detail in this paper. Moreover, we have accurate data on the government role only from western, developed countries, whereas the major change during the last decades was the migration of this industry from these western countries to Asia and South America. Nevertheless, the 1970s and 1980s witnessed an enormous increase in pulp and paper production, massive investments in production machinery, and also leaps in technological development in terms of larger mills and automation.

The emergence of the pulp and paper section within the OEEC/OECD was in the first place motivated by the fact that member countries and their respectable industries needed better international statistics that enabled countries to identify their weaknesses in production and thus emphasize regulation in these particular sectors (if necessary and if politically possible). Investments as such were not the key issue in the discussions surrounding the data gathering, but they can be seen as a long term aim. Through investments it was actually possible to effect better labour productivity, more energy efficient production, and the more sustainable use of raw materials.

In all, one might argue that government support and regulation had an impact on investments, and thus, might have also affected the technological development and change in global dominance in the pulp and paper industry. However, as can be detected from the OECD surveys the ways the government could support and regulate industries in general and pulp and paper industry in particular were many and diverse. Moreover, one might also argue that a number of indirect measures in each country were not even listed in the OECD surveys, especially those that might have affected investment decisions unintentionally.

References

Aydin U (2007) Promoting industries in the global economy: subsidies in OECD countries, 1989 to 1995. J Eur Public Policy 14(1):115–131

Baldwin RE (1969) The case against infant-industry tariff protection. J Polit Econ 77(3):295–305

Bergquist AK, Söderholm K (2011) Green innovation systems in Swedish industry, 1960–1989. Bus Hist Rev 85(4):677–698

Bergquist AK, Söderholm K (2015) Transition to greener pulp: regulation, industry responses and path dependency. Bus Hist 57(6):862–884

Bianchi P, Labory S (2006) Empirical evidence on industrial policy using state aid data. Int Rev Appl Econ 20(5):603–621

Björklund J (1984) From the Gulf of Bothnia to the White Sea: Swedish direct investments in the sawmill industry of Tsarist Russia. Scand Econ Hist Rev 32(1):17–41

Bouwens B (2004) Op papier gesteld. De geschiedenis van de Nederlandse papier- en kartonindustrie in de twintigste eeuw. Boom, Amsterdam

Bouwens B (2012) The paper and board industry in the Netherlands, 1800–2000. In: Lamberg J-A, Ojala J, Peltoniemi M, Särkkä T (eds) The evolution of global paper industry 1800–2050. A comparative analysis. Springer, Dordrecht, pp 191–209

Buigues P-A, Sekkat K (2009) Industrial policy in Europe, Japan and the USA: amounts, mechanisms and effectiveness. Palgrave Macmillan, Houndmills, Basingstoke, Hampshire

Christensen LR, Caves RE (1997) Cheap talk and investment rivalry in the pulp and paper industry. J Ind Econ 45(1):47–73

Clapp RA (1995) Creating competitive advantage: forest policy as industrial policy in Chile. Econ Geogr 71(3):273–296

Foreman-Peck J (2014) European industrial policies in the post-war boom: 'Planning the Economic Miracle'. In: Grabas C, Nützenadel A (eds) Industrial policy in Europe after 1945; wealth, power and economic development in the cold war. Palgrave Macmillan, Houndmills, Basingstoke, Hampshire, pp 13–47

Ford R, Suyker W (1990) Industrial subsidies in the OECD economies

Frey B (1984) The public choice view of international political economy. Int Org 38(1):199–223

Galbraith JR (1983) Strategy and organizational planning. Hum Resour Manag 22:64–77

Ghosal V, Nair-Reichert U (2009) Investments in modernization, innovation and gains in productivity: evidence from firms in the global paper industry. Res Policy 38(3):536–547

Godin B (2004) The new economy: what the concept owes to the OECD. Res Policy 33(5):679–690

Grabas C, Nützenadel A (2014) Introduction. In: Grabas C, Nützenadel A (eds) Industrial policy in Europe after 1945; wealth, power and economic development in the cold war. Palgrave Macmillan, Houndmills, Basingstoke, Hampshire, pp 1–10

Hazley C (2000) Forest-based and related industries of the European Union: industrial districts, clusters and agglomerations. Taloustieto, Helsinki

Hall PA, Soskice D (ed) (2001) Varieties of capitalism: the institutional foundations of comparative advantage. Oxford, Oxford University Press

Iversen MJ, Thue L (2008) Creating Nordic capitalism—the business history of a competitive periphery. In: Fellman S, Iversen MJ, Sjögren H, Thue L (eds) Creating Nordic capitalism: the business history of a competitive periphery. Palgrave Macmillan, Basingstoke, pp 1–19

Jensen-Eriksen N (2007) Läpimurto. Metsäteollisuus kasvun, integraation ja kylmän sodan Euroopassa 1950–1973. Metsäteollisuuden maa, 4. Suomalaisen Kirjallisuuden Seura, Helsinki

Jensen-Eriksen N (2008) A stab in the back? The British Government, the paper industry and the nordic threat, 1956–1972. Contemp Br Hist 22(1):1–21

Jensen-Eriksen N, Ojala J (2015) Tackling market failure or building a cartel? Creation of investment regulation system in Finnish forest industries. Enterp Soc 16(3):521–555

Järvinen J, Lamberg J-A, Murmann JP, Ojala J (2009) Alternative paths to competitive advantage: a fuzzy-set analysis of the origins of large firms. Ind Innov 16(6):545–574

Järvinen J, Ojala J, Melander A, Lamberg J-A (2012) Evolution of pulp and paper industries in Finland, Norway, and Sweden. In: Lamberg J-A, Ojala J, Peltoniemi M, Särkkä T (eds) The evolution of global paper industry 1800–2050. A comparative analysis. Springer, Dordrecht, pp 19–47

King KFS (1974/1975) Forest policies and national development. Unisylva, no.107, pp 31–35

Kuhlberg M (2012) An accomplished history, an uncertain future: Canada's pulp and paper industry since the early 1800s. In: Lamberg J-A, Ojala J, Peltoniemi M, Särkkä T (eds) The evolution of global paper industry 1800–2050. A comparative analysis. Springer, Dordrecht, pp 101–133

Kuisma M (ed) (2008) Kriisi ja kumous. Metsäteollisuus ja maailmantalouden murros 1973–2008. Metsäteollisuuden maa, 5. Suomalaisen Kirjallisuuden Seura, Helsinki

Lamberg J-A, Ojala J (2005) The Nordic forest industry in retrospect: an introduction. Scand Econ Hist Rev 53(1):7–18

Lamberg J-A, Ojala J (2006) Evolution of competitive strategies in global forestry industries: introduction. In: Lamberg J-A, Näsi J, Ojala J, Sajasalo P (eds) The evolution of competitive strategies in global forestry industries. Comparative perspectives. Springer, Dordrecht, pp 1–29

Lamberg J-A, Ojala J, Peltoniemi M (2018) Thinking about industry decline: a qualitative meta-analysis and future research directions. Bus Hist 60(2):127–156

Lima-Toivanen MB (2012) The South American pulp and paper industry: cases Brazil, Chile and Uruguay. In: Lamberg J-A, Ojala J, Peltoniemi M, Särkkä T (eds) The evolution of global paper industry 1800–2050. A comparative analysis. Springer, Dordrecht, pp 243–283

Melander A (2005) Industry-wide belief structures and strategic behaviour: the Swedish pulp and paper industry 1945–1980. Scand Econ Hist Rev 53(1):91–118

Michelsen K-E, Särkikoski T (2005) Suomalainen ydinvoimalaitos. Edita, Helsinki

Nagubadi RV, Zhang D (2008) Foreign direct investment outflows in the forest products industry: the case of the United States and Japan. Int For Rev 10(4):632–640

OECD Economics Department Working Papers 74. OECD Publishing, Paris

Ojala J, Lamberg J-A (2006) The challengers. Kymmene, United Paper Mills, and Metsäliitto. In: Lamberg J-A, Ojala J, Näsi J, Sajasalo P (eds) The evolution of competitive strategies in global forestry industries. Comparative perspectives. Springer, Dordrecht, pp 107–139

Ojala J, Lamberg J-A, Ahola A, Melander A (2006) The ephemera of success: strategy, structure and performance in the forestry industries. In: Lamberg J-A, Ojala J, Näsi J, Sajasalo P (eds) The evolution of competitive strategies in global forestry industries. Comparative perspectives. Springer, Dordrecht, pp 257–286

Pack H, Saggi K (2006) Is there a case for industrial policy? A critical survey. World Bank Res Obs 21(2):267–297

Palo M, Lehto E (2012) Private of socialistic forestry? Forest transition in Finland vs. deforestation in the tropics. Springer, Dordrecht

Ruostetsaari I (1989) Energiapolitiikan määräytyminen. Dissertation, University of Tampere

Scranton P, Fridenson P (2013) Reimagining business history. The Johns Hopkins University Press, Baltimote

Seung-Joon P, Ruuskanen E (2015) Energy supply and regulation. In: Tanaka Y, Toshiaki Tamaki T, Ojala J, Eloranta J (eds) Comparing post war Japanese and Finnish economies and societies. Longitudinal perspectives. Routledge Studies in the Modern World Economy, Routledge, Abingdon, pp 132–154

Succar P (1987) The Need for industrial policy in LDC's-A Re-statement of the infant industry argument. Int Econ Rev 28(2):521–534

Söderholm K, Bergquist AK (2012) Firm collaboration and environmental adaptation. The case of the Swedish pulp and paper industry 1900–1990. Scand Econ Hist Rev 60(2):183–211
Toivanen H (2012) Waves of technological innovation: the evolution of the U.S. pulp and paper industry, 1860–2000. In: Lamberg J-A, Ojala J, Peltoniemi M, Särkkä T (eds) The evolution of global paper industry 1800–2050. A comparative analysis. Springer, Dordrecht, pp 49–80
Turunen O (2012) The paper industry in Germany, 1800–2000. In: Lamberg J-A, Ojala J, Peltoniemi M, Särkkä T (eds) The evolution of global paper industry 1800–2050. A comparative analysis. Springer, Dordrecht, pp 81–99
Wagner M, Van Phu N, Azomahou T, Wehrmeyer W (2002) The relationship between the environmental and economic performance of firms: an empirical analysis of the European paper industry. Corp Soc-Responsib Environ Manag 9(3):133–146
Whitley R (1999) Divergent capitalisms: the social structuring and change of business systems. Oxford, OUP
Zhang D (1997) Inward and outward foreign direct investment: the case of US forest industry. For Prod J 47(5):29–35

Soderbom K, [illegible] A (201[illegible]) [illegible] and [illegible] The case of [illegible]
[illegible] H (2017) [illegible]
[illegible] (201[illegible]) [illegible] In: [illegible] (eds) [illegible]
[illegible] pp [illegible]
Thomas G (2012) The [illegible] 1800–2000. In: Lambert [illegible]
[illegible] The evolution of [illegible] 1800–[illegible]
[illegible] pp [illegible]
Wagner M, Yang R, [illegible] (2017) [illegible]
[illegible]
[illegible]
Whaley [illegible] (2014) [illegible]
[illegible]
Zhang [illegible] (2017) [illegible]
[illegible]

Chapter 6
From Backward to Modern: The Adoption of Technology by the Pulp Industry in Portugal, 1891–2015

Amélia Branco and Pedro Neves

6.1 Introduction

In 2015, Portugal was the third largest European producer of pulp with 7.3% of total output. National production of virgin wood pulp stood at 2,662,000 tons of which 45% was exported to over two dozen countries. Portuguese imports of paper-grade pulp stood at 129,000 tons, a level that makes the country a net exporter of paper-grade pulp. The largest Portuguese pulp producer, Navigator, is also a global producer of uncoated, wood-free printing and writing paper. This picture totally contrasts with that prevailing during the late nineteenth and early twentieth centuries. At that time, there was only one plant in operation that produced chemical pulp, whose output was sold almost entirely to export markets. In 1911 the production capacity of this plant was 3,500 tons and in 1930 it was only slightly above that level (4,343 tons). The main raw material of the Portuguese paper industry was rags and the plants were technologically backward, although some companies had already adopted continuous paper machines. The wood pulp consumed by these paper units was covered by imports, making Portugal a net importer of pulp. All this changed by the mid-twentieth century. At that time, the wood pulp industry in Portugal grew dramatically because it was considered a strategic sector under the state development policy. This factor gave birth to the first plant that processed eucalyptus to produce sulphate pulp, and the enterprise was oriented toward export markets.

A. Branco (✉) · P. Neves
GHES/CSG, ISEG-Lisbon School of Economics and Management, Universidade de Lisboa, Lisboa, Portugal
e-mail: ameliab@iseg.ulisboa.pt

P. Neves
e-mail: pneves@iseg.ulisboa.pt

T. Särkkä et al. (eds.), *Technological Transformation in the Global Pulp and Paper Industry 1800–2018*, World Forests 23, https://doi.org/10.1007/978-3-319-94962-8_6

Taking a long term approach, this chapter[1] aims to address three main topics concerning the life cycle of the Portuguese pulp industry: first, the causes of backwardness for more than half a century, qualifying Portugal as a latecomer to the sector; second, the new circumstances and factors that were present in the industry's emergence phase; finally, the constraints and patterns of expansion that contributed to the industry's maturation. The long-term study of the remarkable development of the Portuguese pulp sector holds a great deal of interest for various reasons. Firstly, the country displayed a significant lag at the beginning of the twentieth century and was among the latecomers to the global pulp sector (Gutiérrez-Poch 2012; Ojala et al. 2012).

A discussion of the causes for this delay represents a contribution towards the broader historiographic debate surrounding the Portuguese economic backwardness (About this debate, see Reis 1993 and Lains 2003). Secondly, following the consolidation of the sector within the Portuguese industrial structure, it became an example of a successful, competitive industry in international markets. A long-term perspective on the Portuguese pulp industry may also shed some light on the international economic integration of a peripheral country with poor endowment of capital and natural resources (Lains 1995; Reis 2000). Finally, because the Portuguese pulp industry received substantial support from the state, this case study may also generate further contributions towards the ongoing debate around industrial policies (On the current state of this debate, see Aiginger and Sieber 2006; Pack and Saggi 2006; Lee et al. 2012; Grabas and Nützenadel 2014).

An analysis of the dynamic interaction between the institutional environment and business strategies contributes towards a broader and deeper understanding of the phases in the sector's life cycle as well as the reasons for its success. The institutional dimension takes on a crucial explanatory relevance in cases of late economic modernisation as is the case with the Portuguese economy (Gerschenkron 1962). The role of the state in the Portuguese pulp sector's path was crucial not only creating conditions for its expansion and growth but also acting as owner of pulp mills. Furthermore, the secular evolution of the industry is anchored in different economic and political regimes. Despite the relevance of institutional forces, private initiative was also fundamental to the success of the sector, specifically by taking advantage of the conditions the state had created. The state and private initiative combined to contribute towards the creation of a comparative and competitive advantage. The forest was crucial for the creation of this advantage, serving the industry well and being a good example of inter-sectoral integration.

In the next section, we set out a long-term view of Portugal's pulp sector and its most relevant features, such as the level of output and exports, product specialisation, business structure, and also the trends in Portuguese forest that allowed for the production of the industry's main raw material. It allows us to identify general trends and phases that shaped the secular evolution of the industry in Portugal. This periodisation will be the roadmap for the remaining sections.

[1]The authors would like to thank the Portuguese national funding agency for science, research and technology (FCT) for financial support (Project UID/SOC/04521/2013).

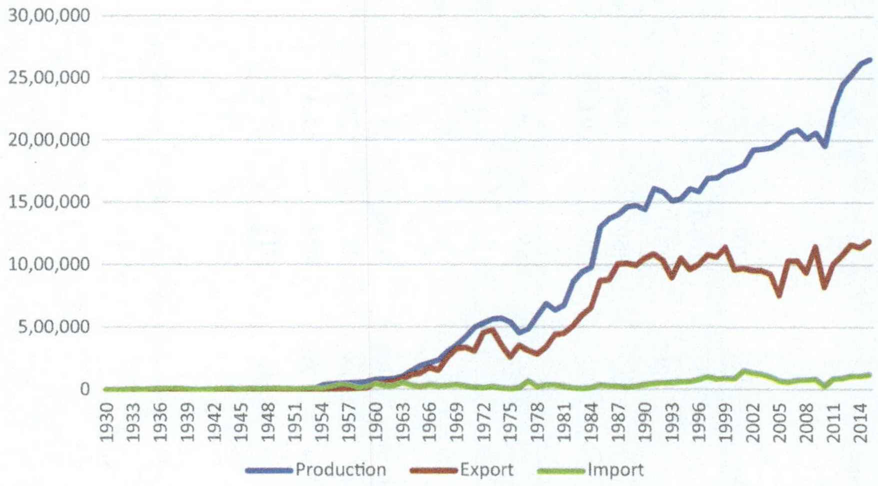

Fig. 6.1 Production and trade of pulp in Portugal, 1930–2015 (tons). *Sources* 1930 to 1974—Branco (2005); 1975 to 2015—FAOSTAT

6.2 Pulp Industry in Portugal: A Long-Term Perspective

The pulp industry in Portugal stands out as a successful case of a latecomer catching up. The wood pulp industry was set up in Portugal in 1891, and after six decades of very slow growth, the increase in productive capacity was both impressive and accompanied by an enormous increase in the volume of pulp being exported (Fig. 6.1).

The remarkable growth in Portuguese pulp production from the 1960s onwards is undeniable. Nevertheless, this broad picture hides different phases through which the pulp industry passed during the period under study. In addition to the level and growth rate of pulp production, other aspects can be taken into account in identifying these phases: export and specialisation trends, business structure, and forest coverage in the country. Institutional framework and corporate strategies are factors that shaped these indicators. Table 6.1 presents the figures for these indicators in several benchmark years for the period under study. Those benchmark years were chosen after taking into account the evolution in the institutional framework in order to assess its impact on the sector.

Up until 1953, pulp production coming from the single plant in operation remained very low, reaching 7,000 tons in that year. The prominent growth in pulp production capacity can be traced back to 1954, when the Cacia plant entered operation and signalled the industry's take-off. The production rose from 7,000 tons in 1953 to 41,600 in 1954. The industry's growth phase spanned the next three decades through the expansion of existing plants and the opening of five new ones. In the mid-1970s the production reached 547,000 tons, thirteen times more than the level in 1954. After a short period of decrease, the high growth pace was resumed by the end of the 1970s. In the second half of the 1980s, the production ramped up again due to

Table 6.1 The growth of Portuguese pulp industry

		1910	1930	1950	1960	1975	1988	2001	2015
Production (10^3 ton)	(1)	3.5	4.3	6.3	86.5	547	1,472	1,806	2,662
Exports (10^3 ton)	(2)	1.9	3.0	4.3	59.4	265	1,017	980	1,194
Imports (10^3 ton)	(3)	8.5	7.2	15.5	53.2	13.8	29	159	126
BEKP to total production (%)	(4)						69	78	94
Market pulp to total production (%)	(5)	100	100	100			83	65	49
No. plants	(6)	1	1	1	3	8	9	7	7
No. companies	(7)	1	1	1	3	7	4	5	3
Forest (10^3 ha)	(8)	1,957	2,332	2,832	2,826	2,956	3,108	3,305	3,155
Forest coverage (%)	(9)	22.3	26.6	32.3	32.2	33.8	35.4	38	35.4
Maritime pine (10^3 ha)	(10)	430	1,132	1,19	1,288	1,293	1,252	978	714
Eucalyptus (10^3 ha)	(11)		8	113	99	214	386	717	812
Pulp firms' forest (10^3 ha)	(12)					45		220	167

Sources [1–3]—same as Fig. 6.1 and Mesquita (1919), Abecassis (1969); [4, 5] CELPA (several years); [8 to 11]; [12]—Baptista (1984), CELPA (several years); Mendes (2007), and ICNF (2013)
Notes in rows 8, 9, 10, and 11 the dates are—1910, 1929, 1950/56, 1963/66, 1968/78, 1980/85, 1995, and 2010

the opening of a new plant; the country's production level rose to 1,300,000 tons in 1985. The industry matured over the next three decades, as no more plants entered operation but pulp production more than doubled to 2,662,000 tons by 2015. This last growth trend did not occur at a steady pace, however, as initially there was a period of slower growth which was followed by another boost in the second decade of the twenty-first century.

The export record also displays considerable dynamism, reflecting the exporting aspect that has been outlined for the sector since its launch. Two phases stand out in terms of the increased exports. First, through to the 1980s, the trends for exports and production were aligned, with the proportion of production exported roughly 70%. Second, after the end of the 1980s, the evolution of exports did not follow the increase in production and its level became stabilised. In 2015 exports were close to the levels in 1999, and today the amount of exported pulp represents about 45% of total production.

In terms of the specialisation pattern in the Portuguese pulp sector, two main features stand out. First, since its very beginning the sector was concentrated in producing chemically bleached eucalyptus pulp, and this specialisation has been reinforced in the last few decades. Second, concerning the forward integration into paper production, there was a significant transformation over the period. Only two pulp plants were originally integrated, and others made investments during their life cycles in order to integrate paper production. The proportion of pulp produced for the open market thus decreased from around 84% in 1990 to 49% in 2015. This was the main reason for the misalignment between the production and export trajectories that was noted above. Concerning the business structure of the Portuguese pulp industry, the evolution of three trends will be analysed: the number of plants, number of firms and ownership of firms. The evolution in the number of plants has already been mentioned previously, and it should be emphasised that the great increase occurred in the 1960s and the last new plant began operating in 1985.

Among others, the institutional framework of the Portuguese economy was a relevant determinant of the changes in the number of firms and in their ownership structure. The first plant was established by a company owned by foreign capital. The second plant was set up by a company with Portuguese shareholders with the state holding a large stake in it. The plants that followed were privately owned, and were both national and foreign. Until the mid-1970s, the dominant pattern remained one company, one plant, with one single exception. The nationalisation process carried out during 1975 affected the ownership of large companies in Portugal but foreign capital was excluded from this process. The pulp industry then became composed of a state owned company controlling five plants and two foreign companies controlling the others. The last plant to go into operation was run by a company controlled by the state but it relied on foreign capital. Since the 1990s the institutional framework changed again, when the international integration and reprivatisation of the Portuguese economy was strengthened by the accession of the country to the European Community (1986). A new restructuring cycle in the pulp industry took place, and it resulted in the state's exit from the ownership of pulp companies. Another relevant feature of this period was the outflow of foreign capital from the industry. The companies that had been in the hands of foreign owners were taken over by national capital and the sector became controlled by two Portuguese business groups. Only one of the re-privatised pulp companies, with only a small share of the total production, ended up in the hands of a foreign business group.

The evolution in the pulp industry traced above is inextricably linked to the dramatic expansion of forest coverage in Portugal. Three aspects can be highlighted about this topic. First, the forest coverage increased from 22% in 1911 to 35% in 2015, with most of this increase taking place prior to the 1950s. Second, there was a change in the Portuguese forest structure in terms of species. The eucalyptus area, the main raw material of the Portuguese pulp industry, was almost non-existent in the country prior to the 1940s. Thereafter, however, it overtook the maritime pine and became the most important commercial tree species. Finally, the pulp firms had an important role in the transformation of the Portuguese forest, not only through buying properties but also renting areas belonging to private and non-industrial owners.

The following sections are structured according to the phases identified above. In section three, we discuss the factors that explain the relatively late investment in the pulp industry. In section four, we analyse the role of the state in fostering the conditions for the pulp industry to take-off. We also highlight the establishment of Companhia Portuguesa de Celulose (CPC), a company that would be a key player in the sector. Section five focuses on analysing the industry's expansion phase and considers not only the business strategies of the firms that came after CPC's establishment but also how the state decisively shaped the sector. Section six maps the evolution of the sector after the nationalisation process (in the mid-1970s) until the end of 1980s. Section seven is dedicated to the most recent decades, when the sector matured, and where we identify the vicissitudes and transformations that resulted from the re-privatisation process. Finally, we present our conclusions.

6.3 A Portrait of Backwardness in the 1930s

The use of wood pulp in the production of paper across the world was increasingly adopted from the mid-nineteenth century onwards, but the pace of development of the wood pulp sector was not equal among countries. This disparity was due to the extent of the forest cover in each nation and also the relative backwardness of its paper industry (Ojala et al. 2012, pp. 345–360). British capital was behind the launch of the first pulp plant in Portugal in the late nineteenth century, and it remained the only plant producing pulp in the country until the mid-twentieth century.[2] The Caima Timber Estate & Wood Pulp Company Limited, known during and after the 1920s as Caima Pulp Company, Limited, had operated since 1891. Its installed production capacity stood at 3,500 tons a year (Abecassis 1969, p. 930). Initially, the firm employed

[2]In fact, during the 1930s, the Companhia de Papel do Prado, the most important firm in the paper industry, produced limited quantities of chemical pulp for its own consumption. However, this activity did not proceed to a larger scale.

the chemical method to make bisulphite pulp from pine. From the 1920s onwards, however, the firm switched to eucalyptus wood but still used the bisulphite method (Radich and Alves 2000, p. 163). Caima's output was sold almost exclusively on the export market, even while Portugal remained a net importer of both chemical and mechanical pulps. This situation raises two interconnected questions. Which factors caused the late emergence of the pulp industry in Portugal, and why did the pulp produced by Caima not get consumed by the domestic paper industry?

The sector's delayed take-off derived from two interwoven causes. The first is more general and underlay the country's overall economic backwardness, and the second is more specific and related to the pulp and paper industry. The broader causes behind the delay in the development of the pulp sector in Portugal involve the shortage of capital for industrial investment and the small scale of the internal market. These two factors explain the difficulties experienced in building up an industry that required high levels of investment in machinery and equipment and that depended on, for its own viability, achieving economies of scale that were only feasible with high production levels. There was also a low level of paper consumption in Portugal (about 1 kg per capita per year in the 1870s and 5 kg in the 1930s), (Gutiérrez-Poch 2012, p. 212; Alves 2000, p. 154) which reflected the country's low rate of literacy, and it greatly reduced the demand for production.

In terms of the factors specific to the sector, the Portuguese paper industry was, in general, small in size and technologically backward (On the paper sector in 1930s, Costa 1946; Alves 2000, 2001; Gutiérrez-Poch 2012). Its scale of production remained very low, with its capacity standing at 22,000 tons per year in the 1930s. Although there were more than seven dozen paper plants, only nine operated continuous production lines. Reflecting the artisanal character of the industry, rags were the most commonly used raw material. Another specific factor that restricted the industry's development was linked to raw materials and the resource endowment of the country. The Portuguese forest cover was dominated by maritime pine. The high level of resin found in this species had to be dealt using chemical products that at that time were not produced in Portugal. Thus, the production costs of producing pulp from pine would exceed those of importing wood pulp from Scandinavian countries (On the discussions around the raw materials used, see Seabra 1943, 1944a, b; Gomes 1969; Branco 2010, 2011).

Concerning the reasons for the non-utilisation of the eucalyptus pulp produced by Caima, it is difficult to arrive to an undisputed conclusion. Paper producers claimed about the whiteness and strength of the paper produced with Caima's pulp and also the higher price in comparison to imported pulp. On the other hand, contemporary studies about the pulp industry in Portugal pointed out obstacles in the use of Caima's pulp related to the technological structure of the Portuguese paper industry, namely the lack of knowledge in using wood pulp and the absence of laboratories that would support the combination of different grades of pulp. When Portuguese paper industry had to consume Caima's pulp, during the Second World War, this mill had to change the raw material mix, through rising the wood pine consumption, in order to respond to the requirements of domestic industry (Seabra 1943, 1944a, b; Gomes 1969; Branco 2010, 2011).

The portrait of backwardness described in the paragraphs above combined with the forecast of an increase in *per capita* consumption of paper and the rise of nationalism, supported the motivation for state intervention in creating conditions for the emergence of the pulp sector in Portugal. There was no declared comparative advantage to produce pulp in Portugal. On one hand, the production costs of pulp were higher in comparison with its international competitors. On the other hand, the initial investment only proved viable when the level of production reached economies of scale, which was difficult to achieve in a small market. Thus, there was a need for the enactment of an industrial policy that rewarded the first companies investing in the pulp sector and also protected it from external competition.

6.4 An Industry Taking-off—Starting Out as a Latecomer

6.4.1 Creating the Conditions for the "Big Push"

Notwithstanding the scenario of backwardness described in the previous section, it was during the second half of the 1930s that the first projects leading to the establishment of the pulp industry in Portugal emerged. In this process, there was a close connection between private interests and the will of the state. The state's role in this process can only be understood through reference to the new political regime that emerged in Portugal during the 1930s. The institutional framework of the new political regime, also known as Estado Novo, was established in 1933 by two important pieces of law: the Constitution and the National Statute of Labour. These statutes legitimised economic intervention by the state and described economic nationalism as a fundamental aim that would be accomplished by an import substitution strategy based on using national capital (On the economic regime of Estado Novo, see Rosas 1986, 2000; Nunes and Brito 1990; Lains 1994; Nunes 1996).

State intervention in the industry had already started through the introduction of Industrial Conditioning (IC) in 1931 (On Industrial Conditioning during Estado Novo, see Brito 1989; Confraria 1992). This measure sought to respond to the Great Depression but became a means to exert close government control over industry during almost the entire dictatorship of Estado Novo. Existing companies also had a say in the process of licensing new installations or expanding the capacities of existing ones. Those industries that depended on importing raw materials were most strongly affected by the auspices of IC as was the case with the paper industry with its great dependence on international supplies of pulp. Under this institutional framework, two questions arise regarding the role of the state in the take-off of the pulp industry. First, what were the reasons behind the state seeking to nurture the pulp sector in the country, and second, what conditions did the state provide for the sector to take-off in Portugal?

In the second half of the 1930s, there was an opportunity for Portugal to expand the productive capacity of its paper industry, which naturally would have led to increased imports of wood pulp (Alves 2000). Following a request submitted to the IC for the reopening of a paper factory, an Industry inquiry took place in 1938. Based on the crucial findings of this report, the authorities recognised the necessity of setting up a pulp industry, recommending several studies about the feasibility of the doing so, particularly with regard to the kind of pulp to be produced, the desire to use domestic raw materials (maritime pine, eucalyptus and cereal straws), and the need to reorganise the paper sector (Direção Geral da Indústria 1939).

In total, this development translated into the beginning of the state's commitment to fostering this sector, and it was reflected in two governmental initiatives that drove the first wave of progress towards launching a national pulp industry. The first began in 1937, when a project for a pulp plant started to be drawn. It was intended to take advantage of the establishment of Soda Póvoa, a plant that produced sodium chlorate, a raw material used in the production of wood pulp. As part of this effort, in 1940 a licence was requested in order to establish in Portugal a mill to produce chemical and mechanical pulps, one that used wood and straw as raw materials. Another potential project was on the drawing board in 1939, and the results of a viability study were presented to the government in 1940. On the initiative of the Secretary of State for Industry, the two projects were merged into a single one that was realised through the establishment of the CPC in 1941 (AAVV 1958; Loureiro 1991).

The economic nationalism was further reinforced by problems arising out of the Second World War, in particular the difficulties the conflict posed to the ongoing operation of Portugal's industrial facilities with their heavy dependence on external suppliers of machinery, equipment and raw materials. The paper industry faced difficulties to import pulp and had to consume that supplied by Caima. Dealing with these difficulties may also largely explain the state's initiative to push for Portugal to develop its own pulp industry. The second governmental initiative would affect the nation's supply of commercial wood. In 1938, through the Reforestation Plan (1938–1968), the state embarked on its pine-based reforestation program. Among the countless objectives enunciated in this Plan, much emphasis was put on the need to supply raw materials to a future pulp industry (Branco 2010, 2011). Additionally, ongoing forestry research took place in order to ascertain the best species for pulp production. To realise this aim, the state set up the Maritime Pine Experimentation Station in Marinha Grande, and the Cork Oak and Eucalyptus Experimentation Station (that later focused only on the cork oak) in Alcobaça. Moreover, in the early 1940s it created the Cellulose Laboratory at the Higher Institute of Agronomy (Seabra 1943; Radish and Alves 2000; Alves et al. 2007).

6.4.2 *Companhia Portuguesa de Celulose (CPC)*

Through the Ministerial Order dated 11.3.1942, CPC was granted a license for setting up a plant to produce bleached and unbleached pulp, mechanical pulp, newsprint and writing paper. In comparison with the licence that had been requested in 1940 and was mentioned above, the license that the government ultimately granted CPC ended up being far wider reaching because the plant in question was an integrated establishment for the production of both pulp and paper. However, it would be necessary to wait a decade for it to go into operation. The delay may have stemmed from the difficulties encountered in raising the capital necessary for this undertaking. The CPC produced several studies after receiving its license that place the emphasis on raising funds especially as the state imposed a share capital requirement of PTE 36,000,000 and that it all come from domestic sources (AAVV 1958). The lack of initial investment capital for the project did not prevent it from being granted the go ahead from state authorities, however. In general, this case reflected the already mentioned problems for a small and peripheral country that was seeking to embark on an economic growth process without any involvement of international capital.

The difficulties in raising capital were attenuated with the Law of Industrial Development and Reorganisation (Law no. 2005 of 11 March 1945) that shaped the industrial policy of Estado Novo. In the field of industrial development, the concept of "core industries" was defined whereby they were to receive different types of support including favourable customs duties and fiscal policies, and involving the direct participation of the state in the company as a shareholder. Law 2005 also went so far as to prohibit any other new plants opening within a ten-year period (Rosas 1986, p. 84).

In 1947, the Portuguese government identified the pulp sector as a "core industry" (Council of Ministers' decision on 24 April 1947) and it became subject to Industrial Conditioning. This development facilitated the financing of the CPC through the participation of the state as a shareholder. In this same year, the company's capital was raised to PTE 32,000,000 through the state's investment of PTE 16,000,000, (AAVV 1978) an amount that was later further boosted through recourse to assistance supplied under the Marshall Plan (Alves 2001, p. 167). By the mid-1950s, only 40% of the capital was private capital with the state holding the remainder through the Treasury, Social Welfare Institutions and the Industrial Development Fund Confraria 1992, p. 85).

In 1953, the CPC plant in Cacia began to produce chemical pulp by making use of maritime pine as its raw material and with an installed production capacity of 114 tons per day of wood pulp and 100 tons per day of paper. The following years saw the launch of the other production lines: newsprint in 1955 and mechanical pulp in 1957 (AAVV 1978). Nevertheless, the project that was initially licensed ended up experiencing deep alterations that would subsequently determine the future of the sector (AAVV 1978, pp. 59–60; Confraria 1992). Even though the CPC's facilities had been set up in order to consume pine wood, in 1957 a fundamental change occurred that had the greatest domino effect in the history of the pulp sector in Portugal: CPC began the production of chemically bleached eucalyptus pulp through the sulphate process. It was the first European plant to produce this type of pulp, and since it first began making this product its output was directed to external markets (Alves 2000). The success of this new production line led to the rethinking of the initial project and the production of newsprint from mechanical pulp ended up being abandoned in the late 1960s (AAVV 1978). Whether due to the fact of CPC's loss of exclusivity, only guaranteed for a decade under Law 2005, or due to its exporting success, the pulp industry in Portugal entered an expansionary phase with the entrance of new players during the 1960s.

6.5 Expanding Productive Capacity: New Entrants and New Investments

Having set down the conditions in the 1930s, state support proved decisive for the start-up of the Cacia plant in the 1940s. The state participated directly in the venture as an investor and granted various benefits to CPC. Additionally, there was a long term plan, for thirty years, implemented upstream that brought about the reforestation of large swaths of the country. The establishment of CPC represented the definitive starting point for the pulp industry in Portugal and its success blazed the trail for the developments that followed in the 1960s. The core foundations were already in place, and the industry entered a phase of expansion. It occurred against the backdrop of the Portuguese economy experiencing modern economic growth coupled with its increasing integration into the international economy and also a shift in industrial policy.

In effect, the 1960s clearly represented a decade of transition. It marked the changeover from an import-substitution based industrial policy, closed to international capital and with the regulation of competition, to a new institutional framework characterised by progressive economic liberalisation. The rising level of international openness constituted one of the main drivers of this new approach, leading to the rejection of the protectionist policies underlying the import substitution model. Also, the main obstacles to the entrance of international capital were removed, opening the country to a wave of foreign investment. The growing integration definitively marked Portugal's growth strategy and allowed its industries to become competitive in external markets (Nunes and Brito 1990; Brito 1993; Lopes 1996).

Concerning its integration into the international economy, Portugal became a member of the European Free Trade Association (EFTA) in 1960 and this development had several consequences for the pulp industry. Chemical pulp fell under the auspices of article 6 of Annex G, benefitting from a certain level of protection as an "infant industry".[3] The transition period for this special regime ended in 1967 and this product entered into the general free-circulation regime and thus became free of any duties and tariffs within EFTA markets.

The growth in the pulp industry took place through the expansion of the productive capacity both in terms of opening new plants and boosting the capacities of existing ones (i.e., CPC and Caima). The state interfered directly in this phase through IC,

[3]Annex G to the Treaty of Stockholm stipulated a special disposition for Portugal in terms of its rights to imports and quantitative restrictions on exports. Annex Article 6 granted the scope for introducing or raising customs duties in order to protect the "infant industries" through to 1 July 1972 and with a broader calendar for the reduction of rights extending through to 1980. See Branco (2005).

regulating not only the number of establishments but also their productive capacity. From the late 1950s and over the course of the 1960s, over thirty requests for pulp licenses were submitted, which reflected the great interest in the sector from private capital (Confraria 1992). Some of these requests gained the technical support of CPC based upon the success and the importance of the technological innovation then implemented by this company (AAVV 1978). This offer of technical support to potential competitors of CPC was a sign of the state's willingness to diffuse innovation. Among the several requests made for new plants, authorisation was granted to Socel, Celbi, Celulose do Guadiana, Celtejo and Celnorte. Two authorisations also went to already existing plants, one to Caima for setting up another production facility[4] and the other to CPC for expanding its installed capacity. Table 6.2 details the requests that were authorised.

The additional capacity created by these projects generated a debate about the risks of continuing to expand production due to concerns over the capacity of Portuguese forests to supply enough raw material. This concern was reflected in the Ministerial Order of 11 July 1966, from the Ministry of Industry, that stipulated the zones in which new factories might be set up and binding their authorisation to the existence of sufficient raw materials to support them. The authorisations would only be granted after the companies had enacted a forestry plan if the existing forests did not guarantee the raw materials they needed. The 27 July 1968 Ministerial Order further reinforced concerns over both the raw materials and competition and postulated the merger of pending projects.

The existing pulp companies took two paths in order to cope with these supply side difficulties. On the one hand, they sought to cut back on the number of intermediaries, forming a cartel for the purchase of raw materials; this had been the case for Caima, CPC, Socel, and Celbi (Paues 1998). On the other hand, they carried out upstream integration through establishing their own forested areas. In this case, the state supported the afforestation plans of the private sector, through the provision of both financing and technical means. From this point forward, the Portuguese forest experienced a growing impact from the expansion of the pulp sector with a steady and sustained increase in the area of eucalyptus plantations (Branco 2010, 2011).

The new mills set up in the wake of CPC were not forward integrated, with the exception of Celnorte. The largest pulp companies acquired part ownership of the major paper manufacturers, as was the case with Companhia do Papel do Prado (with CPC acquiring a stake in 1959) and Inapa (Socel being its founder shareholder). It should be noted that the three plants controlled by foreign capital remained unintegrated. As regards the business structure of the pulp sector, three trends stand out. Firstly, the sector had attracted the interest of the main Portuguese business groups, which participated in setting up the new pulp companies. In some cases, they went into partnership with foreign capital in line with the new openness of the Portuguese

[4]Already in 1949, Caima submitted a project for boosting the productive capacity to 14,000 tons per year. The project was authorised since the production of this company did not call into question the objectives of the CPC.

Table 6.2 Technical features of the authorised requests under IC

Company (Location)	Date		Production system	Raw material	Production capacity (10^3 Ton/year)
	Proposal	Starting			
Caima			Chemical to bisulphite	Eucalyptus	
(Albergaria-a-velha)	1949/1966				14/25
(Constância)	1963				21
CPC (Cacia)	1963		Chemical to sulphate (bleached and unbleached)	Pine and Eucalyptus	119, 5
			Semi-chemical		22, 5
Celulose do guadiana (Mourão)	1952	1953	Semi-chemical pulp and cardboard (Integrated plant)	Straw	20
Socel (Setúbal)	1957	1964	Chemical to sulphate (bleached and unbleached)	Pine and eucalyptus	90
Celtejo (VV Rodão)	1963	1971	Chemical to sulphate (unbleached)	Pine and eucalyptus	60
Celbi (Figueira da Foz)	1963	1967	Dissolving pulp (only until 1969)	Pine and eucalyptus	80
	1968	1970	Chemical to sulphate (bleached)		150
Celnorte (Viana do Castelo)	1963	1973	Chemical to sulphate and paper kraftliner (Integrated plant)	Pine and eucalyptus	100

Sources Abecassis (1969), Gomes (1969), Confraria (1992), Alves (2001)

economy (Santos 1977, 1989). Falling within this scope was the case of Celbi, held by the Portuguese group CUF and the Swedish group Billerud.[5]

In 1973, the state indicated that it would reduce its stake in CPC, thereby marking the beginning of the rolling back of its presence in the ownership of the pulp industry. This retrenchment coincided with the CPC's strategy to rationalise the industry in the early 1970s aiming to merge with Socel and Inapa. This last firm, founded in 1965 with the objective of producing fine quality papers for printing and writing, was built alongside Socel and took delivery of its pulp. This therefore set off a third trend: the concentration of the sector in order to gain scale and efficiency. On the eve of the 25 April 1974 Carnation Revolution, Portugal's pulp industry was made up of seven companies (Caima, CPC, Socel, Celbi, and Celtejo, Celulose do Guadiana, and Celnorte) and eight production facilities with a combined capacity of over 500,000 tons per year, of which about 85% was being exported.

6.6 Consolidating Growth in a Nationalised Economy

As detailed in the previous section, by the mid-1970s the Portuguese pulp sector was experiencing discernible growth with the construction of new plants (and the establishment of new companies), the launch of expansion projects and the emergence of the first signs of business concentration. Following decades of substantial involvement, the state was also planning to withdraw from the sector with the intention of selling the direct stake it held in CPC. This dynamic was disturbed by the impact of the 1974 Carnation Revolution that put an end to the dictatorship of Estado Novo and established a new political and economic regime. In 1976, the main sectors of the Portuguese economy—finance, energy, telecommunications, transportation, and some manufacturing branches—were nationalised. This process correspondingly led to a dramatic increase in the state's involvement in the business sector as it took over direct and indirect ownership of almost 2,000 companies (Nunes et al. 2004; Pintado and Mendonça 1989). This also brought about the downfall of business groups that had hitherto been growing and developing since the 1960s (Silva et al. 2016). The pulp industry was one of the priority economic sectors on the agenda of the nationalisation process. The five companies controlled by Portuguese shareholders—CPC, Socel, Celnorte, Celtejo, Celulose do Guadiana—underwent nationalisation in 1975 and were merged into a single state owned company—Portucel—the following year. Celbi and Caima remained beyond the scope of this process as nationalisation did not extend to foreign capital assets.

Business consolidation, which had already emerged as a trend since the beginning of the 1970s, was driven forward by these nationalisations and this significantly altered the market structure in the Portuguese pulp industry. The eight mills became the property of three firms—Portucel with five, Caima with two and Celbi with

[5]Law 1994, dated 13 April 1943, limited the participation of international capital in national firms. This restriction was lifted in the 1960s.

a single factory. The state became the main player through Portucel, which had the largest installed capacity (53% of the sector's total production in 1977). It was also the largest paper and packaging maker and held controlling stakes in the two longstanding paper producers, Companhia de Papel do Prado and Fapajal, and as well as the new and dynamic Inapa.

This institutional shock associated with the change in the political and economic regime altered the business structure but it did not have a major impact on the industry's trajectory, specifically its rising productive capacity that it had displayed over the preceding decades. On the eve of nationalisation, pulp production in Portugal stood at 578,000 tons. One decade later, the figure had almost tripled to 1,472,000 tons, 81% of which derived from eucalyptus with the remainder being made from pine. In parallel with this rise in capacity, the market structure also changed. On the one hand, the three companies had already adopted different investment dynamics in keeping with the implementation of the expansion projects ongoing since the early 1970s. Pulp production by Portucel doubled in the decade following nationalisation while foreign owned Celbi and Caima increased by only 50% and 30% respectively.

Furthermore, a new project was launched through Soporcel in a process with peculiar origins that resulted in the construction of the largest pulp factory in Portugal, one that made a significant contribution towards the aforementioned rise in capacity. This project traced its origins to Celangol, which, in 1973, received authorisation to produce bleached eucalyptus pulp in Angola with a planned level of output of 210,000 tons. The state held a large stake in the company through an investment bank, Sociedade Financeira Portuguesa, which also proceeded to advance substantial sums in loans. Celangol acquired modern equipment in Europe that was never delivered to Angola as a consequence of the independence process of the former colonies that took place after the Carnation Revolution. The Portuguese government attempted to sell this equipment, but when it could not do so it adopted the solution of installing it in Portugal instead. Soon thereafter Celangol was converted into Soporcel in 1979 with Sociedade Financeira Portuguesa holding a 59% capital stake (Grupo Portucel Soporcel 2014). When its plant went operational in 1984, Soporcel immediately became the second largest Portuguese pulp producer, overtaking both Celbi and Caima.

In 1983, the stake held by Sociedade Financeira Portuguesa in Soporcel rose to almost 100% as a result of converting its loans into share capital. Two years later, the British paper group Wiggins Teape took up a 42.8% capital stake in Soporcel in keeping with the state's goal of getting an international strategic partner involved in the project, especially having in mind the development of the papermaking sector. In 1988, Soporcel produced 387,500 tons of bleached eucalyptus pulp, which represented around 26% of national output and it was almost entirely destined for the international market. This same year saw the launch of the paper manufacturing investment project, which began operating in 1991.

In 1988, the pulp production market shares per company stood at: Portucel 43%, Soporcel 31%, Celbi 17% and Caima 9%. Out of the total pulp produced in 1989 by these four companies, 83% was sold to the market, of which some 87% was then exported. The Portuguese pulp industry thus maintained its international market orientation that had underpinned its growth ever since the 1960s. Only Portucel operated an integrated paper production process, and by the late 1980s sales of paper and cardboard accounted for just over a fifth (23%) of the company's overall turnover.

Control over the supply of raw materials remained high on the agendas of the companies in accordance with the prevailing notion that the forest area of eucalyptus was insufficient to meet the growth in pulp production. The Portuguese Forestry Plan, implemented between 1981 and 1988 with financing from the World Bank, was the first major public intervention program in private forestry since the Forest Service had been established in the nineteenth century. Portucel was the main actor driving the implementation of this project and it was responsible for the afforestation of 60,000 hectares either of its own or leased properties, which represented more than a doubling of the forested area under its control (Mendes and Dias 2001). The other pulp companies also registered growth in their managed areas of forest over the 1980s. The concern over raw materials compelled the pulp companies to reinforce their departments and/or subsidiaries firms that were dedicated to forest management. They furthermore also targeted efforts at fostering research that would leverage productivity gains linked to the supply of raw material.

6.7 Privatisation, Concentration and Specialisation

In the last 25 years, the Portuguese pulp industry has developed in accordance with the following trends. Firstly, production has continued to rise, up from 1,449,000 tons in 1990 to 2,662,000 tons in 2015. This achievement is all the more significant when considering that no new pulp factory opened up during this period. Secondly, there were major changes in the specialisation pattern of the main companies within the scope of the increasing integration of paper production. Thirdly, and finally, the structure of ownership underwent a complete change with both the state and foreign players departing from the sector. The institutional framework faced by the Portuguese economy evolved substantially in the decades following the shock of nationalisation. The course of the accession process to the then European Economic Community, and in particular following membership in 1986, brought with it commitments to the liberalisation of protected sectors and the types of industrial policies for which this membership allowed (Araújo 2010; Confraria 1999). Furthermore, the liberalisation of international trade and capital flows posed new challenges to Portuguese companies.

Within the scope of this chapter, the privatisation process takes on particular prominence given how it drove a sharp decrease in the state's ownership in the business sector and the corresponding emergence of new organisational solutions in the main sectors of the Portuguese economy. The privatisation process began in the late 1980s and featured differences in the timing and the models chosen by each industrial sector (Nunes et al. 2004; Sousa 1995). In some cases, the government considered that the privatisation of state shareholdings should be preceded by a reorganisation in order to solve problems resulting either from high financial liabilities or the economic viability of firms (Amaral and Santos 1995). Pulp was one of these sectors.

At the time of launching the general privatisation process, the two pulp companies under state control, Portucel and Soporcel, were quite different organisations because of their histories. First, they were companies from different eras: the first had decades of existence and the second had only been in existence for a half dozen years. Second, Portucel was wholly owned by the state, while Soporcel had a foreign partner with a significant stake in the company's capital. Third, their product portfolios were also different. Portucel was a multi-plant company with numerous production lines as a result of a longer and more troubled history. The configuration endowed upon its foundation in 1976 remained in effect: production of pulp from eucalyptus and pine and the production of paper, cardboard, and packaging. Soporcel was focused on a single type of pulp, bleached from eucalyptus, and was at that time only starting to integrate paper production. The companies had neither links through their boards of directors nor any evidence of coordinated action.

The privatisation of the state owned pulp sector was conditioned by its restructuring, and Portucel, given both its scale and scope, would be the key actor in this process. Two questions could be raised about the privatisation of Portucel. First, should it be privatised as a single entity or in separate units? Second, was there room to explore associations with other actors in the sector, in particular Soporcel? The first step in the reorganisation took place in 1993 with the transformation of Portucel into a group of companies divided into autonomous business areas. Portucel Industrial, one of the companies emerging from this process, integrated the production of pulp and paper based on bleached eucalyptus pulp, operating the factories in Cacia and Setúbal. Gescartão incorporated the Viana mill's cardboard and packaging operation along with other packaging production units. Portucel Florestal concentrated on the ownership and management of forested properties and Portucel Tejo took over the Vila Velha de Rodão factory and its unbleached pine pulp production. Following the division of Portucel's operations into separate businesses, the privatisation process began. The original idea was to privatise Portucel by sectors with the first step occurring in 1995 when the state sold off 44.3% of its stake in Portucel Industrial. The second operation planned for the same year involved Gescartão but it did not end up happening.

Indeed, the privatisation process was suspended in 1996 and the next steps would have to wait for a "careful" restructuring of Portucel group within the framework of nurturing both critical mass on a global scale and absolute competitive advantages in some of its areas of activity. In the following years, the resumption of the privatisation process remained conditioned by the results of studies on the reorganisation of the sector.[6] This reorganisation extended to the integration of Inapa and Soporcel into Portucel. In turn, the acquisition of the industrial operations of Inapa came about in 2000. Inapa was intimately bound up with the production of the Setúbal pulp factory

[6]Privatisation program 1996/97(Council of Ministers Decision nº 21/96 on 8 February 1996), and Privatisation program 1997/98(Council of Ministers Decision nº 65/97 on 26 March 1996).

(formerly Socel) and had started developing an office paper brand that was gaining acceptance in international markets.

In 2001, the merger of Soporcel into Portucel took place following the former's purchase of the stakes held by the Argo Wiggins group and Papercel and the launching of a takeover bid for the remainder of the firm. The integration of Soporcel proved very important to the overall reorganisation of the sector. In addition to its already large pulp production capacity, the company was engaged in consolidating its position in the paper market with the launch of a second machine and the rising international recognition of its brand. This reorganisation resulted in a company whose main focus was the production of paper based on bleached eucalyptus pulp. In 2001, 54% of pulp manufactured was integrated into the production of paper. The pulp sales of Portucel/Soporcel represented only 25% of its total turnover. The domestic market was residual, accounting for around 5% of total sales.

Following the conclusion of the restructuring process, the privatisation process then advanced through two further phases. The first, in 2004, corresponded to 30% of the Portucel/Soporcel's capital while the second, in 2006, saw the disposal of the remaining stake (25.7%). Outside the Portucel/Soporcel group, Gescartão was privatised in 1999 and Portucel Tejo in 2006. As had been the case with the nationalisations in the mid-1970s, the privatisation process represented an important landmark in the restructuring of the Portuguese corporate sector. In general terms, this was an opportunity for Portuguese capital to invest, grow and/or diversify. The privatisation of state owned pulp companies elicited interest from three private Portuguese groups: Sonae, Semapa and Cofina (later Altri).

Sonae, one of Portugal's largest and most diversified business groups, already had operations in the forest industry (i.e., wood-based panels) and was also a major consumer of paper and cardboard within the scope of its activities as a mass retailer. Sonae targeted Gescartão, Portucel Industrial and Soporcel, in which the group sought to gain a controlling position. This started out with a stake of around 10% in Portucel Industrial. In 2000, in conjunction with the Spanish group Europac, Sonae won the privatisation tender for 65% of Gescartão. Although it was one of the interested parties in the second phase of the privatisation of Portucel in 2003, Sonae ended up withdrawing from the bidding with criticisms about the way the process was being conducted. The group then abandoned the sector following the disposal of its stakes in Portucel in 2004 and in Gescartão in 2005.

Semapa is the holding company of the Queiroz Pereira family, which had been at the very origins of the sector in Portugal in playing a leading role in one of the projects that resulted in the creation of CPC. Semapa acquired a stake in Portucel in 2004 following its winning bid in the second phase of the privatisation tender and the subsequent launch of a takeover bid resulted in it ending up with 67% of the company. The third and final phase saw it slightly boost its holdings.

Cofina unsuccessfully competed in the second phase of the Portucel privatisation process, and later it won the privatisation tender for Portucel Tejo. Furthermore, its presence in the sector was also bound up with the fate of Portugal's foreign owned pulp companies. After nationalisation, Celbi and Caima confined themselves to lower levels of investment in their productive capacity and not integrating their operations into paper production. Their owners ended up selling out to Cofina, which meant the end of any foreign controlling stakes in the Portuguese bleached eucalyptus pulp sector. The first to pull out were the owners of Caima, selling out in 1998. Stora, heir to Billerud, which had founded Celbi in 1967, alienated its holding in 2006.

Semapa and Cofina concentrated the entire Portuguese industry bleached eucalyptus pulp, and made strong investments in the last decade, but followed very different paths in doing so. Portucel maintained its growth strategy by increasing its production of paper, specifically by making the major investment to install a third paper machine that started up in 2009. Currently the company produces 1,400,000 tons of pulp, but sales of pulp represents only 9% of its turnover. The core business of Portucel is the production of uncoated wood-free paper, and it is the global leader in the premium office paper segment with its Navigator brand.

In contrast, Altri's strategy focused on the production of pulp. Between 2008 and 2009, it converted production at its Celtejo plant from unbleached pine and eucalyptus pulp to bleached eucalypt pulp, and it doubled the productive capacity of the Celbi plant. Furthermore in 2011 Caima started producing dissolving pulp. At present, Altri produces about one million tons of pulp, 93% of which is sold abroad.

6.8 Conclusions

Portugal is a country with scarce natural resources and a small-scale domestic market, and it predominantly specialises in low value added and labour intensive products. Nevertheless, it was able to become an international leader in producing several products, namely pulp and paper, that have a major technological component. The success of Portugal's pulp industry is indebted to the creation and fostering of comparative and competitive advantages, which stemmed from the coordinated efforts between the state and the country's firms. It was based in some key areas—product innovation, export orientation, business concentration, raw material supply, and forward integration into paper production—that have emerged over the 60 years since the industry first took off.

The state played an important role in all the phases of this industry's development. During the early years of Estado Novo, it created conditions for the launch of the sector. Concerning upstream measures, its forest policy aimed to supply the raw material base, first by direct afforestation and then by supporting private forest owners. The pulp industry was considered strategic under Estado Novo's economic development policy, and thus benefitted from the state's financial support. His regime took several other measures to facilitate the industry's vertical integration and were directed to a specific company.

The history of the pulp industry in Portugal is inextricably linked with that of Companhia Portuguesa de Celulose. The state chose a "national champion" and "picked its winner" by using an import substitution model and providing aid to an "infant industry", and this was the result of merging industrial policy measures with a protectionist trade policy. The strategy of CPC was fundamental to the future of the industry and nurturing its competitive advantage. Initially its project was designed to utilise pine wood, but it ended up relying on eucalyptus wood. Similarly, the objective was to produce mechanical pulp and newsprint to supply the internal market. This goal was soon abandoned, however, and CPC was the first company to produce chemical pulp from eucalyptus using the sulphate process, and most of its production was exported. In the end, positioning pulp within the import substitution policy allowed this industry to be well placed to capitalise on the subsequent export promotion policy, which came to define the sector. As a symbol of a company that achieved success in global and competitive markets, other firms mimicked CPC. This led to an expansionary phase during which the state regulated the market under a scenario of potential scarcity of raw materials, choosing who the new players were going to be.

The strong state commitment to the sector was not jeopardised with the political transformation of mid-1970s. Nationalisation led to the concentration of ownership in the hands of the state. Over the course of twenty-five years the state controlled the industry and supported its development. The creation of Soporcel, which proved to be an enormous success, was an example of that commitment. Notwithstanding the state's role as entrepreneur, it was also front and centre in designing the privatisation scheme. In this regard, the government transformed the sector's landscape in the last quarter of a century, in what could be considered its maturity phase. Today's business structure, with two Portuguese business groups controlling the six mills that produce bleached eucalyptus pulp, reflects the decisions taken under that process.

Just as it did fifty years ago, the state forced the creation of a champion, shaping the course of privatisation to the reorganisation of the sector in order to create a company that could compete in global markets. This goal was achieved through the growing integration of pulp and paper and focusing on the production of a certain type of paper. As a result of this process, Portucel, heir to Companhia Portuguesa de Celulose, became a completely different company. Significantly, the company whose name has always alluded to pulp has recently changed its name to the paper brand with which it leads the premium segment of office paper—Navigator.

Ensuring the industry was well supplied with raw materials was a constant concern. The actions of the state, forest owners and pulp companies led to a radical transformation in the country's forest cover, with eucalyptus moving from being a species that was virtually non-existent to being the country's most important type of tree.

References

AAVV (1958) Companhia Portuguesa de Celulose. [s.l.]. Companhia Portuguesa de Celulose

AAVV (1978) 25 anos Celulose Cacia. Portucel-Centro de Produção Fabril, Cacia

Abecassis JLC (1969) Problemática da Indústria Portuguesa de Pasta de Celulose à base de Madeira e sua Expansão. Boletim Semanal da Direcção-Geral dos Serviços Industriais:198–200

Alves AM, Pereira JS, Silva JMN (2007) A introdução e a expansão do eucalipto em Portugal. In: Alves AM, Pereira JS, Silva JMN (eds) O Eucaliptal em Portugal: Impactes Ambientais e Investigação Científica. ISA Press, pp 13–24

Alves JF (2000) A estruturação de um sector industrial: a pasta de papel. Revista da Faculdade de Letras: História, III série 1:153–182

Alves JF (2001) Indústria e pasta e do papel em Portugal: o grupo Portucel. [s.l.]: Portucel SGPS, Edições Inapa

Amaral LM, Santos AD (1995) Privatização e Reestruturação dos Sectores Industrial e Energético. Direcção-Geral da Indústria, Lisbon

Aiginger K, Sieber S (2006) The matrix approach to industrial policy. Int Rev Appl Econ 20(5):573–601

Araújo JFFE (2010) Policy-making for industrial competitiveness in Portugal: patterns of change in a traditional bureaucracy. Public Manag Rev 3(2):255–269

Baptista FO (1984) A Política Agrária (anos trinta-1974). Dissertation, Instituto Superior de Agronomia, Lisbon

Branco A (2005) O impacto das florestas no crescimento económico moderno durante o Estado Novo (1930–1974). Dissertation, ISEG—Lisbon School of Economics and Management, Lisbon

Branco A (2010) Was the Portuguese forest policy a contribution towards economic modernization? The case of the paper pulp industry during the Estado Novo (1930–1974). Revista de Historia Ind 44, XIX(3):69–96

Branco A (2011) El Plan de Repoblación Forestal (1938–1968): una medida del Estado Novo por evaluar. In: Freire D, Lanero D (eds) Agriculturas e Innovacíon Tecnológica en la Península Ibérica (1946–1975). Ministerio de Medio Ambiente y Medio Rural y Marino, Madrid, pp 135–166

Brito JMB (1989) A industrialização portuguesa no pós-guerra (1948–1965): o condicionamento industrial. Publicações Dom Quixote

Brito JMB (1983) Rumos velhos e novos da industrialização portuguesa. Evolução Recente e Perspectivas de Transformação da Economia Portuguesa. ISE, Lisboa, pp 2535–2561

Confraria J (1992) Condicionamento Industrial - Uma análise económica. Direcção-Geral da Indústria, Lisbon

Confraria J (1999) Portugal: industrialization and backwardness. In: Foreman-Peck J, Federico G (eds) European industrial policy. The twentieth century experience. Oxford University Press, Oxford

Costa AP (1946) A Indústria do Papel em Portugal. Separata do Boletim da Direção Geral da Indústria (2.ª série), pp 5–126

CELPA (several years) Boletim Estatístico da Indústria Papeleira Portuguesa. CELPA – Associação da Indústria Papeleira

Direção Geral da Indústria (1939) Indústria do Papel - Parecer do Conselho Superior da Indústria sobre o Inquérito à Indústria do Papel e Despacho de S. Ex[a]. o Ministro. Boletim da Direção Geral da Indústria - Ministério do Comércio e Indústria, Lisbon

Gerschenkron A (1962) Economic backwardness in historical perspective: a book of essays. Belknap Press of Harvard University Press, Cambridge, MA

Gomes AMA (1969) Fomento da arborização nos terrenos particulares. Fundação Calouste Gulbenkian, Lisbon

Grabas C, Nützenadel A (2014) Introduction. In: Grabas C, Nützenadel A (eds) Industrial policy in Europe after 1945; wealth, power and economic development in the cold war. Palgrave Macmillan, Houndmills, Basingstoke, Hampshire, pp 1–10

Gutiérrez-Poch M (2012) Is there a Southern European model? Development of pulp and paper industry in Italy, Spain and Portugal (1800–2010). In: Lamberg J-A, Ojala J, Peltoniemi M, Särkkä T (eds) The evolution of global paper industry 1800–2050. A comparative analysis. Springer, Dordrecht, pp 211–242

ICNF (2013) IFN6—Áreas dos usos do solo e das espécies florestais de Portugal continental. Resultados preliminares. Instituto da Conservação da Natureza e das Florestas, Lisbon

Lains P (1994) O Estado e a Industrialização em Portugal. Análise Social, 128:923–958

Lains P (1995) A Economia Portuguesa no Século XIX. Crescimento Económico e Comércio Externo. Imprensa Nacional, Lisboa
Lains P (2003) Os Progressos do Atraso. Uma Nova História Económica de Portugal, 1842–1992. Imprensa de Ciências Sociais, Lisboa
Lee J, Clacher I, Keasey K (2012) Industrial policy as an engine of economic growth: a framework of analysis and evidence from South Korea (1960–96). Bus Hist 54(5):713–740
Lopes JS (1996) A Economia Portuguesa desde de 1960. Gradiva, Lisbon
Loureiro JA (1991) Economia e sociedade: a indústria no após-guerra anos 50 e 60. Edições Cosmos, Lisbon
Mendes AMSC (2007) The Portuguese forests. Working Papers de Economia (Economics Working Papers) 13, Católica Porto Business School, Universidade Católica Portuguesa
Mendes AMSC, Dias RARS (2001) Financial instruments of forest policy in Portugal in the 80s and 90s. Faculty of Economics and Management, Portuguese Catholic University, Porto. http://www.metla.fi/eu/cost/e19/finincport.pdf. Accessed 31 Jan 2017
Mesquita EM (1919) Tese sobre os meios a recorrer para atenuar a importação de pasta vegetal para o fabrico de papel. Boletim do Ministério da Agricultura
Nunes AB, Brito JMB (1990) Política Económica, Industrialização e Crescimento. In: Rosas F (coord) Portugal e o Estado Novo (1930–1960). Editorial Estampa, Lisbon, pp 306–351
Nunes AB, Bastien C, Valério N (2004) Nationalisations et dénationalisations au Portugal (XIXe-XXe siècles): une évaluation historique. Entreprises et histoire 37:104–119
Nunes AB (1996) Control and regulation in the Portuguese economy (1945–1973). GHES Documento de Trabalho/Working Paper 2. Instituto Superior de Economia e Gestão
Ojala J, Voutilainen M, Lamberg J-A (2012) The evolution of the global paper industry: concluding remarks. In: Lamberg J-A, Ojala J, Peltoniemi M, Särkkä T (eds) The evolution of global paper industry 1800–2050. A comparative analysis. Springer, Dordrecht, pp 345–356
Pack H, Saggi K (2006) Is there a case for industrial policy? A critical survey. World Bank Res Obs 21(2):267–297
Paues N (1998) A história da Stora Celbi. [s.l.]: [s.n.]
Pintado MR, Mendonça A (1989) Os novos grupos económicos. Texto Editora, Lisbon
Radich MC, Alves AAM (2000) Dois Séculos da Floresta em Portugal. CELPA, Lisbon
Reis J (1993) O atraso económico português em perspectiva histórica, 1850-1930. Imprensa Nacional, Lisbon
Reis J (2000) How poor was the periphery before 1850? The Mediterranean versus Scandinavia. In: Williamson JG, Pamuk S (eds) The Mediterranean response to globalization before 1950. Routledge, London, New York, pp 17–44
Rosas F (1986) O Estado Novo nos anos trinta (1928–1938). Editorial Estampa, Lisbon
Rosas F (2000) Salazarismo e o Fomento Económico. Editorial Notícias, Lisbon
Santos AR (1977) Desenvolvimento monopolista em Portugal (fase 1968–73): estruturas fundamentais. Análise Social XIII(49):69–95
Santos AR (1989) Abertura e bloqueamento da economia portuguesa. In: Reis A (coord) Portugal Contemporâneo, vol V. Publicações Alfa, Lisbon, pp 109–150
Seabra LSV (1943) O Problema Nacional da Celulose: Utopias e Realidades. Direção Geral dos Serviços Florestais e Aquícolas
Seabra LSV (1944) Perspectivas Nacionais na Produção de Celulose. Revista Agronómica 32(1):1–15
Seabra LSV (1944) Celulose. Matéria-prima Florestal. Revista Agros 3–6:206–214
Silva AF, Amaral L, Neves P (2016) Business groups in Portugal in the Estado Novo period (1930–1974): family, power and structural change. Bus Hist 58(1):49–68
Sousa F (1995) O processo de privatizações em Portugal. Associação Industrial Portuense, Oporto

Chapter 7
Natural Potential, Artificial Restraint: The Dryden Paper Company and the Fetters on Adopting Technological Innovation in a Canadian Pulp and Paper Sector, 1900–1950

Mark Kuhlberg

7.1 Introduction

Beginning in the late 1800s, Canada's massive storehouse of wood and water power resources became the foundation upon which a multitude of pulp and paper mills was built. Although the industry arrived relatively late—it was still in its infancy at the turn of the twentieth century, it enjoyed truly extraordinary growth once it had set down roots in the country. For example, in 1890 the industry represented a total capital investment of roughly CAD$7.5 million, and twenty-five years later this figure was approximately CAD$132 million, an increase of nearly 1,700% in only one generation. Moreover, the newsprint sector grew from barely 50,000 tons annual capacity in 1900 to well over 3,500,000 thirty years later; already by that time Canada was the world's largest newsprint maker, a title it still holds. During this period, the industry seemed perpetually driven by the adage that "bigger is better", and as a result it eagerly embraced practically each new technological innovation in an effort to improve efficiency and profitability (Kuhlberg 2012a, b).

Despite a steadily rising demand for all the products Canadian pulp and paper makers manufactured in the first half of the twentieth century, not all of these enterprises aggressively adopted the latest technology, specifically in terms of economies of scale. Technology exists within a matrix of many other forces, upon which it acts and which act upon it. Although logic would suggest that a business is driven to maximize profitability, and that doing so often entails adopting ever-more efficient machinery (assuming there is a rising demand for its product), there are a few glaring exceptions to this rule. For example, resource-rich areas of the globe, of which Canada is a classic example, can be averse to developing domestic technologies because the "staples" upon which they rely for economic growth are readily exportable with prac-

M. Kuhlberg (✉)
History Department, Laurentian University, Sudbury, ON, Canada
e-mail: mkuhlberg@laurentian.ca

T. Särkkä et al. (eds.), *Technological Transformation in the Global Pulp and Paper Industry 1800–2018*, World Forests 23, https://doi.org/10.1007/978-3-319-94962-8_7

tically no processing (Innis 1930).[1] Furthermore, the implementation of nationalist government policies, such as protective tariffs, in the producer's potential markets can fundamentally shape the desirability and speed of technology transfer. Similarly, in jurisdictions in which governments control access to raw materials, their cooperation and support can be crucial to fostering an industry's growth. So, too, can a company's culture and business designs. For example, an enterprise operating in a sector in which it controls its marketplace, by virtue of either its propitious location, control over patents or some other mechanism, has little incentive to expand.

Conducting a case study of the pulp and paper mill in Dryden, Ontario, demonstrates that more than merely abundant forests and waterfalls were needed to compel a firm to take advantage of the latest technology and thereby expand its productive capacity by leaps and bounds. Although the plant in Dryden had a rated capacity of over 100,000 daily tons by the time of the Korean War, for most of the previous four decades it had remained remarkably small-scale even though the technology existed for much of this period for it to have expanded dramatically.

A number of factors, operating on the provincial, national and international levels, caused this lengthy developmental delay in the case of the Dryden mill. Initially, its very existence was thwarted by technological challenges. Later, its early commitment to expanding was stifled by the Ontario government's refusal to support this agenda, and by the existence of stiff American tariffs and fierce foreign competitors. A few years later, a change in ownership fostered a new corporate ethos, one that focused not on growth but instead on controlling its market and operating according to the principles of monopoly capitalism. As a result, even though revolutionary new technology was developed during the Great Depression and aggressively adopted by most kraft producers to grow their operations exponentially, the mill in Dryden chose not to participate in this movement. When it changed its perspective during the Second World War and sought to expand its capacity, the conflict made it virtually impossible to do so. Ultimately, all these factors resulted in this facility representing an anomaly in the Canadian industry. Although it was located amidst a prodigious supply of natural resources, for nearly forty years it remained relatively tiny because a complex set of factors prevented it from adopting the technology needed to expand its operation in a meaningful way.

[1] Harold Innis is most often credited with having given rise to the "staples thesis" to explain Canada's development. It contends that capital, most often foreign, was drawn to exploit a succession of the nation's natural resources (first cod and fur and later timber), which could be readily exported with very little domestic processing.

7.2 The Context and Early Challenges to Establishing a Pulp and Paper Mill

7.2.1 Ontario's Dominant Place in Canada

From the moment that Canada was founded in 1867, Ontario was the country's most dominant province. At the time of the nation's birth there were four provinces in total, and roughly 43% of Canada's approximately 1,500,000 residents lived in Ontario. Over the next few decades, as the young country grew and the number of provinces increased, the proportion of Canada's population living in Ontario rose slightly and edged ever closer to 50%. The province's preponderance in this regard was reflected politically in the strong voice it was accorded in the national government; Ontario received 24 of 72 (33%) seats in 1867 in the Senate and 82 of 181 seats (45%) in the House of Commons. The province's dominance also extended to the economy. Ontario boasted a robust agricultural sector that produced large surpluses for export markets because the province was blessed with such productive farmland and favourable climate. These dynamics allowed it to generate significant capital that was then available for investment in diversifying the agricultural sector (i.e., into dairy farming, and cheese and butter production) and expanding Ontario's industrial capacity (McCallum 1980). All these forces propelled the province forward and gave observers ample reason to nickname it "Empire Ontario" by the end of the nineteenth century.

By the turn of the twentieth century, however, Ontario's hegemonic position in the country was slipping, and slipping quickly. Canada's vast swaths of vacant prairie lands were attracting millions of settlers from the United States and Europe, and the extraordinary pace at which this region was being populated threatened to end Ontario's dominance. Although the province remained the most populous in the country, the percentage of Canadians living in it had slipped to 36% by 1911 and 33% ten years later. During this same period, Ontario's voice in national politics had also weakened considerably. By 1921 its representatives occupied only 25% of the seats in the Senate and 35% of those in the House of Commons. Moreover, Ontario's economic foundation had been built upon a strong farming sector, but by the mid-nineteenth century the province's best land—in its southern reaches—was all settled. Thereafter, the government pushed agricultural development into the central part of the province, but this campaign had lagged badly due to this region's limited growing season, generally poor soils and the enormous investment of energy needed to clear its trees. This "colonization" push suffered even further during the late nineteenth century when the aforementioned allure of Canada's prairies drew wave upon wave of settlers out west, many of whom were drawn from Ontario. If these forces continued to act in this way, the province would soon be usurped from its predominance over Canada.

Several forces combined around the turn of the century to give the province's guiding lights hope that they might realize their dream of transforming the vast land base that made up northern Ontario, which represented roughly nine-tenths of

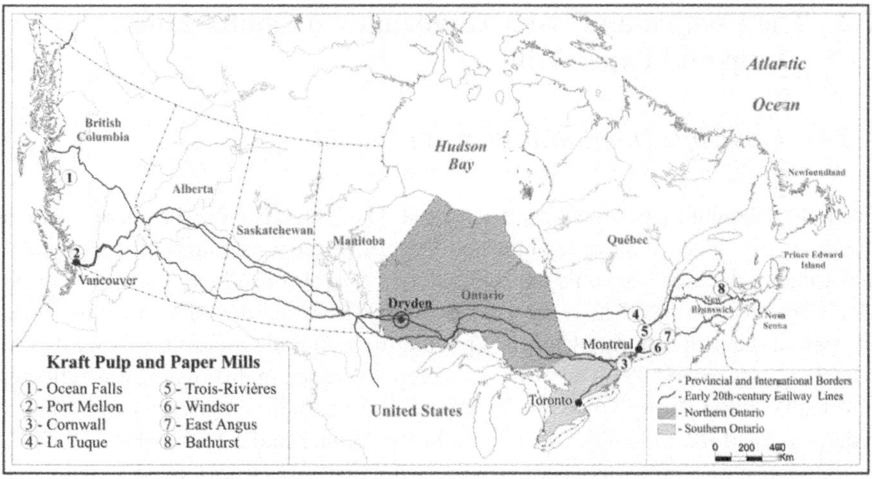

Fig. 7.1 Location of kraft pulp and paper mills built in Canada between 1907 and 1932

the province's size, into a massive new farming region (Fig. 7.1). The provincial government funded a major propaganda campaign to convince potential settlers that much of northern Ontario was as good for growing wheat as the prairies because both areas experienced virtually the same climate. In an effort to entice homesteaders north, the government also distributed virtually free land in the area. Although a string of experts contended that commercial grain farming was a non-starter in the region (much of the area was composed of swamp, rock or poorly drained soils, and its growing season was extremely short), the government viewed promoting this chimerical notion as a means of permanently connecting thousands of settlers to the hinterland (Kuhlberg 2014).

Achieving this goal only became possible because North America's pulp and paper industry was undergoing a fundamental reorientation around this time. Its transition from using wood instead of rags as its raw material was a boon to northern Ontario's colonization program because this part of the province was covered in huge volumes of black spruce (*Picea mariana*) and white spruce (*Picea glauca*), the most valuable pulpwood species during this period (it also boasted vast swaths of jack pine, *Pinus banksian*, which would soon become a valuable pulpwood). The provincial government quickly realized that the only way to colonize the hinterland was to create a robust market for the trees—especially small-diameter spruce pulpwood—that the settlers would harvest as they cleared their homesteads and worked as cutters in the nearby forests. To realize this aim, the government denied nearly every pulp and paper mill in northern Ontario enough pulpwood to sustain its operations, thereby forcing them to rely for their existence on the purchase of large supplies of settlers' timber on the open market. Moreover, although the provincial government had implemented an embargo on exporting pulpwood from Ontario in 1900, thereafter the politicians went to great lengths to foster this trade. In fact, during the first three decades of the twentieth century, Ontario was the only province in Canada that controlled substantial supplies of pulpwood and from which the export of it continued to climb, and it did so dramatically (Kuhlberg 2015).

Dryden was established within the context of this state-sponsored colonization impulse. The construction of the Canadian Pacific Railway through northwestern Ontario during the early 1880s opened the area's resources to economic exploitation and helped reveal much about them (Fig. 7.2). Although logging began at that time in the vicinity of present-day Dryden, it took more than another dozen years for the area to get a major development boost. In 1895, John Dryden (1840–1909), Ontario's Minister of Agriculture and a zealous proponent of the north's suitability for pastoral pursuits, was apparently on a train trip out west to examine the Canadian prairies when he noticed a thick crop of clover growing around a watering stop—at which boxcars had been washed down—near the shore of Lake Wabigoon. Convinced that the area could support farming he created an experimental farm there and optimistically christened the site New Prospect. He also ordered immediate surveys of two local townships and opened them to settlement, and laid out the boundaries for the town, which was soon renamed in his honour. Thereafter, "Dryden" benefited from the Ontario government's major investment in pushing settlement in the north. In doing so, the provincial politicians repeatedly pointed to the success of this effort in the area around Dryden as proof positive that farming was indeed possible in Ontario's farthest reaches (Stewart 2003; Wice 1967; The Pioneer Farm and the Wabigoon Country 1896; The Wabigoon Country 1897, 1898).

7.2.2 Government Hinders the Development of the Industry

Northwestern Ontario was blessed with more resources than simply soil, and entrepreneurs soon began exploiting the ones around Dryden. By the late 1890s, dozens of local mining claims were being worked and yielding significant volumes of gold. Around the same time, British capital began eyeing the region's timber and water power resources. At the turn of the twentieth century, several major pulp and paper firms that were based in England began looking to Canada as a potential site for their operations, including Edward Lloyd, Limited, which acquired a new mill in northeastern Ontario at this time (Carruthers 1947). In 1903, another group of British investors incorporated a new firm, the Dryden Board Mills Company, to establish a pulp and paper plant in the town of the same name. To support the project, its backers entered into an agreement with the Ontario government for the "Dryden and Wabigoon pulpwood concession (i.e., limit)" (approximately 666 square miles). To be valid, however, the contract required the approval of the Ontario Legislature. The deal obliged the firm to build a pulp and paperboard mill—the enterprise was thus known as the Dryden Board Mills Company—in exchange for which it was given both the privilege of cutting the tract's pulpwood and a lease to develop the local water power site on the Wabigoon River near Dryden. The project was delayed by the chaos and uncertainty that erupted after Ontario's first pulp and paper mill went into

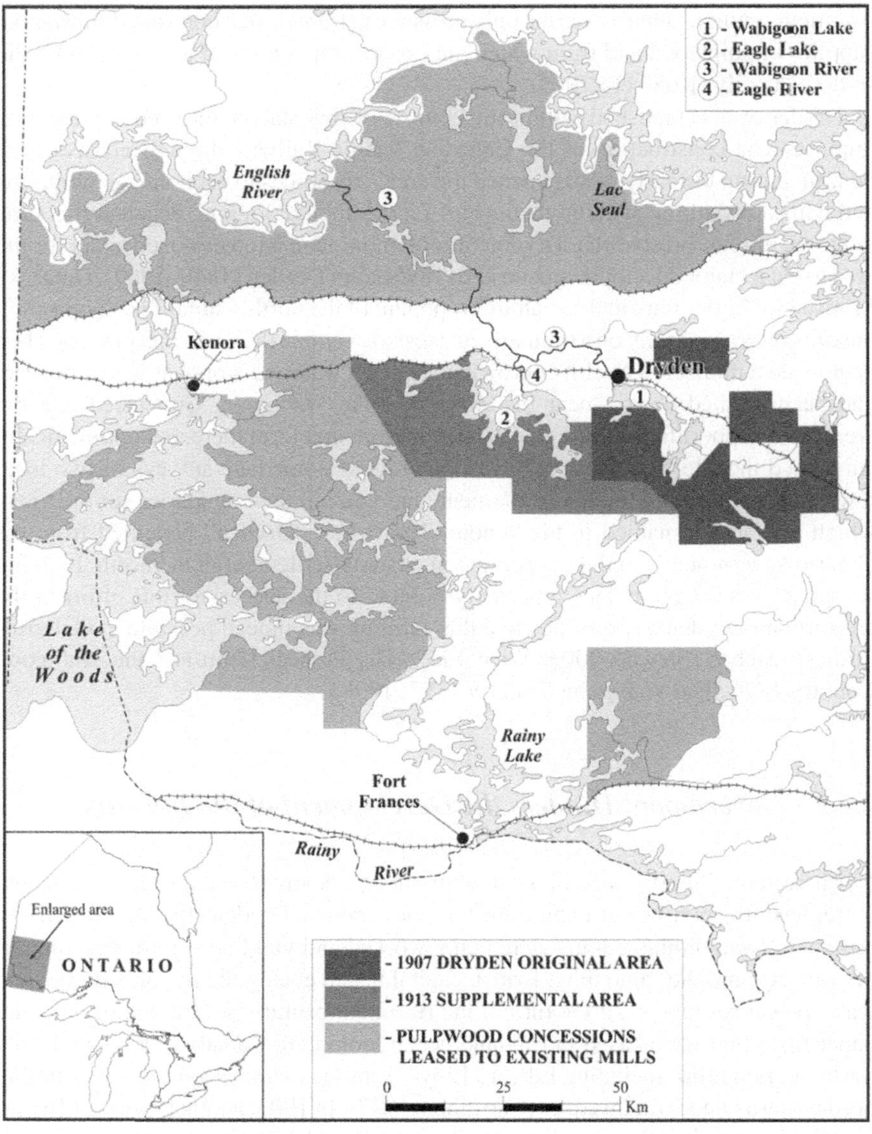

Fig. 7.2 Timber situation and waterways in the vicinity of the pulp and paper mill in Dryden, Ontario

receivership, and the provincial government thus re-set the deadline by which Dryden Board was to build its new facility (Pulp and Paper Magazine of Canada [hereafter PPMC] 1904a, b, 1905a and b; Archives of Ontario [hereafter AO], RG1-E-6, 6 January 1906; Mail and Empire 1906).

But political power changed hands in 1905 in Ontario, before the Legislature had approved the firm's pulpwood agreement with the government, and the new Conservative regime proved unwilling to assist the pulp and paper industrialists in realizing their aims.[2] The Tories refused to ratify the original pulpwood lease into which the Liberals had entered with Dryden Board and instead re-tendered the timber in 1906 (AO, RG1-246-3, 6 June 1916).[3] They did so under terms that practically precluded it from being used to support the construction of a pulp and/or paper mill. First, they set a preposterously high diameter limit on the trees that could be cut. Even though timber estimators considered spruce and jack pine pulpwood to be mature when it was between four and eight inches (10.1 cm–20.3) in diameter, and the previous licence to the local timber had granted the licensee permission to harvest trees above six inches (15.24 cm) in diameter, the tender in 1906 restricted the winning bidder from cutting timber that was under eight inches in size. By significantly decreasing the volume of pulpwood that was being sold at this time, however, the government ensured a high demand for the timber the settlers cut as they cleared their land. Incidentally, the government attached this provision to all the pulpwood limits it tendered at this time, and not surprisingly the result saw not one ton of capacity being added to Ontario's pulp and paper industry (AO Library 1906).

In addition, the Conservatives also refused to offer for lease the nearby waterfalls that could be harnessed to meet a new pulp or paper mill's demand for energy. They permitted Dryden Board to retain the rights to this crucial potential energy source, the lease to which it had obtained in 1903, but refused to award the firm the pulpwood the government tendered at this time. Instead, the Tories sold the timber to Robert McLaughlin, a contractor from southern Ontario, for the measly sum of CAD$6,000. Although McLaughlin's pulpwood agreement with the government, which he signed in 1907, obliged him to construct a pulp mill within three years and spend at least CAD$200,000 on the project, he would never do so and yet he would be permitted to tie up this timber for another half decade (Fig. 7.2) (AO, RG75-57, OC56/108 and OC56/236; Wice 1967).

---

[2]For nearly 40 years, historians have been arguing that provincial governments in Canada, particularly in Ontario, did all they could to facilitate the establishment and growth of pulp and paper mills within their jurisdictions. The best known example of this perspective is Nelles 1974. Recent research has revised this understanding and demonstrated that the story was much more complex (Kuhlberg 2015).

[3]Technically, the original agreement was only provisional and required the approval of the Ontario Legislature, ratification that the new Conservative government refused to give (AO, 5520, 15 April and 6 June 1916).

7.3 A Pulp and Paper Mill Is Built in Dryden Despite Government Hurdles

7.3.1 Dryden's Kraft Pulp and Paper, 1913

In the meantime, the party—consisting of the Gordon brothers—that had originally sought to construct a pulp and paper mill in Dryden pushed ahead with its venture, and its efforts would eventually lay the foundation for this goal to be realized. In 1908, they built a major sawmill near Dryden, and began cutting timber to supply it from the timber concession McLaughlin had leased in 1906 from the government. The next year, the Gordons began constructing their pulp mill, but when a fire destroyed their partially completed structure and dealt them a devastating financial blow, they decided to cut their losses. They sold their power lease, which they still held, and other assets to the Dryden Timber and Power Company, a firm that was led by veteran British and Canadian pulp and paper makers and whose financial backing (i.e., in the form of bond purchases) again came largely from the United Kingdom; Dryden Timber also acquired McLaughlin's pulpwood agreement at this time. Over the course of 1911–1913, Dryden Timber succeeded in building a 40-ton kraft pulp and paper mill, a project that included constructing a small hydroelectric generating station on the Wabigoon River, about which more will be said below. By the early part of 1913 the town of Dryden was feting the opening of this new enterprise, which stood as the first pulp and paper mill erected in northwestern Ontario (Wice 1967; Hall 1992; PPMC 1913a, b; AO, RG75-57, OC70/361).[4]

Although the provincial government's policy had played a hand in delaying the realization of this project, so too, had other factors. The local environment was one. The Canadian pulp and paper industry enjoyed meteoric growth during the early 1900s, and its existence had been based on producing almost exclusively sulphite and groundwood pulps and newsprint (made from spruce). Naturally, early plans for the mill in Dryden had entailed manufacturing these same products. Around 1910, however, the decision had been made to focus on producing primarily sulphate (or kraft) pulp and paper (made from jack pine), and only small volumes of groundwood pulp. This decision reflected a sensible evaluation of prevailing local conditions. First, it spoke to the unique nature of the energy and fibre resources available to the project. There was no local water power that would have been large enough to

[4]The relationship between McLaughlin and the Gordon brothers is unclear, as is how the former's pulpwood concession lease was transferred to Dryden Timber. It is known, however, that the Gordon brothers were associated with Samuel G. Nesbitt, who was president of their enterprise. Nesbitt was a Conservative Member of Provincial Parliament from 1908 to 1919. Considering the preeminent role patronage considerations played in timber transactions during this period, logic suggests that the Gordons and Nesbitt were able to use their political connections to secure access to the pulpwood on the local concession even if the lease was not in their name.

support the production of newsprint on the economies of scale that were required at the time,[5] and the local forest was dominated by jack pine instead of spruce.

Technology had also played a major role in delaying the realization of the mill project in Dryden, specifically in terms of developing the means by which the local jack pine could be processed into a valuable product. In the early 1900s, the means for producing kraft or sulphate pulps and papers had not yet been transferred from Scandinavia to North America, and it arrived initially in Canada and only in 1907. At that time, the Brompton Pulp and Paper Company opened the continent's first kraft pulp mill in East Angus, Quebec, and the first kraft paper mill was built three years later in the same province (Fig. 7.1). The first kraft pulp mill was not constructed in the United States until the eve of the First World War. This technology was still very new, however, and it took several years to improve it to the point whereby other industrialists felt confident investing in utilizing it (PPMC 1907, 1908; AO, RG75-57, OC59/424 and OC67/490; Lockwood's 1912; Toivanen 2004; Carruthers 1947).

7.3.2 Government Fetters Development: Timber

Although the townsfolk in Dryden were celebrating by the eve of the First World War the existence of their mill, a facility that was committed to capitalizing on the latest technology, the Ontario government still proved reluctant to assist it in resolving the challenges it faced in terms of natural resources. The first concerned its fibre supply. The mill now controlled over 600 square miles of forest, but it was only legally permitted to harvest pulpwood that was over eight inches in diameter even though many of the mature trees on its woodlands were much smaller. Dryden Timber's managing director, E.W. Bonfield, raised this issue with the government even before the company had started production, and in doing so he stressed the major damage that this provision was inflicting on the firm's financial standing. Writing to the Ontario government in mid-1912, Bonfield explained that his company had already invested well over one million Canadian dollars in the operation but was now in need of raising more capital—CAD$200,000—to complete it. The company's British bondholders had refused to provide the funds, however, because they were now convinced that the enterprise lacked a long-term fibre supply, which was the *sine que non* of any successful pulp and paper venture. They had every reason to be particularly dubious at this point. Dryden Timber's original prospectus, which had been the basis upon

[5] A study in 1911 of the water powers in Canada had estimated that the rapids on the Wabigoon River could be harnessed to produce roughly 1,000 hp (100 hp was needed for every ton of newsprint production), and it had not even included those available on the Eagle River, which was roughly 20 miles west of Dryden (Commission of Conservation 1911). A subsequent, more thorough study fourteen years later found that the total potential of the water power sites in the vicinity of Dryden was roughly 8,500 hp (Ontario Department of Lands and Forests 1925). By the mid-1920s, newsprint mills in Ontario were averaging well over 200 tons, and several were being built across eastern Canada whose capacities were over 500 tons.

which the British financiers had initially supported the venture, had broadcast that the mill controlled more than 4,300,000 cords (app. 15,500,000 m^3) of pulpwood. When the bondholders investigated the state of the company's woodlands in 1912, however, they had learned that its timber limits held just over one-eighth of this total (AO, RG1-246-3, 1424 Vol 1A 28 December 1917). They thus asserted that 'any security coming behind the existing Bonds would have little value'. As Bonfield explained the situation, 'the development by the Company should have timber ahead to last for forty years in order to protect the investment and ma[k]e the industry a solid one.' He added that 'the situation to-day with regard to the lumbering operations is that being limited to logs of not less than eight inches on the stump there is not sufficient timber on the concession to last at a liberal estimate for more than ten years.' The financiers thus felt that they were not 'warranted in undertaking to raise the additional CAD$200,000 without securing additional territory which will make the operations of the Company reasonably safe and sure.' As a result, Bonfield asked that the Ontario government grant his firm an additional timber limit that covered roughly 500 square miles of the drainage basin of the nearby Eagle Lake (Fig. 7.2) (AO, RG75-57, OC71/571; AO, RG1-246-3, 1424 Vol 1A, 14 July 1917).

The Ontario government's senior bureaucrats empathized with Dryden Timber's position and were crucial to convincing the elected officials to grant it. Aubrey White, Ontario's veteran Deputy Minister of Lands, Forests and Mines, presented a précis of the situation for his minister. He agreed wholeheartedly with Bonfield's assessment of the matter and reiterated that the financiers were willing to continue sinking their money into the venture but 'there is this fact standing in their way:- they are unable to show that they have sufficient timber on the territory covered by their present Concession to keep their plant running until the capital expenditure is wiped out and pay interest from year to year.' White added that there was thus good reason to grant the firm the privilege of cutting pulpwood from another 500 square miles of nearby forest, a directive upon which the minister acted in mid-1913 (AO, RG1-246-3, 1424 Vol 1A 10 December 1912).

But the mill in Dryden was still short of fibre, and this was because the Ontario government continued to favour the northern settlers over the pulp and paper makers (AO MU1311 ca1911).[6] While the firm slipped into receivership in August 1913 due to production problems at its mill and the recession that gripped Canada at the time, it continued to operate under a receiver and did so with increasing success. In attempting to procure its annual wood supply, however, it continued to bump up against the eight-inch diameter limit that applied to the pulpwood concessions it leased from the government and greatly limited the volume of timber available to it. In late 1916, Dryden Timber took the matter up with the government, and literally pleaded for consideration. The bureaucrats prepared a report on the matter that summarized the diameter limits that applied to all of Ontario's pulpwood concessions. In doing so, they revealed how arbitrarily this measure had been applied; there was no logical

[6]The Conservative government refused to resolve the Dryden mill's fibre problems at this time even though the politicians boasted at election time that the enterprise's very existence demonstrated the efficacy the government's northern development strategy.

reason why some licensees were subject to a six-inch limit whereas others operated under a nine-inch limit. Nevertheless, the deputy minister recommended Dryden's diameter limit be lowered by only one inch, and the government did so. When the firm reapplied two years later for this cutting restriction to be lowered from seven to six inches, however, and again the government recognized the logic of the company's request, this time the Tories refused to budge. This meant that Dryden Timber would continue to be forced to purchase much of its fibre supply from local settlers (AO, RG1-246-3, 1424 Vol 1A, 14 July 1917, 14 December 1916 and 10 October 1918).[7]

Consequently, the mill, which was renamed Dryden Paper Company in 1920 when it was purchased by a few veteran pulp and paper makers from Montreal and its ownership switched from being British to Canadian (PPMC 1918 and 1920), took aggressive steps to conserve its own timber supply as much as possible.[8] It maximized the volume of wood it purchased from local settlers (just as the politicians had wished it would do). It also purchased small timber licences that the provincial government occasionally sold in the Wabigoon area and entered into agreements to cut pulpwood from nearby "Indian Reserves" (AO, RG1-E-4-B, Box 9, Timber Sales Book—Volume 9, W-8-583; LAC 1909–1930). The upshot saw the company obtain between one-half and three-quarters of its annual fibre requirements on "the open market", and barely harvest any wood at all from its own timber limits during some years.[9]

7.3.3 Government Fetters Development: Hydroelectricity

Similarly, the Ontario government limited Dryden Paper's ability to increase its production after the First World War by tightly restricting the amount of hydroelectric power the pulp and paper mill could generate from the local waterways. Wabigoon Lake is drained by a river of the same name, and Dryden Falls is found just north of its mouth (Fig. 7.2). The Canadian government had built a dam there in 1897 to raise the water level of Wabigoon Lake to assist with local navigation. Ten years later the

[7]The Conservative government agreed to lower the diameter limit in mid-1919 but did not before it lost that year's election. The diameter limit remained at 7" until 1929, when Dryden Paper renewed its leases (AO, RG1-246-3, 6306 Vol. 1, 1919–1920, all correspondence).

[8]The politicians remained surprisingly hostile to the company in general. When the government's own forest rangers were organizing a timber cruise of the firm's woodlands in 1917, for example, and Dryden Timber asked to send a representative to be part of the investigating team, G. Howard Ferguson, Ontario's Minister of Lands, Forests and Mines, responded tersely to 'refuse permission' (AO, RG1-246-3, 1424 Vol. 1A, 7 June 1917). Similarly, J. B. Beveridge, a veteran and successful pulp and paper industrialist and general manager of the Dryden mill, applied in 1917 for another moderately sized pulpwood limit northeast of his existing plant to support a new mill he committed to constructing at the Lakehead (now Thunder Bay). The government refused to entertain his proposition and did so in an inexplicable manner (AO, RG1-246-3, 13773, 1917).

[9]Like many firms in Ontario that were faced with fibre shortages, Dryden Paper began looking at implementing progressive forestry measures to maximize the efficiency with which it managed its timber supply (LAC 1921–1922; Kuhlberg 1999, 2001).

Ontario government had given a lease to develop the hydroelectric potential of Dryden Falls to the Gordon brothers, who sought to establish a pulp and paper mill in the town. They built a power station and dam just downstream from the existing government dam, and the Gordons' was higher (i.e., it could raise the lake to a higher level). The Gordons transferred their control and ownership of the hydroelectric installation to Dryden Timber in 1911, at which time Dryden Timber asked for permission to raise Wabigoon Lake above the high water mark that the Ontario government had set. The latter agreed to this request, but only on the condition that the company promise to do so in a manner that would 'not injure any public or private interest.' After Dryden Timber constructed its pulp mill over the next few years, it was keen to increase the amount of energy available to the facility. As a result, it used its dam at Dryden Falls to raise the level of Wabigoon Lake to an unprecedented level, a move that touched off a protest among some local settlers whose properties were being flooded by this action (AO, RG1-246-3, 11916 vol. 1, 22 December 1910 and 21 August 1947).

In adjudicating this conflict, the Ontario government showed Dryden Paper no sympathy despite the company's manifest interest in investing in the latest technology.[10] Over the course of 1918–1920, for instance, the firm installed a new paper machine, thereby raising Dryden Paper's capacity to 60 tons of kraft pulp and 40 tons kraft paper (sheathing and wrapping papers) (Fig. 7.3) (Lockwood's Directory 1915; Post's Paper Mill Directory 1914; Canadian Pulp and Paper Association 1920). Naturally, the new equipment required additional power to run (particularly during the low-water winter season), and so the company pleaded for favourable consideration from the Ontario government. In making its case, its representatives stressed that the local homesteaders did not farm on the land near the lake, and thus their interests were hardly being impaired. Moreover, Dryden Paper offered to compensate all *bona fide* complainants for any damage to their property. Nevertheless, over the late 1910s and into the early 1920s, officials from the Ontario government were uninterested in these concerns, and repeatedly ordered Dryden Paper to lower the lake's level. J.B. Beveridge, the company's general manager, continually begged for the government to reconsider the matter, adding that not keeping the water stored in the lake for use in the winter would 'jeopardize the operation of this mill' (AO, RG1-246-3, 11916 vol 1 1917–1922a).

The Ontario government continued to turn a deaf ear to such pleas, even as the reasons for assisting Dryden Paper grew stronger. In an effort to buttress its lobby, the firm beseeched the government to investigate the situation for itself, which the latter did during the early 1920s. In the meantime, the local municipal politicians joined the chorus of voices asking for the provincial government to assist their town's mill. And once the field officials from the Ontario government had examined conditions on the ground, they, too, added their support to Dryden Paper's cause. 'No doubt,' L. V. Rorke, who was Ontario's Director of Water Surveys and personally visited the site at this time, declared emphatically to his superiors 'that the Company would be

[10]The Ontario government took practically the same stand during this period in its dealings with another mill that was established in northeastern Ontario, namely the Mattagami Pulp and Paper Company in Smooth Rock Falls (Kuhlberg 2015).

Fig. 7.3 Dryden paper company, ca. early 1920s. *Source* Photo courtesy of Charlie Rankin

seriously handicapped if they were not allowed to hold the water in order to secure efficient operation.' Furthermore, D. A. Hutchinson, the person who had been chosen to represent the settlers' interests in this dispute, urged the government to allow Dryden Timber to regulate the water level in the firm's interests even if it would result in further damage to the shoreline. As Hutchinson put it, recent experiences had poignantly demonstrated 'to me and to the other settlers the necessity of the Company being allowed to keep the level of the lake as high as [Dryden Paper felt was necessary]. … On this new level, I, personally would be more damaged than all the other settlers put together, on Lake Wabigoon, and I am willing to concede the Company the level of water that they need for the reason that I do not want to do anything which will jeopardize the operations of their mill, and I sincerely hope that your Government will give the Company the necessary permission.' (AO 1922b).

7.4 Dryden Paper Company in the 1920s

7.4.1 Difficulties Due to Technology, Global Forces and Market Trends

Although the Ontario government would finally accede to Dryden Paper's wishes in September 1923, by then it was too late.[11] The politicians' insistence over the preceding few years that the mill lower the level of Wabigoon Lake to a point at

[11] Dryden Paper had taken steps to resolve its energy issues by applying for a lease to develop the small water powers on the nearby Eagle River, some twenty miles east of the mill, and building a transmission line to deliver the electricity to Dryden. It applied for a lease to those sites in 1918, and although the Conservatives refused to grant its wish, the new United Farmers of Ontario government did so in 1920 (AO, RG1-246-3, 6306 vol. 1, 1918–1920; AO, RG75-57, OC135/232 and OC152/433). In addition, Dryden Paper acquired the lease to a local power source at Wainright

which it could not generate sufficient energy for its plant had prevented it from operating at capacity. Moreover, the provincial government had insisted that the firm pay compensation to any and all claimants, including those whose cases rested on the flimsiest grounds. These factors caused the firm to bleed cash at a time when it could not afford to do so, and it slipped into receivership in 1923 (AO, RG1-246-3, 11916 vol. 1, 1922–1923; The Globe 1923a, 1923b and 1924).

This downturn in Dryden Paper's fortunes was also the product of an unfortunate intersection of technology with international factors and market trends. When the new owners had taken over Dryden Paper in 1920, they had clearly done their homework in terms of the enterprise's potential and endeavoured to capitalize on it. For example, they recognized that they had little hope of competing in the American market with kraft pulp sold by Scandinavian (largely Swedish) producers either in terms of price or quality; by the late 1920s kraft pulp producers—particularly International Paper's (IP) new subsidiary, Southern Kraft Corporation—in the southern United States would represent further competitors on this front (Heikkinen 1999; Heinrich 2001). As a result, Dryden Paper's owners had decided in 1920 to invest in upgrading their machinery in order to produce higher value, "finished" paper products. Specifically, they had chosen to specialize in manufacturing boxboard and selling it in the United States. Not only was North America's consumer society growing rapidly, but also the First World War had greatly accelerated the trend toward using paper instead of wood for packaging (Dryden and District Museum 1920).

While Dryden Paper was well positioned to meet this demand, its strategy was sideswiped by the sweeping Republican victory in the American election of 1920. The party had gained power on a pledge to return "normalcy" to the United States, and the newly minted president, Warren G. Harding, was fervently committed to a pro-business agenda that his backers believed had been sorely lacking in the White House for nearly two decades. This perspective resonated perfectly with that held by Harding's Secretary of the Treasury, Andrew W. Mellon. The upshot saw the US implement "emergency" tariffs in 1921 and the Fordney-McCumber Tariff the following year. Among other things, these measures subjected imports of kraft paper—the very product Dryden Paper was keen to export to the US—to staggering duties (i.e., 30% ad valorem) (Lamm 1927; AO, RG1-246-3, 6306 Vol. 1, 25 February 1922 and 3 December 1923; AO, RG75-57, OC135/232). Concomitantly, the wake of the First World War had also touched off a boom in the construction of new pulp and paper capacity in Canada. Naturally, the mills were anxious to begin production as quickly as possible. As a result, many of them began marketing their groundwood and chemical pulps long before they were able to convert these products into higher value paper products. The result was a flooded pulp market and declining prices in much of northeastern North America, a reality that had pushed a string of Canadian mills into receivership (Dryden and District Museum 1921; Pulp and Paper Magazine of Canada 1920 and 1923; The Globe and Mail 1922).

Falls that was owned by the municipal government in exchange for which the firm was to deliver cheaper power to the town (AO, RG1-246-3, 31460 vol. 1–2, 1918–1930).

7.4.2 Americans Take Control: Monopoly Capitalism Delays Expansion, 1928

Nevertheless, Dryden Paper endured a remarkably successful 5-year receivership (1923–1928), and the manner in which it emerged from this period had a fundamental impact on its development—or more specifically, its lack thereof—over roughly the next twenty years. Although few details are known because the corporate documents do not survive, this much is clear. In late 1928 St. Regis Paper Company, a major pulp and paper maker in the United States, purchased a controlling interest in Dryden Paper.

This transaction was enormously important because of St. Regis Paper's activities at the time in the North American industry. Born in 1899 in upper New York State strictly as a newsprint maker, over the next few decades St. Regis gradually increased its productive capacity through acquisitions and by enlarging its existing mills. Its two driving forces, Roy K. Ferguson and F. L. Carlisle, used their investment firm, F. L. Carlisle and Company, to control and finance their expanding papermaking business. In 1921, Carlisle entered the hydroelectric industry in a major way, overseeing the aggressive acquisition of a string of power companies in the northeastern United States, all of which were organized under the Niagara-Hudson Power Company umbrella by the end of the decade. By this time, St. Regis had begun diversifying its paper production away from newsprint and into higher-value products, including paper bags. Its first step in this direction involved the formation in 1924 of Taggart Corporation as a holding company for a variety of paper bag makers. Four years later St. Regis literally cornered the market for 'multi-wall valve bags', which were used to package things such as cement, when it purchased the Bates Valve Sack Company (the exact date on which St. Regis acquired control over Bates is unclear, but it had done so by 1929 at the latest in an attempt to escape prosecution under American anti-trust laws) (United States District Court 1930). Bates had commercialized the technology for filling closed bags through an opening that would seal automatically when the bag was filled; this advance dramatically improved safety in industries such as cement and grain, particularly the former, in which the workers had been suffering crippling respiratory illnesses (Drasner 2004). Bates held the monopoly on the machinery both for making and filling the bags, and it was also the de facto sole supplier of them to the firms that used this technology. In 1929 St. Regis officially acquired the International Bates Valve Company, thereby solidifying its hold on this sector of the world's market. Part of this strategy entailed operating in Canada, which it did through its plant in Trois-Rivières, Québec (Drasner 2004; Amigo and Neuffer 1980; Maunder and Ross 1977).

Another element in this plan involved gaining a controlling interest in Dryden Paper's mill. Although no official announcement would be made, those moving in Canada's most powerful financial circles had learned by mid-1928 that St. Regis had been keen to acquire a plant that was conveniently located to supply the growing multiwall bag market on the Canadian prairies[12]; Dryden Paper's proximity to this region allowed it to enjoy a virtual natural monopoly there. St. Regis's subsidiary, Bates Valve, would soon begin supplying this demand by constructing a new plant in the town of Dryden, one that would use paper from Dryden Paper and be built adjacent to the latter's existing plant. In addition, Dryden Paper entered into a contract to supply Bates Valve's plant in Trois-Rivières with paper. Together, these new contracts would utilize nearly all the kraft paper Dryden Paper produced, and the mill would continue selling its surplus pulp on the open market in the American Great Lakes states. With these clearly defined markets allocated to it, the mill in Dryden would show little interest in expanding its operation even though the Canadian kraft pulp and paper industry enjoyed a significant growth spurt during this period. Between 1928 and 1932, three new large kraft pulp and paper mills were built in eastern Canada: one in Bathurst, New Brunswick (1928), another in Windsor, Quebec (1929) and a third in Cornwall, Ontario (1932) (Fig. 7.1) (AO, F1056 1928; The Globe 1928a).

7.4.3 Government Now Supports the Firm

Remarkably, St. Regis's involvement with Dryden Paper completely changed the Ontario government's attitude towards the latter firm's operations, an about-face that was consistent with the politicians' strong sympathy for American monopoly capitalism at this time. During the mid- to late 1920s, the Ontario government had surreptitiously assisted IP, which sought to control the North American newsprint industry, in acquiring a string of pulpwood limits in the province that were urgently needed by Ontario's own mills. More importantly, the provincial politicians forced those same domestic mills to adhere to a price-fixing and tonnage-sharing arrangement that both practically bankrupted them and left IP's mills operating at capacity (Kuhlberg 2015). Furthermore, the Ontario government gave IP's hydroelectric subsidiary, Gatineau Power, an egregiously lucrative contract to supply the province's public utility with energy. The agreement permitted IP to subsidize its newsprint operations in a manner that strengthened its control over the market. Although numerous governments facilitated the formation of export cartels to protect their domestic industries, in Ontario during these years the politicians consistently supported a cartelization plan that was ruinous to the province's own newsprint makers (Heikkinen 1999; Jensen-Eriksen 2013; Report of Commissioner 1945; Kuhlberg 2012).

It was hardly surprising then, that once the Dryden Paper Company became part of St. Regis Paper's monopoly in the multi-walled bag industry, the Ontario government extended a friendly hand to the operation. It did so through its control over the natural resources in the vicinity of Dryden Paper's mill, specifically through the latter's contracts for the local timber and water power supplies. Paramount in this regard

[12]Two decades later, a small article appearing in *The Globe* confirmed that St. Regis owned 'some 61,000 shares' in Dryden Paper and the latter company had only issued 150,000 shares in 1928, when insiders described how it had gained a controlling share of Dryden Paper (The Globe 1948).

was Dryden Paper's agreement with the government for pulpwood (AO, RG75-57, OC171/360 and OC172/335). It had been signed initially in 1907, and was set to run for 21-years; it thus expired in 1928. In re-negotiating its contract for this timber, Dryden Paper was able to secure all for which it could have asked, and more. Its new pulpwood agreement allowed it to cut trees of any diameter (no longer was it hamstrung by an absurd "diameter limit"), and it was required to pay a mere fifty cents per cord (app. 3.6 m^3) for the main species it sought (i.e., jack pine) and far below market values for the spruce pulpwood it cut. Most importantly, the Ontario government signed a string of new pulpwood concession agreements at this time with a number of pulp and paper mills that were much larger than the one in Dryden, and those contracts ran for merely one 21-year term. These mills and the financiers who backed them railed against this provision, arguing that it profoundly weakened their enterprises' control over their fibre supplies and ability to finance their operations. For the same reason they also protested against the "employment provision" the government inserted into their contracts, whereby they were obliged to employ a specific number of persons each and every year of the contract and that failure to do so would void the agreement. For a cyclical industry like pulp and paper, this clause essentially rendered the pulpwood agreements worthless as collateral to raise financing. Although the government was unmoved to amend these clauses that Ontario's newsprint-makers loathed, it granted Dryden Paper practically unlimited tenure to its pulpwood supply and released it from the objectionable "employment provision".

In fact, what was *not* included in Dryden Paper's new contract was just as important as what *was*. Again, practically all the other pulpwood concession agreements the Ontario government signed with pulp and paper firms at this time demanded that they dramatically expand their mills according to a very tight timetable even though practically everyone connected with the industry recognized that the market for the product that they all produced—newsprint—was flooded (Kuhlberg 2015). In sharp contrast, the deal Dryden Paper signed with the Ontario government in 1929 obliged the company to increase the scope of its operations such that it consisted of 'sulphate and groundwood pulp mill daily capacity of 60 tons … of which a minimum of 25 tons to be manufactured into paper or some other finished product'. The astounding part was that the mill had been operating since the early 1920s with a rated capacity of 75 tons of sulphate pulp, 40 tons of groundwood, and roughly 50 tons of kraft papers (Canadian Pulp & Paper Association 1923; Financial Review 1924).[13] In other words, in exchange for getting long-term tenure to a very cheap fibre supply, the Ontario government had required Dryden Paper to "expand" its operations to a size that was well below its existing capacity.

Even more perplexing was Dryden Paper's decision *not* to ask the government for a large, supplementary source of fibre at this time, a move that strongly suggests it did not intend to expand its operations over the next few decades. By 1928, the

[13]The figures for these two documents are slightly different in terms of the Dryden mill's capacity to make specific classes of pulps and papers, but its overall capacity is roughly the same in each case.

company had cruised the roughly 1,100 square miles of timber limits it leased from the Ontario government, and the data were most disappointing. They confirmed that there were only about 500,000 cords (app. 1,800,000 m^3) of pulpwood, a supply that would last the mill merely 15 years at its present capacity. At the time, mill managers considered a 60-year supply of wood the bare minimum for believing that they controlled a 'perpetual' supply of fibre. Moreover, Dryden Paper had publicly declared years earlier that there was plenty of pulpwood available for lease near its mill. On one occasion, its management team had even stated that, in addition to the timber limits Dryden Paper leased, there were '2,000 square miles, containing some 4,000,000 cords [app. 15,000,000 m^3], available to the company in the Dryden district, according to the report of the company's forestry experts' (Fig. 7.2). Nevertheless, in renegotiating its pulpwood agreement with the Ontario government in the late 1920s, the company bafflingly did not even raise the issue of acquiring a significant, additional tract of timber (AO, RG1-E-3-B, A-19 Dryden Paper Company, Licence Reference Sheet; The Globe 1920b).[14]

Thereafter, Dryden Paper flourished for a few years, just as it had begun to do during its receivership. Although the company had not released publicly its annual statements since 1923, when it had gone into receivership, these data were revealed when the firm was re-financed in early 1929. They indicated that Dryden Paper had lost money until 1925, but then it had begun churning out net earnings of nearly CAD$100,000 the following year and over CAD$210,000 in 1928. The next year the firm's profit soared to over CAD$330,000, and it remained profitable until the end of 1931 (The Globe 1928a, b, 1929a, b, c, 1931; Financial Review 1924–1929).

7.5 Dryden Paper in the Great Depression and War

7.5.1 Status Quo During the 1930s

By this time, however, Dryden Paper—just like nearly all pulp and paper makers around the world—was already enduring a very difficult operating environment. The deep and widespread depression that gripped the world economy during these years deleteriously affected its business. Whereas the mill ran practically at full capacity during 1929, two years later it was only operating half time. Over the next few years, sometimes it ran for merely a handful of days a month (AO RG1-E-5 1929–1934; Johnston 2014).

But in relatively short order, considering the malaise that was afflicting Ontario's and Canada's newsprint industry during the 1930s, the Dryden Paper Company rebounded. In 1934, the company was nearly breaking even, and the following year it generated a modest profit of nearly CAD$7,000. Thereafter, the enterprise's balance sheet improved dramatically, with net earnings rising to over CAD$215,000 by

[14]Dryden Paper's new agreement added a relatively tiny block to its timber holdings and the government deleted the diameter limit from its contract, but the mill still did not control anything approaching a perpetual supply of pulpwood.

1937 (The Globe 1935, 1937a–d; Johnston 2014).[15] Despite facing strong headwinds in selling its surplus kraft pulp in the United States, a market in which it competed against the 'Scandinavian product and southern [US] crops', Dryden Paper continued to prosper (The Globe 1938a–b; Toivanen 2004, 2012). In fact, when F. A. Sabbaton, the firm's vice-president, reflected on Dryden Paper's performance during the Depression, he vaunted the company's remarkably positive results. After describing the series of challenges the firm had faced, particularly in terms of successive crop failures on the Canadian prairies, 'which is the natural market for its products', he boasted that 'the company has been able during this ten year period to make a sufficient profit to meet its bond interest, pay sinking fund, now amounting to something over CAD$200,000.00, and to write off a reasonably large percentage of what should be a regular depreciation charge on its plant' (The Globe and Mail 1939b).

There was a curious aspect to the financial success Dryden Paper enjoyed during the Depression. It was the result of the company simply maintaining its typical modus operandi; Dryden Paper did not tap new technology or expand its operations even though North America's kraft pulp and paper industry experienced a virtual revolution during the 1930s. The development of the Thomlinson recovery boiler allowed for the industry to increase substantially its economies of scale and tap the relatively cheap and plentiful supplies of pine in the southern United States. The upshot was a dramatic expansion in the industry's capacity across the continent, but not in Dryden (Toivanen 2004, 2012).

7.5.2 *Keen but Unable to Expand During the Second World War*

Ironically, the Second World War would finally compel the company to demonstrate a genuine commitment to incorporating new technology to expand its capacity, but the conflict would fetter its ability to do so. Back in 1935, Canadian kraft producers had reached an agreement with Scankraft, the Scandinavian cartel, to divide the western world's markets. Thereafter Scankraft would regard Canada as reserved for Canadian mills while Canadian mills would not obtrude into Europe except for Great Britain, to which market Canadian producers could ship only agreed tonnages at prices that were uniform with Scankraft's. Then came the Second World War, which significantly increased the demand for kraft products in Canada and virtually cut off Skancraft's access to North American and European markets. The latter factor allowed Dryden Paper to compete more effectively in the former and expand its share of sales in the latter, particularly in Britain (Commissioner 1945). The upshot saw Dryden Paper hit new record levels of production and profitability despite the Canadian government's decision to control the domestic prices at which it sold its wares (The Globe and Mail

[15]During this period the mill certainly diversified the range of bag products it produced, but this had little impact on its total production.

Fig. 7.4 Dryden paper company, ca. 1940s. *Source* Photo courtesy of Eagle River Historical Society, Ontario, Canada

1939a, 1940, 1942, 1943a–b and 1944). At the same time, however, the Canadian government had taken control over nearly all aspects of the country's economy, and this included the supply of critical war materials such as heavy equipment and steel. This meant that, although the demand for Dryden Paper's products outstripped its ability to supply these goods, it was unable to procure the machinery it needed to expand its capacity at this time.

As a result, in 1945 the mill in Dryden was barely larger than the operation that had existed there some 25 years earlier. In the early 1920s, Dryden's mill had a capacity of roughly 60 tons of pulp and 40 tons of kraft papers. As the conflict in Europe was ending in the spring of 1945, the mill was rated at roughly 100 tons of pulp and 50 tons of kraft paper. Moreover, the facility was using roughly the same volume of pulpwood—roughly 50,000 cords (app. 181,000 m^3)—in the mid-1940s as it had back in the early 1920s (Fig. 7.4).[16]

[16]To be fair, Dryden Paper invested in improving its operation in 1944–45 and thus raised its pulp capacity to 125 tons, but this effort did not produce results until the latter part of 1945 (Dryden and District Museum undated; The Globe 1945).

7.6 Dryden Paper at Mid-Century: Dramatic Expansion

7.6.1 International Impetus to Expansion, Late 1940s

Only in the late 1940s—some three and half decades after the original pulp and paper mill had been built in Dryden—did international and local forces drive this enterprise to begin investing in more and better machinery to capitalize fully on the area's natural resource potential. Canada benefited enormously both from having ramped up its industrial potential during the Second World War and by having avoided the destruction that the conflict had wrought across much of Europe and Asia. It was thus ideally positioned to exploit the post-war demand for goods. While Dryden Paper had installed a new paper machine over the course of 1945–1948, the 1950s saw the firm truly stretch its wings. The first stage in the process saw it install a new evaporator set and chemical recovery boiler, and more digesters. By 1953, its capacity was four times what it had been in 1940, and the firm was acquired by the Anglo-Canadian Pulp and Paper Company that same year (Johnston 2014; AO, RG1-A-I-10, 7 November 1951).[17]

The change in ownership touched off a growth spurt that occurred at dizzying speed, and again international factors were largely responsible. The post-war period saw an explosion in the demand for bleached kraft pulp, and this dynamic drove a major reorientation of the entire Canadian pulp and paper industry. While newsprint production continued to dominate, the country's capacity to produce bleached kraft pulp experienced spectacular growth, and Dryden Paper was well positioned to exploit this opportunity (Kuhlberg 2012a, b). By the early 1950s the mill's new owner had already discussed with the Ontario government its strategy for growing exponentially the size of both the town and its mill. Whereas the enterprise produced a total of 58,000 tons of pulp and paper (that required nearly 96,000 cords [app. 350,000 m^3] of pulpwood) in 1953, within a few years Anglo-Canadian would increase this figure to 128,000 tons (that would require over 211,000 cords [app. 765,000 m^3] of pulpwood). The undertaking also diversified its operations in a major way by adding a bleached kraft line to tap into the exploding demand for this product. Anglo-Canadian envisioned one day requiring 1,000,000 cords (app. 3,625,000 m^3) of pulpwood annually to supply a "super-mill", an augmentation that would necessitate the population of the town of Dryden growing to over 10,000 (the census of 1951 recorded it as having just over 2,600 residents and by 1961 its population had more than doubled to over 5,700) (AO, RG3-23, 16 December 1953; The Globe and Mail 1951).

[17]At practically the same time that the Dryden Paper Company was sold to Anglo-Canadian (i.e., St. Regis Paper lost its supply of pulp and paper), St. Regis Paper was opening a brand new kraft pulp mill in Hinton, Alberta (Bott and Murphy 1997).

7.6.2 *New Owners and Government Support for Huge Expansion, 1950s*

The support of the Ontario government was crucial to Anglo-Canadian succeeding in augmenting the mill in Dryden in the post-war era. During this period, the provincial politicians zealously pursued an agenda that saw economic development as its touchstone; the exploitation of the province's northern resources was thus a policy priority. Moreover, the government had jettisoned in the midst of the Depression its commitment to colonizing northern Ontario; industrial mining, forestry and hydroelectric projects were what mattered most now. In this context, the politicians who controlled the resources that these enterprises sought were generally inclined to assist in their development, an approach that occasionally led to dire human and environmental consequences (Shkilnyk 1985). The upshot saw the Ontario government shower Dryden Paper in largesse. It granted the company additional pulpwood limits, and also permitted it to acquire several local sawmills and the timber licences they controlled. Whereas in 1940 Dryden Paper had legal title to barely 1,100 square miles of forest, roughly one decade later that total had doubled, and by 1970 it was managing over 6,600 square miles of woodlands (Graham 1990; Johnston 2014; AO, RG1-E-10, 26 September 1956 and 2 January 1957).

7.7 Conclusion

While the presence of bountiful natural resources that are generally in high demand does not always translate into immediate and strong economic development for a region, the combination of forces that retarded the growth of the Dryden Paper Company's mill in northern Ontario for nearly four decades was particularly complex. On the eve of the twentieth century, the town of Dryden was born amidst a wide swath of virtually unexploited timber and waterfalls, and it was also blessed with good transportation links to western Canada and the mid-western United States. In other words, the community was favoured with many of the prerequisites needed to support a burgeoning pulp and paper industry.

And burgeon is exactly what the industry did in Canada during the early to mid-1900s. Between 1920 (the first year for which reliable comparative data are available) and 1945, for example, the production of newsprint—in which Canada led the world—rose from just over 938,000 tons to more than 3,592,000 (Fig. 7.5). During the same period the production of non-newsprint grades of paper increased roughly threefold (from 339,000 to 1,035,000 tons), the production of all pulps soared from 1.96 million tons to over 5.6 million, and the production of sulphate pulp rose from 188,000 tons to over 478,000 tons (Fig. 7.6). In the process, pulp and paper became Canada's most important manufacturing activity, and the industry's products represented the country's second most valuable export behind wheat.

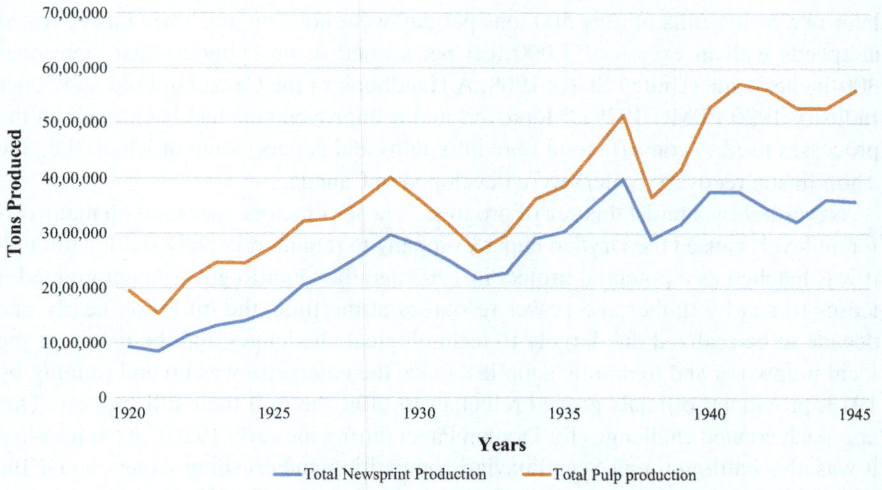

Fig. 7.5 Total production of newsprint and pulp in Canada, 1920–1945. *Sources* Canadian Pulp and Paper Association, Annual Newsprint Supplement, 1920–1945; Canadian Pulp and Paper Association, Reference Tables, 1955, Table 23

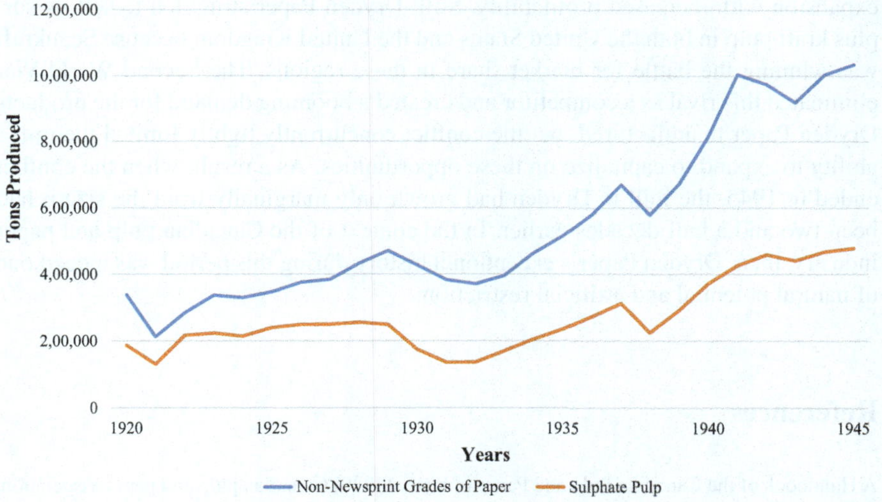

Fig. 7.6 Total production of sulphate pulp and non-newsprint grades of paper in Canada, 1920–1945. *Sources* Canadian Pulp and Paper Association, Reference Tables, 1955, Tables 29 and 64

The explosive growth that occurred across virtually all sectors of Canada's pulp and paper industry during this period was a result of transformational technological advances. Whereas in the early 1900s typical mills had produced in the range of 30–50 tons per day of newsprint and run paper machines at speeds of 300–400 feet per second using cylinders that were in the range of 100-inches wide, thirty years later newsprint mills of over 500 tons per day were housing machines that operated at speeds well in excess of 1,000 feet per second using cylinders that were over 300-inches wide (United States 1908; A Handbook of the Canadian Pulp and Paper Industry 1920 PPMC 1929). Moreover, major improvements had been made in the processes used to convert wood fibre into pulps and papers, some of which (i.e., the Thomlinson recovery boiler) were developed in Canada.

Nevertheless, amidst this sea of progress, a host of factors operating on many different levels caused the Dryden Paper Company to remain relatively staid. Although it was hatched as a potential project in 1903 and the Ontario government granted it leases to nearby timber and power resources at the time, the mill took nearly one decade to be realized due largely to technological challenges and the nature of the local pulpwood and hydraulic supplies. Once the enterprise was up and running by 1913, provincial officials proved reluctant to offer the mill their full support. This approach created challenges for Dryden Paper during the early 1920s, at a time when it was also battling tough Scandinavian competition and crushing American tariffs. When the firm emerged from its half-decade receivership in 1928, it had been incorporated into a continental—and indeed international—paper bag monopoly whose market on the Canadian prairies was protected as much by patents as it was by geography. The mill's new corporate culture thus had good reason not to associate major expansion with increased profitability. Still, Dryden Paper struggled to sell its surplus kraft pulp in both the United States and the United Kingdom because Scankraft was winning the battle for market share in these regions. The Second World War eliminated this rival as a competitor and created a booming demand for the products Dryden Paper manufactured, but the conflict concurrently tightly limited the mill's ability to expand to capitalize on these opportunities. As a result, when the conflict ended in 1945, the mill in Dryden had grown only marginally from the size it had been two and a half decades earlier. In the context of the Canadian pulp and paper industry, then, Dryden Paper's exceptional history during this period was indeed one of natural potential and artificial restriction.

References

A Handbook of the Canadian Pulp and Paper Industry (1920) Canadian pulp and paper association, Montreal

Amigo E, Neuffer M (1980) Beyond the adirondacks: the story of St Regis Paper Company, Greenwood Press, Greenwood, Connecticut

Archives of Ontario (AO), F1056 (1928), MU8655, RL Hearn Correspondence, 1929, Apr 1928, CM Lynn to RL Hearn and 18 Dec 1928, Hanson Bros Incorporated to RL Hearn

AO, MU1311 (ca1911), Northern Ontario: its progress and development under the Whitney government

AO, RG1-246-3, 1424 Vol 1A (10 Dec 1912), A. White to minister of Lands, Forests and Mines, from which the citations are taken; (14 December 1916), FJ Niven re: application of DL Mather on behalf of Dryden Timber & Power Company to be allowed to cut … down to 7" instead of 8" as per present agreement; (7 June 1917), HRM to Minister of Lands, Forests and Mines; (14 July 1917), Bondholders' Memorandum for the Minister of Lands and Forests; (28 Dec 1917),

Memorandum for the Minister of Lands, Forests and Mines from A Grigg re: Dryden …; (10 Oct 1918), WB Raymond to A Grigg
AO, RG1-246-3, 5520 (15 Apr 1916), TA Gordon to Deputy Minister; (6 June 1916), TJ Niven, Memorandum regarding application by TA Gordon …
AO, RG1-246-3, 6306 vol 1 (1918–1920), all correspondence; (25 Feb 1922), JB Beveridge to B Bowman; (3 Dec 1923), WA Black to J Lyons
AO, RG1-246-3, 11916 vol 1 (22 Dec 1910), copy of Order-in-Council re: transfer of power agreement; (1917–1922a), documents between 1917 and 1922, with the citation taken from 17 July 1919, JB Beveridge to A Grigg; (1917–1922b), documents between 1917 and 1922, with the second citation from 26 June 1922, LV Rorke to Deputy Minister, ODLF and the third from 23 Oct 1922, AR Hutchinson to WC Cain; (26 June 1922), LV Rorke to Deputy Minister, ODLF; (23 Oct 1922), AR Hutchinson to WC Cain; (1922–1923), all documents from 1922–1923; (4 July 1923), HH Bullard to WC Cain; (21 Aug 1947), LB Rose, Memorandum for file regarding water power on the Wabigoon River at Dryden
AO, RG1-246-3, 13773 (1917), all documents
AO, RG1-246-3, 31460 Vols 1–2 (1918–1930), documents surrounding granting a lease to develop Wainwright Falls to the town of Dryden and then the latter assigning it to Dryden Paper Company
AO, RG1-A-I-10 (7 Nov 1951), Box 7, Advisory Committee … Management (Timber), Nov/51–Jan/52, EL Goodall to HR Scott
AO, RG1-E-3-B, A-19 Dryden Paper Company, Licence Reference Sheet
AO, RG1-E-4-B, Box 9, Timber Sales Book—Volume 9, W-8-583
AO, RG1-E-5 (1929–1934), vol 19, Dryden Paper Company Limited, Returns for 1929–1934
AO, RG1-E-6 (6 January 1906), vol 3, 186–191, F. Cochrane, Memorandum re: Cancelling pulp concessions and agreements
AO, RG1-E-10 (26 September 1956), vol 78, 2-1-1 Timber Management—Minister of Lands and Forests, 1956–60, vol 1, EM Little to C Mapledoram
AO, RG1-E-10 (22 Jan 1957), vol 78, 2-1-1 Timber Management—Minister of Lands and Forests, 1956–60, vol 1, Mapledoram to Anglo-Newfoundland Development Company Ltd
AO, RG3-23 (16 Dec 1953), Box 91, Department of Lands and Forests 1951–6, FA MacDougall to LM Frost
AO, RG75-57, OC47/104; OC55/222; OC56/108; OC56/236; OC59/424; OC67/490; OC70/361; OC71/571, in which is enclosed 29 November 1912, E.W. Bonfield to Department of Lands and Forests; OC135/232, enclosing 4 March 1924, L.V. Rorke to Minister of Lands and Forests; OC152/433; OC171/360; OC172/335
AO Library (1906) General Conditions with respect to Pulpwood Areas—Offered for Lease by Tender—10 Feb 1906
Bott R, Murphy P (1997) Living legacy: sustainable forest management at hinton. Weldwood of Canada, Alberta, Edmonton
Canadian Pulp and Paper Association (1920) A Handbook of the Canadian Pulp and Paper Industry. Canadian Pulp and Paper Association, Montreal
Canadian Pulp and Paper Association (1923) Pulp and Paper Mills of Canada. Bulletin Number 37, 15 May 1923
Carruthers G (1947) Paper in the making. Garden City Press Co-operative, Toronto
Commission of Conservation (1911) The Water-Powers of Canada. Commission of Conservation, Ottawa
Commissioner—Combines Investigation Act (1945) Canada and International Cartels: An Inquiry into the Nature and Effects of International Cartels and Other Trade Combinations—Report of Commissioner, Combines Investigation Act, Ottawa, 10 Oct 1947
Drasner R (2004) The history of bags and baggers. Dissertation, Langston, Alabama
Dryden and District Museum (undated), Historical notes on operation of the pulp and paper mill in Dryden
Dryden and District Museum (1920) Dryden, Ontario, Dryden Paper Company Meeting Minutes, Nov 1920

Dryden and District Museum (1921) Dryden, Ontario, Dryden Paper Company Minutes, 21 Nov 1921
Financial Review (1924–1929) Canadian Annual Financial Review. Dryden Paper Company, Houston's Standard Publications, Toronto
Graham R (1990) Old Man Ontario: Leslie M. University of Toronto Press, Toronto, Frost
Hall K (1992) The sweet smell of success: a study of the origins and development of the pulp and paper industry in Dryden, Ontario. Dissertation, Lakehead University, Thunder Bay, Ontario
Heikkinen S (1999) Paper for the World: The Finnish Paper Mills' Association—Finnpap, 1918–1966. Otava, Helsinki
Heinrich T (2001) Product diversification in the US pulp and paper industry: the case of International Paper 1898–1941. Bus Hist Rev 75(3):467–505
Innis H (1930) The Fur Trade in Canada: an Introduction to Canadian Economic History. University of Toronto Press, Toronto
Jensen-Eriksen N (2013) A potentially crucial advantage: export cartels as a source of power for weak nations. Revue économique 64(6):1085–1104
Johnston G (2014) 100 years and beyond—Dryden mill. For Chronicle 90(3):338–346
Kuhlberg M (1999) We have 'Sold' forestry to the management of the company: Abitibi's forestry initiatives in Ontario, 1919–29. J Can Stud 34(3):187–210
Kuhlberg M (2001) We are the pioneers in this business: Spanish River's forestry initiatives after the great war. Ont Hist 93(2):150–178
Kuhlberg M (2012) An accomplished history, an uncertain future: Canada's pulp and paper industry since the early 1800s. In: Lamberg J-A, Ojala J, Peltoniemi M, Särkkä T (eds) The evolution of global paper industry 1800–2050. A comparative analysis. Springer, Dordrecht, pp 101–133
Kuhlberg M (2012) The Invisible Hand: International Paper's Invasion of Ontario, 1920–1930. Paper presented at the Annual Business History Conference, Philadelphia, PA, Mar 2012
Kuhlberg M (2014) 'Pulpwood is the Only Thing We Do Export': they myth of provincial protectionism in Ontario's forest industry, 1890–1930. In: Smith A, Anastakis D (eds) Smart globalization: the Canadian business and economic history experience. University of Toronto Press, pp 59–91
Kuhlberg, M (2015) In the power of the government: the rise and fall of newsprint in Ontario, 1894–1932. University of Toronto Press, Toronto
Library and Archives Canada [hereafter LAC] (1921–1922) RG39, correspondence between the Dominion Forest Service and the Dryden Paper Company, vol 329, p 45928
LAC (1909–1930) RG10 documents concerning cutting on the Wabigoon Lake and Eagle Lake Reserves, vol 7850, pp 30130–30136
Lamm LM (1927) Tariff history of the paper industry of the United States 1789 to 1922—Prepared for the American paper and pulp association. The American Paper and Pulp Association, New York
Lockwood's Directory of the Paper, Stationery and Allied Trades (1912) Lockwood Trade Journal Co., New York
Lockwood's Directory of the Paper, Stationery and Allied Trades (1915) Lockwood Trade Journal Company, New York
Mail and Empire (1906) 9 February 1906
Maunder ER, Ross JR (1977) Evolution of a paper company: the Carlisle-Ferguson Years at St Regis. Forest History Society, Santa Cruz, California
McCallum J (1980) Unequal beginnings: agriculture and economic developments in Quebec and Ontario until 1870. University of Toronto Press, Toronto
Ontario Department of Lands and Forests (1925) List of Water Powers in the Province of Ontario, 1925. Clarkson W James, Printer to the King's Most Excellent Majesty, Toronto
Post's Paper Mill Directory for 1915 (1914) Post Publisher, New York
Pulp and Paper Magazine of Canada (PPMC) (1904a) March; (1904b) September; (1905a) January; (1905b) November; (1907) May; (1908) July; (1913a) March; (1913b) May; (1918) October; (1920) September; (1923) October; International Number (1929)

Report of Commissioner—Combine Investigations Act (1945) Canada and International Cartels: An Inquiry into the Nature and Effects of International Cartels and other Trade Combinations. Combine Investigations Branch, Ottawa

Shkilnyk A (1985) A Poison Stronger Than Love: The Destruction of an Ojibway Community. Yale University Press, New Haven

Stewart, IM (2003) Dryden, John. In: Dictionary of Canadian Biography, vol 13, University of Toronto/Université Laval. http://www.biographi.ca/en/bio/dryden_john_13E.html. Accessed 22 Feb 2018

The Globe and Mail (1920a) 7 September 1920; (1920b) 9 September 1920; (1922) 21 December 1922; (1923a) 5 October 1923; (1923b) 17 December 1923; (1924) 30 January 1924; (1928a) 4 October 1928; (1928b) 11 December 1928; (1929a) 15 January 1929; (1929b) 24 January 1929; (1929c) 19 December 1929; (1931) 10 December 1931; (1935) 2 December 1935; (1937a) 26 May 1937; (1937b) 20 July 1937; (1937c) 26 October 1937; (1937d) 2 December 1937; (1938a) 16 July 1938; (1938b) 13 December 1938; (1939a) 14 November 1939; (1939b) 9 December 1939; (1940) 16 May 1940; (1942) 19 July 1942; (1943a) 8 January 1943; (1943b) 15 January 1943; (1944) 17 March 1944; (1945) 20 June 1945; (1948) 22 December 1948; (1951) Dec 1951

The Pioneer Farm and the Wabigoon Country, Rainy River District (1896). Ontario Department of Agriculture, Toronto

The Wabigoon Country, Rainy River District (1897) Ontario Department of Agriculture, Toronto

The Wabigoon Country, Rainy River District (1898) Ontario Department of Agriculture, Toronto

Toivanen H (2004) Learning and corporate strategy: the dynamic evolution of the North American pulp and paper industry, 1860–1960. Dissertation, Georgia Institute of Technology

Toivanen H (2012) Waves of technological innovation: the evolution of the US pulp and paper industry. In: Lamberg J-A, Ojala J, Peltoniemi M, Särkkä T (eds) The evolution of global paper industry 1800–2050. A Comparative Analysis. Springer, Dordrecht, pp 49–80

United States District Court for the District of Delaware (1930)–39 F.2d 162 (D. Del 1930), March 11 1930, United States v. Bates Valve Bag Corporation et al, No 705

United States—Select Committee of House of Representatives (1908) Pulp and paper investigation hearings, vol 1. Government Printing Office, Washington

Wice G (1967) Carved from the wilderness: the intriguing story of Dryden. Canada Confederation, Dryden

Chapter 8
The Endless Sheet: Technology Transfer and the Papermaking Industry in Spain, 1800–1936

Miquel Gutiérrez-Poch

8.1 Introduction

The industrialisation process in the nineteenth century was based on the transfer of technology from pioneering countries to countries that had not yet participated in the process. Southern Europe in general and Spain in particular were very late in experiencing this modernisation wave.[1] Technology transfer is crucial for developing countries to increase their wealth by raising productivity (Coe and Helpman 1995). Nevertheless, the transfer is seldom an easy or simple process. Jeremy (1991) states that 'it is complex. It is the product of many ingredients' (p. 2). Many factors explain why this is the case, including the myriad ways in which the transfer can occur. Keller (2004) identifies some of them: foreign trade (especially imports), FDI and geography, to name just a few (pp. 765–773). At a micro level some other factors can be analysed, including industrial espionage, technical press, skilled labour emigration and trips abroad. In addition, the mechanisms for the transfer can vary according to the degree of knowledge codification, with the "person-to-person" means being the most crucial tacit one. Dasgupta and David (1994) maintain that 'the information contained in patents, blueprints and other codified forms of knowledge often are insufficient for successful implementation of the technical innovations they purport to describe' (p. 494). To solve this problem, it is often the case that skilled labour arrives with the machinery to form the package of technology (Bruland 1989).

[1] This investigation is part of the research project HAR2015-64769-P, "Industrial crisis and productive recovery in the history of Spain, 1686–2018". Spanish Ministry of Economy (MINECO) and European Regional Development Fund (ERDF).

M. Gutiérrez-Poch (✉)
Department of Economic History, Institutions, Policy and World Economy, The Research Centre in Economics and Economic History 'Antoni de Capmany', University of Barcelona, Barcelona, Spain
e-mail: mgutierrez@ub.edu

T. Särkkä et al. (eds.), *Technological Transformation in the Global Pulp and Paper Industry 1800–2018*, World Forests 23, https://doi.org/10.1007/978-3-319-94962-8_8

In the technology transfer process, both the place from whence it originated and the destination at which it will arrive are important. Regarding the former, the assimilation process is much easier when the economic structure and cultural milieu of the country transferring the technology is very similar to that of the nation that is receiving it. With regard to the destination, access to and the use of the new technology do not guarantee the successful adoption of it (Keller 1996). The characteristics of the receptor condition the process, with the skill level of labour and the institutional framework playing crucial roles. Abramovitz (1986) pointed out the significance of an adequate national social capacity of absorption and the technological congruence to achieve positive effects. Similarly, Berg and Bruland stress (1998) how 'institutions and cultural frameworks set the terms for the diffusion of technologies across regional and national boundaries' (p. 11).

The institutional framework and its stability can accelerate or restrain the transaction costs and determine the impact of the new technology. The absorption capability (Cohen and Levinthal 1990) would be, therefore, a crucial consideration in choosing the most appropriate technology, decoding the information, and incorporating and adjusting it to local needs (the so-called technological capacity). Furthermore, the labour skills and the institutional framework of the reception area are very important. The existence of a previous tradition in the considered sector (path dependence), or simply in the industry as a whole, facilitate the assimilation of the technological change. In areas with a high density of firms, the technological spillovers will happen more easily because of the intra-industry relations. This reasoning refers to A. Marshall's industrial district concept. However, it would be wrong to focus on a one-sector reality and ignore diversified economic structures, because the technology transfer can transgress a single industry affecting other sectors. These links highlight the importance of inter-industry relations in overlapping districts (Scherer 1982).

The study of technology transfer during the Industrial Revolution and the Second Technological Revolution has been a recurrent theme in economic history. This subject becomes relevant in cases such as the Spanish one, in which foreign technology fed industrialisation. This investigation explores the technology transfer in the Spanish case[2] by examining the papermaking sector. In this regard, it follows in the footsteps of Nelson and Winter (1977), who pointed out that technological change can be better understood by analysing industries instead of models of aggregated production (p. 46). Spain began the change with a considerable handmade paper manufacturing sector, and it was modernised slowly with machines and foreign skilled labour. With its nascent maturity in the late nineteenth century this dependence decreased, allowing a better adaptation of the technology to local needs. The improvement in labour training and the establishment of an active engineering sector contributed decisively to this modest qualitative advance.

The chapter consists of three main sections. The first analyses international papermaking engineering. The second section is devoted to the mechanisation of Spanish papermaking until 1880. The final one deals with the technology transfer from that year until the outbreak of the Spanish Civil War in 1936.

---

[2]About the technology transfer in Spain see Inkster and Calvo (2010).

8.2 An International Perspective on Papermaking Engineering

The only options to increase output with handmade paper technology were multiplying the units of production (the vats) and/or increasing the amount of work performed by the labour force. The demand dynamism boosted production using these tools in the last third of the eighteenth century,[3] thereby showing the limits of the traditional technology. The answer was mechanisation, symbolised by the continuous machine; it allowed paper to be made in one endless sheet instead of many small, individual ones. The first prototype was conceived in France in 1798 by Louis-Nicolas Robert, and it was first utilized commercially by Bryan Donkin in the United Kingdom in 1804. The development of the French prototype occurred in an English workshop owned by John Hall and commissioned by the Fourdriniers, a family of stationers. At that time, Donkin was Hall's employee. Eventually, he was supported by the Fourdriniers to set up his own workshop. The international presence of Donkin's machines was remarkable. A second device was invented by the Englishman Joseph Bramah between 1797 and 1805 and developed by John Dickinson, who registered it in 1809. It resulted in a sheet of paper being formed by a cylinder, and thus the device was called a "round machine". The diffusion of the continuous machine in the United Kingdom accelerated after 1820 (Coleman 1958, p. 195). There were only 42 working machines in 1822 but fifty years later there were 471. The first engineering firms to follow Donkin's path were also British: Tidcombe from Watford; George and William Bertram and James Bertram & Son both from Edinburgh; and Redfern, Bentley & Smith and Walmsleys Ltd., both from Bury, Lancashire. These workshops had an important exporting bias. The skilled labour mobility was central to the diffusion of the new technology and frequently when a factory began its productive life it was managed by British engineers. Soon enough, however, locals were trained and took over these jobs. The technical press,[4] the technical handbooks, and World Exhibitions were usual paths of diffusion at this time. The state was an essential mechanism of innovation, but it could both lead[5] and restrain the process.[6] Other institutional factors that could slow down the diffusion process were political instability and the nascent system of patents.

The international diffusion of the new technology allowed modern papermaking to achieve a certain level of maturity by 1840 in a few countries (e.g. France, Belgium, and Germany). Its diffusion in peripheral Europe (i.e., Scandinavian countries, Eastern and Southern Europe) occurred later and more slowly, arriving only during the 1830s. The former countries became centres for the diffusion of the new technology. The sale of machinery, the emigration of British technicians and

---

[3]In 1798 there were 1,061 vats in France, and the number of mills in the UK grew from 100 in the late seventeenth century to 400 in the late nineteenth century.

[4]The main journals were *Paper Mills Directory*, *Papermakers Circular and Rag Prices*, *Papemakers' Monthly Journal* and *Paper Mill Directory*, all of which were first published in the 1860s.

[5]The Swedish State extended credit to Sven Summerdahl of Klippan in 1830 to buy a paper machine.

[6]For example, the Papal States forbade in 1847 the import of a machine.

the licence system led to the appearance of new papermaking engineering centres. During the early 1840s, the most important outside France were in Belgium, Switzerland,[7] and Germany.[8] Norway and Sweden also had some modest papermaking engineering from the mid-nineteenth century. Even the United States began to have an important papermaking machine building sector (e.g. Merrill and Houston and Pusey and Jones). The means through which the new technology was diffused from these second prime movers were the same as in Britain (skilled labour[9]; technical press[10]; technical handbooks[11]; and national and World Exhibitions).

The French machine shops merit deeper attention because of their importance for the Spanish case. These workshops, located in Paris (e.g. Chapelle and Sanford et Varrall), Mulhouse (e.g. Biesler frères & Dixon and Koechlin), Angoulême (e.g. Alfred Motteau and Trousset et Duveau et Cie.), and Isère (e.g. L'Huillier-Jouffray and Allimand), began to compete with the British workshops beginning in the mid-1830s. The Parisian papermaking workshops were characterised by their high degree of specialisation. For example, by 1839 Stanislas Casimir Chapelle had already built 37 machines for the domestic market and 19 for abroad. Sanford et Varrall, set up in 1834,[12] had already built 35 machines by 1842 for France and 16 for export. The Alsatian workshop by André Koechlin, set up in 1826, built 25 machines from 1831 to 1844, and it had exported 12 to Switzerland and Germany. The rapid mechanisation of southwestern France allowed for the rise in Angoulême of firms such as Alfred Motteau, founded in 1835, whose first machine was built in 1840, and Trousset et Duveau et Cie., in 1845. In the mid-1850s a new centre was established in Isère. L'Huillier-Jouffray, located in Vienne, had already delivered machines to more than 30 factories by 1855. Allimand, created in 1849 in Rives, built 34 machines from 1856 to 1864 (André 1996, pp. 157–160, 343–348).

---

[7]Escher Wyss from Zürich had already delivered 20 machines by 1852 (André 1996, p. 336). It set up a branch in Ravensburg, Germany in 1856, where it started to build papermaking machines. Its 104th machine was built in 1875.

[8]The most important papermaking engineering district was located in Württemberg. Johann Widmann built the first German Fourdrinier in Heilbronn in 1830. By the mid-nineteenth century J. M. Voith, H. Füllner and Gottschald, Nötzli & Steiner had been born.

[9]For instance, the first Polish machine was operated by Gabriel Planche from France. Amedée de Montgolfier, also from France, managed a factory in Naples.

[10]The most important French technical journals were *Journal des fabricants de papier* (1854), *Moniteur de la Papeterie Française* (1864), *Annuaire Général de la Papeterie française et étrangère* (1865), and *Revue de la Papeterie* (1867). Some prominent technical journals were published in Germany, including *Zentral-Blatt für die Deutsche Papier-Fabrikation*, which was first published in 1849, *Wochenblatt für Papierfabrikation* (from 1869 onwards) and *Papier Zeitung* (from 1876 onwards).

[11]The most important technical handbooks were *De l'industrie de la papeterie* by Gabriel Planche (1853), *Manuel du contremaître et du chef d'atelier de papeterie* by Louis Piette (1861) and *Guide pratique de la fabrication du papier et du carton* by Albert Prouteaux (1864). The German Carl Hofmann began to publish a technical handbook in the early 1870s in the United States, where he worked at the time. Its German version, *Praktiques Handbüch für Papier-Fabrication*, emerged as the era's leading technical handbook.

[12]William Varrall and Henry Sanford had arrived to France in 1829 and 1827 to work for Bryan Donkin (André 1996, pp. 108–109, 156).

A new scenario appeared with the introduction of wood pulp. The wood grinder was developed, using an earlier prototype, in the 1840s and the 1850s by Heinrich Voelter from Heidenheim, Germany. A milestone in its diffusion was the Paris World Exhibition of 1867, when between 80 and 100 grinders were operational; by 1875 that number had risen to 400. The growing markets and the expiry of the Voelter patent stimulated the setting up of alternative centres of production (e.g. Voith, Bell, and Escher Wyss).

The technological advances in papermaking machinery were practically continuous beginning in the late nineteenth century. The innovation process also affected the existing machines because of the modular nature of the new technology. In addition, there were also changes in the pulping process. The improvements allowed for wider and faster machines and they thus boasted capacities. This required machines that were conceived to produce specific types of paper. The international market for papermaking machinery was transformed from 1870 onwards as a result of its dominance by the German, Swiss, and Americans. The qualitative leap by German workshops was boosted by their specialization in wood pulp grinders and the increasing vertical integration of both activities. The reliability of the German equipment and its low prices ensured its success. J. M. Voith continued to focus on auxiliary machines, wood pulp technology and turbines.[13] Other German workshops were Maschinenbauanstalt Golzern, H. Füllner, Gustav Toelle, I. W. Erkens, and Bruderhaus. In Switzerland, Escher Wyss & Co. and Théodore Bell still occupied a hegemonic place. The papermaking engineering sector in the United States sold mostly in the domestic market but also in Canada. The most important workshops were the Beloit Iron Works, successor of Merrill and Houston, and Bagley & Sewall.[14] The Scandinavian countries also developed an active engineering industry specialised in papermaking.[15] On the contrary, the United Kingdom, France and, to a lesser extent, Belgium, lost ground in the late nineteenth century and in the first third of the twentieth. The most prominent British firms were those from Bury and Edinburgh. These workshops mainly supplied the domestic market although, just before the First World War, they retained some export orientation. In France, practically all the pulp and paper plants used domestic machines. However the international orientation of the French paper engineering was diminishing. The centre of this machine-building sector was located in Isère, while the main Belgian workshops were Thiry et Cie (formerly Dautrebande et Thiry) and J. & E. Chantrenne, both of which had an important international presence.

In addition, this period saw dramatic improvements in the production of more refined pulp using chemical processes. Until 1880, the production of chemical pulp

---

[13]By 1905, Voith had already built 2,000 turbines (Nerheim 1991, p. 337). Its first papermaking machine dated back to 1881.

[14]Beloit Iron Works was set up in 1885 and it produced its first machine in 1887. From 1889 onwards, Bagley & Sewall, which had previously focused on wood grinders, started to build high capacity, continuous paper machines.

[15]In Sweden, there were Nydqvist & Holm from Trollhättan and Hellefors Bruks Co, Ltd., in Norway there was A. S. Myrens Verkstad, and in Finland there was Wiborgs Mekaniska Verkstad and Karhula Works.

was very limited due to its high cost. Advances in sulphite pulping and the introduction of the sulphate process led to the diffusion of the use of chemical pulp. The methods conceived by A. Mitscherlich and Carl D. Ekman crystallised around 1880. The German engineer C. F. Dahl introduced the sulphate or kraft method in 1844, although its widespread use began only in the twentieth century.

8.3 The Genesis of the Spanish Market of Papermaking Engineering

8.3.1 Foreign Machines and Mechanisation

The Spanish industrialisation process in the nineteenth century was backward, poor, and based on foreign technology. Nadal (1988) maintains that 'Spain in the nineteenth-century [...] has had a dependence almost complete on European technology' (p. 33). The papermaking sector was a prime example of this situation. In 1862, a Spanish government official emphasised that, in reference to continuous papermaking machines, 'in our country they are not built'.[16] The institutional framework and the political instability made it difficult for this technology to be diffused. The patent (or privilege) of introduction is dated 5 July 1836 and two persons applied for it at the same time, namely Mariano de la Paz García and Juan Sans, and it prevented the installation of any other device for the following five years.[17] The aim was 'to introduce from England an endless papermaking machine'. However, the machine was not imported immediately, so Juan Sans obtained in 1838 an extension of the privilege. It is not known how Tomás Jordán, a printer and the owner of a paper warehouse in Madrid and who came from the hand-making paper village of Alcoy, acquired it. The machine began to produce in late 1838 or early 1839 and was still the only one working in 1841. The finalization of the prerogative and the end of the First Carlist War (1833–1840),[18] jointly with a new tariff regime[19] and some dynamism from the demand side, allowed for the modest development of the modern papermaking industry. By 1845, there were 15 working machines and in 1879 that number had grown to 50. In addition, there were about 10 paper factories, which enjoyed only an ephemeral existence, and they were proof of the difficulties involved in the technology transfer.

---

[16]Archivo del Congreso de Diputados (ACD), Sección General, Legajo 112, Exp. 3º, Proposición de ley sobre la introducción de papel extranjero.

[17]Oficina Española de Patentes y Marcas (OEPM), Privilegios de invención e introducción (1826–1878), Expediente 120.

[18]In 1846 it was said that 'as soon as, [...] finished civil disturbances and the horrifying war [...] fourteen factories of continuous paper [...] were built' (*Semanario de la Industria*, nº 12, 23 May 1846, p. 94).

[19]The import of continuous paper was forbidden from 1840 onwards. Soon after the ban had been enforced, it was stated that 'due to the ban the industry has grown extraordinarily' (*Semanario de la Industria*, nº 11, 16 May 1846, p. 88).

Until 1880, the most important modern papermaking centres were located in the vicinity of Madrid, in the Catalan province of Girona and in the Basque province of Guipúzcoa. The most dynamic area was the last mentioned followed by the Catalan province. Both areas enjoyed the advantages of agglomeration that facilitated the technological spill-over. The weak industrial tradition in inland Spain inhibited the development of papermaking and caused the stagnation of the factories surrounding Madrid.[20]

The machines in use in Spain were mainly French but also British and Belgian. The most important suppliers were Chapelle, Sanford et Varall, Alfred Motteau and Trousset and Duveau et Cie. A contemporary newspaper claimed in October 1841 about La Gerundense (located in Girona) that it was 'moving to this town one of the four or five big continuous paper factories located in Bordeaux'.[21] La Esperanza from Tolosa was inspired by the factories of Angoulême 'but on a smaller scale'.[22] The pioneering machine located in Manzanares el Real (close to Madrid) was Belgian.[23] It probably arrived due to the contacts between Tomás Jordán, the owner of the paper mill, and the machine builders from Belgium.[24] Jordán was from Alcoy, an important centre of woollen textiles, which already used Belgian machines. The presence of British engineering was minimal, although there is some information about machines by Bryan Donkin & Co.[25] Some Swiss machines were set up in the late 1870s.[26] The auxiliary parts of the machine, such as wire and felts, were also imported, mainly from France.

The French supremacy is largely explained by previous existing paths for technology transfer, which had already been established in the late eighteenth and early nineteenth centuries. Likewise, French technology was used widely in the Spanish textile industry, in steam engines, in railways and in mining. Furthermore, the French machines were best suited to meeting the demands of the Spanish market. It was characterised by its low demand, which necessitated installing machines that were versatile enough to easily change from producing small volumes of a wide range of papers.

---

[20]The pioneering Candelario factory (in Castile) had 'a very primitive technology', 'a low capacity' and it was 'run by people not well versed in mechanics' (Díaz y Pérez 1880, pp. 115–116).

[21]*El Espectador*, I, nº 68, 7 October 1841, p. 1.

[22]*Documents sur le commerce exterieur. Espagne. Faits comerciaux nº 2, 3ª sèrie, nº 282*, August 1845, p. 77.

[23]Tomás Jordán had set up to his first mill 'a continuous machine from Belgium' (*Gaceta de Madrid*, nº 4,778, 14 October 1847, p. 3).

[24]The machine and a part of the auxiliary machinery set up in Morata de Tajuña (in Madrid province) in 1879 came from 'H. Dautrebande et F. Thiry' (*Crónica de la Industria*, nº 152, 31 January 1880, p. 18).

[25]The machine, which was in operation in the 1850s in Benalmádena, Málaga, was from Donkin (*Diario Oficial de Avisos de Madrid*, nº 915, 25 December 1862, p. 4). Clapperton (1967) also refers to a machine by Donkin in Spain (p. 209).

[26]The machine set up by Josep Flores in Sarrià de Ter came from Théodore Bell (Brugada and Vila 1999, p. 28). The glazing machinery of Morata de Tajuña had the same origin (*Crónica de la Industria*, nº 122, 31 January 1880, p. 18).

Wood pulp production in Spain was very limited at this time. In 1880, there were only two factories: Felip Flores in Sarrià de Ter, Girona and Vda. Ribed e Hijos in Villava, Navarra. Flores, previously involved in the pioneering papermaking factories in Girona, had learned about the Voelter grinder at the World Exhibition in Paris in 1867. He ordered a replica of it to be built by Porredón, Comas y Cia, a local workshop. Flores registered a privilege on 17 January 1870 for a 'machine and methods to grind wood mechanically [and] reduc[e] it to paper pulp'.[27] The factory was operational by the following April. The manufacturing plant in Navarra opened on 1 August 1872, and it was integrated with the existing papermaking factory under the same ownership. The grinder came from the franchised workshop by Voelter in Munich, being the first to be sold in Spain.

8.3.2 The Means of Transfer: The Importance of Direct Contact

The transfer sources of the new technology were various, and they included foreign skilled labour, travels abroad, imitation effect, technical press and commission agents. Each of them had an effect to varying degrees. An essential element in the transfer of knowledge, especially the tacit one, was foreign skilled labour. In this regard, the interaction between the foreign and local workers was crucial to adopting the technology to the particular needs of the local market. The pioneering machines were assembled and started up by foreign workers. In 1862, it was said that 'the machine operator and the most important workers have had to come from abroad'.[28] The preferential origin was French (Table 8.1).[29] Some firms even had French participation in the ownership. In Brunet, Guardamino, Tantonat y Cª the majority of partners were French or they had links with Bordeaux and Bayonne.[30] Similar circumstances were present in the second factory in Tolosa: Echazarreta, Larión y Aristi, with partners from Bayonne.[31] In this case, the contacts with French partners greatly

[27]OEPM, Privilege nº 4,689.

[28]ACD, General Section, File 112, Exp. 3º, Law proposal on the imports of foreign paper. See André (1996), pp. 219–221.

[29]The skilled workers of La Esperanza in Tolosa were recruited from Angoulême and Montfourrat (*Documents de Commerce Extérieur publiées par le Département de l'agriculture, de commerce et des travaux publics. Espagne, Faits commerciaux, nº 2*, nº 282, p. 77).

[30]Among the founders of the firm, there were only two Spaniards. Brunet was from San Sebastián and Ramón Guardamino from Madrid, although at the time the latter was the Spanish consul in Bayonne. Juan Tantonat was from Bayonne, Juan Conte from Bordeaux and Julio Larion from Riberac (Archivo Histórico de Protocolos de Guipúzcoa de Oñate [AHPGO], File III-111, f. 141).

[31]The investors from Tolosa were Juan Gregorio Echazarreta, Pedro José Aristi and Julio Larion (from Tolosa, but originally son of a French founder of La Esperanza). Adam Weidemann and Rodriguez y Salcedo, both living in Bayonne, were also part of this group (Archivo General de Guipúzcoa [AGGT] Public notary files of Tolosa, José Mª de Furundarena, 28-X-1850, f. 505).

facilitated the technology transfer. In contrast, British[32] and German workers[33] were in the minority. Many skilled French workers and engineers settled in Spain, and would go on to start papermaking dynasties, such as Limousin, Vignau, Larion, Duras, and Grelon. In this family atmosphere, the first French technicians trained their descendants.[34] In a context of precarious formal technical training, with merely a few engineering schools that enjoyed only poor financial support, the knowledge transfer through the person-embodied skills and "learning by doing" were essential.[35] As early as the 1850s, some factories were already being managed by Spanish employees.[36] Some of them even participated in designing new factories. Examples of this dynamic include José Canalejas,[37] Nemesio Uranga,[38] and Luis Anitua,[39] and all three of them trained in Liege. Another Spanish technician was Benigno Yrazusta,[40] although he spent his entire career in Belgium. Many of the first Spanish technicians had previous training in mills that made paper by hand.

Travels abroad were also a source of technology transfer, especially of explicit knowledge. Visits to World and National Exhibitions were integral to this process. The French exhibitions exposed the Spanish papermakers to the first advances in modern production. For example, the machines built by Chapelle were the focus of attention at the exhibition of Paris in 1844. La Aurora, a factory of Girona, bought

[32]In 1848, in Villalgordo, there was 'a renowned English engineer'. A partner of the owner of a factory in Rascafría was the British engineer William Sanford, who had a workshop in Madrid. Sanford was committed to 'to get ready, to run and to implement the necessary works to set up the factory' (Archivo Histórico de Protocolos de Madrid [AHPM], Public notary files 24,965, fols. 571r.-584v).

[33]Fernando Wilhemi, born to a papermaking family in Heilbronn, purchased a paper factory in Granada. His link with Spain dated back to 1866. Arthur Tischer, a German engineer, supervised the setting up of a wood grinder in Villava.

[34]Francisco Limousin Claussure trained his son Ernesto Limousin Trijasson, born in Tolosa, to papermaking business. The latter mentioned passed the profession to his son, Francisco Limousin Mugerza. Both were managers of La Tolosana.

[35]William Sanford, a partner in the Rascafría factory, was obliged during four years 'to attend to the factory, at least three months yearly, not continuously […] to teach and to train the workers without keeping any secret' (AHPM, Public notary files 24,965, fol. 575).

[36]It was said about Gárgoles that 'for a long time the management was in charge of French papermakers, and nowadays is carried out by […] the same owner or by a former employee […], who substituted him with success' (Memoria 1851, p. 400).

[37]Canalejas worked in the workshop owned by Valentí Esparó in Barcelona, and he studied engineering in Barcelona and Liège. He said in 1862 that 'I have visited the main factories abroad and I have helped to plan some Spanish factories' (ACD: General Section, File 112, Exp. 3º, Law proposal on the imports of foreign paper, minutes of the session 26 April 1862).

[38]Uranga directed the modernisation works of his family-owned paper mill in Tolosa (Anduaga 2011, p. 53).

[39]Anitua graduated in Liège in 1867. Years later he became the manager of the Papelera Vizcaína in Arrigorriaga, which had been in operation since the early 1890s.

[40]Yrazusta, native of Tolosa and member of a papermaking family, obtained an engineering degree from the École centrale des arts et manufactures in Paris in 1878. After his graduation he began to work in a paper factory called De Naeyer in Willebroeck, Belgium. The same Belgian firm made investments in the 1890s in La Papelera Vasco-Belga in Rentería, the Basque Country.

Table 8.1 French presence in Spanish paper mills, 1841–1875

Name	Job/Skill	Year	Mill (location)
Constantino Lancelieu	Director	1841	Candelario (Salamanca)
Jean Lami	Carpenter-mechanic	1842	Manzanares el Real (Madrid)
Juan Conte	Director-partner	1842	La Esperanza (Tolosa)
Julio Larion	Partner	1842	La Esperanza (Tolosa)
Francisco Limousin Claussure	Technician-assembly operator	1842	La Esperanza (Tolosa)
Juan Conte	Director-partner	1846	Villava (Navarra)
Francisco Limousin Claussure	Technician-assembly operator	1846	Villava (Navarra)
Agustin Montgolfier	Director	1846	La Zaragozana (Villanueva del Gállego) y Villarluengo (Teruel)
Napoleón Paquier	Director-partner	1847	La Barcelonesa (Barcelona)[a]
Victor Fourvel	Director	1849–1851	La Esperanza
Lesvignes	Director	1851	La Gerundense (Girona)
Jean Joseph Grelon	Industrial partner	1854	Borgonyà (Girona)
Ernesto Autreux	Director	1855	El Catllar (Tarragona)
Claudio Viaud y Trovillart	Director	1858	La Providencia (Tolosa)
Louis Duras	Director	1862	La Confianza (Tolosa)
Alfred Motteau	Industrial partner	1862	La Cristina (Lavadores, Pontevedra)
Charles de Villedeuil	Director	1864	Villalba (Madrid)[a]
Francisco Limousin	Director	1873	La Tolosana (Tolosa)
Auguste Kaindler	Director	1870–1875	Villalgordo de Júcar

Source Collected by the author from technical journals and handbooks of the time
[a]Failed project

in 1845 'a machine like those which the owners saw last summer in the industry exhibition in Paris'.[41] José Canalejas and other Spanish observers visited the London World Exhibition of 1862, focussing their attention on Donkin and Bertram's machines. The best example of the impact of the World exhibitions is Paris in 1867

[41]*Gaceta de Madrid*, 28 January 1845, nº 3,789, p. 2.

where Felip Flores discovered, in the 'Württemberg hangar', the Voelter grinder.[42] In the Spanish context, the industrial exhibitions at both the national and regional levels spread the knowledge of the new technology. The visits to foreign papermaking factories[43] and workshops[44] were also a source of technological transfer.

The imitation effect was also important. Occasionally, the leaders of a pioneering project quickly became involved in a second one. Viuda de Ribed e Hijos in Villava began operating in 1846 and could be considered an example of this pattern; it had among its founders Juan Conte, previously involved in La Esperanza from Tolosa. Likewise, Felipe Flores and Félix Pages, shareholders of La Gerundense, which began operations in August 1843, established a second factory in Girona, La Aurora, roughly two years later. In both cases, these were the first steps in establishing important industrial districts and they also served as good examples for others to copy.

The tacit knowledge supplied by the workers was complemented by other explicit sources, and in this regard the technical press played a central role. The absence of a Spanish papermaking journal was compensated for by the circulation of foreign ones, especially from France (e.g. *Moniteur de la Papeterie Française* and *Journal des fabricants de papier*). The general interest technical press also contributed to filling this void with its own articles and translations of foreign papermaking journals. Examples include *El Semanario de la Industria* (1846-1848), *La Gaceta Industrial*,[45] *El Porvenir de la Industria*, and *Crónica de la Industria*. Journals that were focused on graphic arts, such as *La Tipografía* (1866), also published news concerning papermaking engineering. Moreover, more general interest journals such as *Gaceta de Madrid*,[46] *El Imparcial*, and *Mercurio de España* acted as a collective transmission belt of information. *La Gaceta Industrial* published in the mid-1870s a handbook titled *El papel y sus aplicaciones*. Likewise, the foreign technical handbooks were on the shelves of Spanish papermakers.

Another means of diffusion was the commission agent. Hiring foreign commission agents allowed Spanish papermakers to purchase imported equipment much faster because of the formers' direct contact with foreign machine builders. For example, A. Kaindler, a former director of a Spanish factory, was a partner in Everling & Kaindler with headquarters in Paris in the late nineteenth century. Some of the most

[42]OEPM, Privilege 4,689.

[43]Felip Flores visited in 1867 various factories close to Grenoble. Ribed travelled to Germany to see working of the Voelter grinder.

[44]Luis Oseñalde, promoter of a factory located in Rascafría, had to travel with William Sanford 'to France and to England to buy the required machines' (AHPM, Notary files 24,965, fols. 574v.-575r). Eliseo Frois, a partner of La Salvadora of Tolosa, travelled in 1870 to Angoulême to find machines for his factory.

[45]His editor-in-chief, José Alcover published articles on Voelter grinder and the World's Exhibition held in Paris in 1867.

[46]Proof of that was an article published in 1823 on the origins of the papermaking machines and a pioneering factory in Berlin (*Gaceta de Madrid*, nº 123, 11 December 1823, p. 462).

important French commission agents kept close contacts with the Spanish markets. For instance, Louis Piette[47] advertised paper machines in the Spanish press in 1860.

The Spanish network of commission agents was mainly located in Barcelona and Madrid, and it was also very important to the technology transfer. Those from Barcelona were linked to textile machinery, illustrating the importance of inter-industry relations. The Frenchman Juan Pedro José Canal, who specialized in wool machines, in 1842 offered 'machines according to the most modern and renowned system for making paper'.[48] For instance, Canal contacted Donkin using a French acquaintance.[49] Oller Chatard y Cª, from Paris and focused on textile machinery,[50] also had representation in Barcelona and it offered machines for making continuous paper. The Centro Científico-Industrial y Agencia Comercial by Martín Ziegler[51] had in the 1860s among his activities the representation of papermaking machinery. Other agents included Estanislao Malingre, an engineer, and Miguel Cheslet y Hermano, a commission agent in Madrid for French and Belgian firms. He acted as the liaison in the late 1870s in the purchase of the new machinery built by Dautrebande et Thiry for a factory located near Madrid.

8.3.3 *National Workshops: An Essential Piece in Technological Transfer*

The new machinery required a national technical structure to maintain it and adapt it to local conditions. The pioneering factories and their foreign technicians, added to the skills acquired in other sectors and in handmade paper, were the basis of these national workshops. However, papermaking was only as a secondary activity due to the poor development of this industrial sector. Nevertheless, maintaining the machinery gave them the practical know-how to build some parts of the machines. The machine shops with higher degree of development were completely integrated in the papermaking district dynamics. The two best examples were Tolosa and Girona. Tolosa had an iron-making tradition, and this supported its growth as a papermaking centre. In fact, it led to the creation of small workshops and some foundries.

[47] Piette was the author of *Manuel du contremaître et du chef d'atelier de papeterie* and the editor of *Journal des fabricants de papier*.

[48] Diario de Barcelona, 25 February 1842, pp. 7–9.

[49] Canal received a letter in May 1844 from Donkin's workshop in Bermondsey that described a papermaking machine. The contact between Canal and Donkin was established through John James Hopwood from Boulogne sur Mer, where he owned a linen factory (founded in 1839) (A copy of the original letter sent by Richard Donkin to the author).

[50] A Catalan, who had settled in Paris in 1837 after his family woollen firm had gone bankrupt, set up Oller Chatard y Cª. The firm supplied machinery in the early 1840s to La España Industrial, one of the most important cotton factories in Barcelona.

[51] Ziegler was a chemist, who worked in a calico factory in Alsace. He was employed in 1847 as a technician for La España Industrial where he worked until 1852 (Gutiérrez i Medina 1997, pp. 217–220).

Fossey y Cía in Lasarte, Guipúzcoa was founded in 1853 by the British engineer Edward Fossey. It built the most important part of the turbines and hydraulic wheels of the first papermaking factories in Guipúzcoa.[52] The first steps to develop the papermaking district in Girona were linked to workshops in Barcelona, such as the one of Valentí Esparó's specialising in textile machinery.[53] Girona soon had papermaking engineering workshops. Planas, Junoy, Barné y Cª was created in 1857 and specialised in producing turbines (Nadal 1992). Planas had as a second specialisation 'cylinders and calenders for papermaking' (Martínez Quintanilla 1865, p. 308). In 1870, it offered a continuous machine, claiming that it was the first Spanish workshop to have built them. Another workshop in Girona was Porredón, Comas y Cia (later Porredon, Claret y Cía.), which in 1872 offered, among other types of machinery, 'machines and devices for papermaking'.[54]

Some workshops were set up in other places in Spain that also supported modern papermaking activity. Many of them were run by foreigners, especially from France. For example, Nicolás Cardhaillac founded Cardhaillac y Aldea in 1842 in Valladolid.[55] The next year he got involved in setting up a new papermaking factory in this Castilian town. The same French mechanic also got involved in the emerging Castilian flour industry (Moreno 1998, pp. 248–250), and enjoyed major development from 1856 onwards with Félix y Aldea y Cía. Aragón was another important Spanish paper machine building centre with French links. La Maquinista Aragonesa SA in Zaragoza, set up in 1853, was partly owned by Villarroya, Castellano y Cia., a firm with an interest in modern papermaking. The firm was composed of a group of French engineers (Germán 1994, p. 77) that had arrived in Spain in the early 1840s.[56] Antonio Averly set up in 1863 his own works (Sancho 2000), and its specialities included complete factories for the production of white and straw paper. The emergence of modern papermaking in the vicinity of Madrid boosted the presence of workshops connected to this activity. William Sanford settled in Madrid in 1837 to set up Bonaplata, Sanford y Cia. He was also the industrial partner of the papermaking factory in Rascafría, operational in 1842, but he set up his own workshop in 1846. He was described in 1865 as 'a builder of all kind[s] of papermaking machines'. In the Valencian region, an area with a strong papermaking tradition, there were two centres of engineering with some specialisation in this sector: Alcoy and Valencia. The workshops in Alcoy were given a boost with the modernisation of the wool industry. The pioneering workshop, founded in 1845, was established by Antonio Boronat Payá, who was a member of a papermaking family. El Vulcano Alcoyano-Rodes Hnos., a new machine shop, was set up in 1870. Until 1880, this

[52]Fossey remodelled the papermaking factory located in Irura, close to Tolosa in 1852 and supplied turbines to La Tolosana.

[53]Esparó cooperated in 1854 with Varrall of Paris to assemble a machine of La Gerundense.

[54]La Provincia, I, 33, 31 October 1872, p. 8.

[55]Cardhaillac was a mechanic from Toulouse, who built a machine in 1836 for a French factory (André 1996, p. 163). He went to Valladolid in the late 1830s to assemble machinery for the Castile Channel.

[56]Goybet, a member of the papermaking family of the Montgolfier, arrived to Zaragoza in 1841 to help his uncle Agustín Montgolfier. Goybet lived in Zaragoza until 1863, and later directed the Science and Industrial Arts School La Martiniere of Lyon.

firm worked only for mills that made paper by hand. In Valencia, Fundición Primitiva Valenciana, which in the late 1870s increased its papermaking activity,[57] benefited from setting up factories in its vicinity.

A very important characteristic of the Spanish papermaking industry was the continuity, even dynamism, of making paper by hand. It needed its own engineering workshops, as illustrated by the Alcoy example. An opportunity for its consolidation was the Hollander diffusion, which arrived quite late in Spain, in the second half of the nineteenth century. This mechanical device substituted the traditional hammers that ground the rags. These workshops were given an important boost with the adoption of the Picardo paper machine, which was the answer from the mill owners to a series of strikes in the mid-1870s. A prototype of this round machine, which made paper sheet by sheet and came from Italy, was operational in early 1877 in Gelida, Barcelona and in August a group of Catalan firms patented it in Spain. The machine was built by Lerme y Cia., a workshop located near Barcelona and focused on textile machinery. The machine shop set up by Isidro Soteras in 1860 and known, first, by its Hollanders and later by its Picardo machines was located in Capellades, a papermaking district on the outskirts of Barcelona and one of the most important hand-made paper areas in Spain. There were examples of this same trend in other handmade paper Catalan centres such as La Riba in Tarragona, which was also supplied by the dynamics of the neighbouring towns of Valls and Reus (two of the most prominent industrial Catalan towns).

8.4 Foreign Technology and Spanish Engineers: The Improvement of Technological Capacity

The Spanish papermaking industry began to grow both qualitatively and quantitatively in the late nineteenth century. By 1900, there were more than 100 working machines in the country. The new capacity competed for a limited but growing domestic market, and the problem of overproduction triggered the merging of the most important factories into La Papelera Española (LPE). The growth of the Spanish paper industry accelerated in the first third of the twentieth century, and by the eve of the Spanish Civil War there were over 200 Fourdriniers and round machines in use. During this period, the decline of papermaking in inland Spain contrasted with the growth pattern in the Basque Country, Catalonia and Valencia region.

---

[57] La Fundación Primitiva Valenciana was announced in 1879 to be a builder of 'improved machines to cut rag, and iron cylinders for papermaking factories; factories of continuous paper' (Almanaque de la Gaceta Industrial para 1879, p. 58).

Table 8.2 Imports of papermaking machinery into Spain by country, 1925–1935 (%)

Country	%
Germany	73.5
Belgium	5.1
France	18.0
United Kingdom	1.9
Other	1.5

Source Estadísticas de Comercio Exterior (ECE)

8.4.1 The German and Swiss Supremacy

The majority of the new machines still came from abroad, but the Spanish presence increased in terms of the production of auxiliary equipment. However, some of them, especially those in big factories, were imported (for example, the Hollanders of La Papelera Vizcaína were German and Swiss). According to the papermaking census of 1943, the percentage of foreign machines was higher in the Basque Country (65–70%) whereas it was marginal in the Valencian (5–15%) and the Catalan mills (20–30%).[58] A papermaking journal stated in 1909 that 'paper mill machinery was imported chiefly from Switzerland, Germany, France and the United Kingdom'.[59] Between 1925 and 1935, German-made machinery dominated in Spain, representing 73.5% of the total (Table 8.2). Even though the French presence had diminished from the late nineteenth century onwards, French machines still represented in the mid-1920s between 15 and 20% of the Spanish imports, falling to 5% in the first half of the 1930s.[60] Some British machines were imported in the early twentieth century[61] although they disappeared after the First World War. The Belgian paper machines remained important in the Spanish market in the late nineteenth century[62] and the early twentieth century.[63] They were more prevalent in the Basque Country, largely as a result of the links between the Basque engineers and Belgium (Anduaga 2011).

[58]Ministerio de Industria y Comercio. Dirección General de Industria (1944).

[59]The World's Paper Trade Review (hereafter WPTR), vol LII, no 5, 30 July 1909, p. 184.

[60]Examples include paper machines of Neyret Beyler, imported in 1925, and Allimand, imported in 1936. Both machines were for the cigarette-paper producer Miquel y Costas & Miquel.

[61]Sociedad Española de Papelería purchased a Bentley & Jackson in 1912. *La Vanguardia* of Barcelona bought a Bertram in 1913.

[62]Camilo Gisbert Terol de Alcoy purchased a Thiry in 1885. The machine was defined as 'of good size and is promising to be more productive in silk paper than the locally built machines (*El Serpis*, IX, nº 2,701, 23 October 1886, p. 1). Papelera del Cadagua had a Belgian machine in the early 1890s, and another machine that was assembled in the late nineteenth century was a J. et E. Chantrenne.

[63]Portu Hermanos y Compañía of Villabona purchased a Thiry in 1910. In the mid-1930s, Papelera de Arzabalza planned to purchase a Thiry, but the plans were disrupted by the outbreak of the Spanish Civil War in 1936.

The German hegemony began in the late nineteenth century and it was firmly established by the early twentieth century.[64] It was a function of the need to utilize machines with bigger capacity. The most prominent firm was J. M. Voith, which supplied 13 machines to Spain between 1911 and 1935, mainly to the biggest factories (eight of them to La Papelera Española). The second most important country of origin was Switzerland. Escher Wyss built machines for Papelera del Cadagua. This trend was accentuated in the early twentieth century due its growing sales of hydraulic turbines.[65] Wyss built machines for Torras Hermanos, La Gelidense (in 1904) and Miquel y Costas & Miquel (in 1916).

The development of the pulp industry in Spain during this period was still modest, however. Around 1900, for instance, there were only six wood pulp mills in the country, and they were focused mainly on producing mechanical pulp. There was only one integrated chemical pulp and paper mill, which used a patent obtained in 1893. The grinders and the other machinery came mostly from Voith, Escher Wyss and Bell. The increasing integration between pulp and paper production reinforced the hegemony of German and Swiss machines, especially in the biggest factories.

8.4.2 *The Transfer Mechanisms: The Growing Importance of Networks*

From the late nineteenth century onwards, the capacity for technological absorption of the Spanish papermaking sector had increased notably. Proof of that was the reduction of Spain's dependence on foreign skilled labour and the proliferation of more specialised workshops. In addition, some traditional channels of technology transfer, such as World Exhibitions,[66] lost importance. In contrast, the contact between the papermakers and the machine builders increased in significance. The preferential mechanism of acquisition was through the network of agents, because the industrial growth allowed the main papermaking engineering firms, using their other specialisations, to have a permanent presence in Spain. Especially important was the stimulus coming from the process of electrification beginning around the turn of the twentieth century. These agents (i.e. representatives) assured that spare parts could be readily obtained and so could news related to advances in the industry. The best two examples of this dynamic were J. M. Voith and Escher Wyss.

Voith had permanent representation in Spain from the early twentieth century onwards, due to its specialisation in making turbines at the height of the electrification process (Table 8.3). In April 1901, Voith looked for an 'agent in Spain and Portugal'.[67]

---

[64] Some of these machines could have come from the German branch of Escher Wyss.

[65] From 1901 to 1924 Spain was the third largest market for the Swiss firm, after Switzerland and Japan, with 208 turbines (Nadal 1992, p. 80).

[66] The Escher Wyss machines exhibited in the Paris Exhibition of 1900 impressed Joaquim Jover, who bought one in 1904 for his La Gelidense factory. The Spanish technical press paid special attention to the machines built by Füllner and Escher Wyss.

[67] El Heraldo de Madrid, XII, nº 3,808, 18 April 1901, p. 3.

Table 8.3 Commission agents of Voith in Spain, 1900–1935

Agent	Town	Period
Ermano Schilling	Madrid	Early twentieth century
Ahlemeyer. Compañía Anónima	Bilbao and Madrid	Early twentieth century
Ricardo Zaragoza	Barcelona	1910–1920
Sotomayor y Compañía	Bilbao	Early 1920s
Emilio Ziegler	Madrid	Early 1920s
Ewald Gutensohn	Barcelona	Mid-1920s
A. L. Richard	Tolosa	Mid-1930s

Source Collected by the author from technical journals and handbooks of the time

Its agents also generally represented firms that made electrical supplies or German machinery. In terms of relations with the papermaking sector, its agent was shared by the group Echevarría-Zuricalday with interests in La Papelera Vizcaína. In the mid-1930s, the Voith papermaking business was channelled through A. L. Richard, a French engineer, with headquarters in Tolosa, the centre of Spanish papermaking. Just before the outbreak of the Spanish Civil War, customers dealt directly with Voith's German headquarters because the firm did not have an agent in Spain. Golzern in 1908 was represented in Barcelona by Emilio Guillermo Schierbeck, an engineer focused on electrical matters.

The two main Swiss papermaking engineering firms (Escher Wyss and Bell Maschinenfabrik) also specialised in making turbines. D. R. Zerbone represented the first firm in Spain in the early twentieth century.[68] Later, the engineer Francisco Vives Pons (with headquarters in Barcelona, Madrid and later in Bilbao) was in charge of it. Vives's office opened in 1908[69] and he was still working in the mid-1920s, although in 1930 his activities were limited to Barcelona. The delegation in Madrid was led by the Swiss engineer Rodolfo Liner. Bell Maschinenfabrik had a travelling sales agent throughout Spain in the early 1880s. In the early twentieth century it was represented by the R. de Eguren firm, which had been set up in 1906 and focused on electrical supplies and lifts.

Other channels, which allowed papermaking engineering firms to stay in contact with Spanish costumers, were graphic arts, timber and textile representatives, links that highlight the centrality of inter-industry relations. Some German engineering firms used commission agents linked with graphic arts. For example, H. Füllner in

[68]El Liberal, 29 April 1901, p. 4.

[69]Vives managed the Planas workshop of Girona and its follower Construcciones Mecánicas y Eléctricas SA, until he decided to establish his own workshop. His works in the papermaking sector included an esparto grass factory, which was located in Andalusia in 1918 and operated machines supplied by Escher Wyss.

the early twentieth century was represented by Richard Gans,[70] a German owner of a foundry located in Madrid. In the 1930s, Füllner was represented by Pablo Weeber, also the agent in Spain of Vomag, a builder of graphic arts machinery. The Scandinavian machines arrived through the channels of commercialisation of wood, wood machinery, and wood pulp. For instance, Agencia Imex SA was set up in 1933 in Tolosa and it was focused on selling Nordic papermaking machines. Imex shared headquarters with The Northern Pulp Co. S. A., an importer of Scandinavian wood pulp. Svenska Alliance Co., with an office in Barcelona, imported both machines and wood pulp from the Nordic countries in the 1920s. Also, the German firm Füllner was represented in 1927 by Thorwald Schiott, a pulp commission agent in Bilbao. In Catalonia, many of the papermaking machine builders were still using the distribution network of textile engineering. Freíd-Krupp, a German firm specialising in armaments and a builder of components for papermaking machines (such as calenders), had representation in Barcelona from the early twentieth century. Its agent was the German Ernest Leonhardt, who was also devoted to textile machinery from his country and other papermaking engineering firms. Isidore Dietlin, mainly focused on textile machines, also offered papermaking machinery. John Sumner y Cª, which imported machinery to Barcelona from 1856, had offices in Manchester, close to Bury, the centre of English papermaking engineering.

The Belgian engineering firms were represented by commission agents specialised strictly on papermaking machinery. Ernesto Limousin from Tolosa was acting in 1892 as 'the representation of a very renowned Belgian papermaking machinery builder'.[71] Limousin had worked in the factories of Tolosa, advertising himself as a specialist in studies for system and reform of papermaking factories. Floro Izaguirre, also from Tolosa, represented Thiry in 1912.[72] The most important firms bought the machines after visiting the workshops. The process of building them could take a couple of years and often involved the intense exchange of information and mutual visits. The growing complexity of the technology and the building "à la carte" required an intense exchange of information and feedback.[73] The Spanish papermakers also received

[70]Gans arrived in Spain as a commercial agent of a Belgian papermaking firm. In the early twentieth century the firm made 'all kind of machinery for making paper and wood pulp' (Anuario de electricidad, 1903, p. 477).

[71]Museo Molino Papelero de Capelladas (MMPC), Fondo Cal Violant (FCV), Correspondencia 1892.

[72]Anuario Técnico e Industrial de España, año 2, 1912, p. 418. Izaguirre had previously been a commission agent of the German chemical firm Badische Anilin & Soda Fabrik.

[73]A Flores family representative visited Grenoble in 1889 to buy some components for a machine. Nicolás Mª de Urgoiti, manager of La Papelera del Cadagua, bought few machines after traveling to Germany. Benito Portu, a Basque papermaker, visited Belgium in 1910 together with an agent of Thiry in Spain to order a new machine. Tomás Costa, an engineer of the paper factory owned by the Catalan newspaper *La Vanguardia*, visited in 1924 many papermaking factories and engineering workshops in Germany. Costa finally chose a Voith machine. Executives of Miquel y Costas & Miquel travelled to Paris and Grenoble to sign an order of a machine in 1923 and 1924. The representatives of the same firm travelled to Paris and Rives from 1933 to 1936 to buy another machine.

advice from foreign engineering consultants,[74] who were crucial to facilitating access to the new technology using their knowledge and contacts with foreign workshops.[75]

8.4.3 The Subsidiary Role of District Workshops

From the late nineteenth century onwards the supply of Spanish made machines increased, especially the supply of auxiliary machines. Nevertheless, continuous machines were built only sporadically and were limited to those of small capacity. Import substitution was, therefore, modest. The main activities of the Spanish workshops were the adaptation of foreign machines to local needs and the maintenance of them. For most of these workshops, the production of papermaking equipment was a secondary activity. Three areas experienced the most important growth in this field, namely the Basque Country, the Region of Valencia, and Catalonia. In contrast, the decline of the mills located in inland Spain caused the end of papermaking activity in some of this region's workshops.[76]

The most important papermaking engineering concentration was in Tolosa in Basque Country, and it was essential to feed the district's industrial dynamics. The first example of intensive papermaking specialisation was the workshop owned by Félix Yarza (and followed by Javier Luzuriaga) founded in 1884. Yarza's and Luzuriaga's firm finally became Talleres de Tolosa SA in 1918,[77] whose Board of Directors was formed by members of the Basque papermaking oligarchy. Other important papermaking workshops in Tolosa were Gorostidi y Gozategui, founded in 1915, Pedro Pasabán, founded in 1928,[78] and Ramón Basagoitia. Beasaín, which is close to Tolosa, saw the founding in 1892 of La Maquinista Guipuzcoana, which boasted some dedication to papermaking engineering (building some Hollanders and wire ends). Outside the district of Tolosa, during the first productive years of LPE when the factories were restructured, the Spanish papermaking emporium strengthened its mechanical workshop in Aranguren, Biscay, the Basque Country.

Alcoy was another papermaking engineering centre that was increasingly dynamic,[79] and its workshops had as a main specialization wine, oil, and textile

[74]For example from Frenchman called Édouard Héry, who was the director of a technical journal called *Le Papier*. Miquel y Costas consulted Héry in 1923/24 for his plans to purchase a machine made by Neyret Beyler.

[75]Miquel y Costas & Miquel contacted E. Héry in late 1926 to acquire information about the use of Millspaugh's cylinders in the French factories.

[76]In Aragon, Averly, before suspending their operations, supplied some machines or machine parts in the late nineteenth century.

[77]Talleres de Tolosa S. A. was focused on 'setting up complete paper and board machines as well as modifications […]; and in general, on all machines used in the papermaking industry' (Commercial Catalogue of Tolosa, undated).

[78]Pasabán had worked in LPE and involved in the setting up of Talleres de Tolosa.

[79]In 1908, it was aid that 'papermaking machinery is made at Alcoy' (WPTR, 3 April 1908, vol XLIX, no. 14, p. 577).

machines. In terms of papermaking equipment, these works focused on auxiliary machinery and on small capacity machines that could make cigarette and silk papers, which were the district's most important paper products by the early twentieth century. The most important workshops were La Maquinista Alcoyana. Hijos de José Boronat (founded in 1904), Jorge Serra, Tomás Aznar Hermanos,[80] Francisco Blanes (founded in 1910), El Vulcano Alcoyano-Rodes Hnos., (which had already built complete papermaking lines in the 1880s), and José Pérez y Cia. In the papermaking district of Valencia, the regional capital, two firms (La Primitiva Valenciana, and La Maquinista Valenciana) began building papermaking machinery in the late nineteenth century, and their range of products included Hollander beaters, presses and even some paper machines.

The high density of papermaking capacity in Catalonia explained the existence of a network of workshops in three of its zones: Barcelona, Girona and the other areas with a strong papermaking tradition. Those located in Barcelona included, for example, Marcelino Vilarasau, successor of Lerme y Cia.; Puig y Negre, founded in 1882 and builder of auxiliary papermaking machines; and Joan Trabal Casanella, which set up in 1929 a workshop focused on the same activity. Planas, Flaquer y Cª in Girona specialized in papermaking machinery and moved its works to Barcelona. In the early twentieth century, however, it increasingly focused on producing turbines and electrical equipment. The tradition it had established of manufacturing papermaking equipment in Girona was sustained by Talleres Alberch (founded in 1902 and whose first papermaking machine was built in 1935) and by Construcciones Mecánicas Sarasa (operational from 1923). The workshops linked with the production of handmade paper acquired a new dimension with the introduction of the Picardo machine. Isidre Soteras from Capellades, after the patent had expired, began to build and sell them throughout Spain. These workshops would play an auxiliary but essential role when modern papermaking was later adopted.

Also, there was a long history of import substitution in terms of the auxiliary parts of the continuous machines. The wire used to form the sheet of paper was purchased mainly from France and Germany until the last quarter of the nineteenth century. Thereafter, producers of Spanish wire gained an increased share of the market. However, the most important makers were French[81] and these firms dominated the Spanish market.[82] Moreover, the felts were imported mainly from France and the

---

[80]Tomás Aznar built in 1877 a papermaking machine for La Clariana (in Onteniente), which special feature was that it was capable of making 'cigarette paper which look like handmade' (El Graduador, III, 1,105, 27 November 1877, p. 2). The machines by Aznar were characterised by their 'narrow dimensions' and their design 'to make exclusively cigarette paper' (El Serpis, IX, 2,701, 23 October 1886, p. 1).

[81]Pérot had worked as a commercial agent of Laboreaux et L. Lescure after visiting Spain from 1870 to 1874 and noticing the potential of the wire market. It was said in 1888 that he made 'Endless wires. This kind of wire is to be used in the making of continuous papermaking' (Serrate 1888, pp. 112–113).

[82]José de Garaizábal from Valladolid used wires made by Rivière and pointed out that 'today I have the pleasure to tell you that the wire is of my liking, and if the following supplies are as good, I am not going to purchase any more from France and Germany' (Reforma 1890, III, p. 71).

United Kingdom. Also, some firms from Alcoy made them.[83] La Papelera Española had its own workshop to make this kind of felts, which were also sold to other firms in Spain.

8.4.4 *Skilled Labour: More Spanish Technicians*

In the last quarter of the nineteenth century, skilled labour became increasingly Spanish. Foreign technicians declined in importance and remained only in the biggest factories[84] and when a new product or process was introduced.[85] As a result, they were the exception by the early twentieth century.[86] During the late nineteenth century, informal learning systems, such as families, were still important to the process of technology transfer in Spain. Examples of this dynamic were present among many of the country's leading papermaking families. Nevertheless, the industry's growing technological complexity boosted the demand for formal training and the creation of specialised educational institutions. In 1899, for instance, a papermaking school was proposed for Tolosa, but it failed to materialise. As a result, different types of engineers compensated for this lack of specialised training. Many papermaking technicians were civil engineers. Also, the opening of new industrial engineering schools in Spain was decisive to the nationalization process of the technicians. They would emerge as the new technical staff for Spanish factories. For example, during this nascent period in the country's industrial development, several native technicians held chemistry degrees from local institutions. Forestry engineers also had an important presence in papermaking firms. The innovative spirit of these engineers was reflected in the different patents they registered.

These years also saw it become typical for highly skilled workers in Spain's papermaking industry to pursue their higher education in foreign schools that specialized in papermaking studies. It is symptomatic that the German Ingenieurschule of Altenburg advertised in 1923 and 1924 in the Spanish newspapers to recruit students for its papermaking courses. The main destinations were in France and Germany (Table 8.4). During their training, Spain's papermakers often spent internships in French or German factories before joining their family businesses. The personal relations established during this training period with teaching staff and other students were also essential later to enjoying quick and effective access to the latest technical information. The foreign training centres, equipped with better laboratories and

[83]The factories of Alcoy in 1908 consumed local felts (WPTR, 3 April 1908, vol. XLIX, no 14, p. 577).

[84]The production manager of La Guipuzcoana in 1884 was German-born. The installation work of a round paper machine in Alcoy in 1886 was also supervised by a German.

[85]Hippolite Piqués, a papermaker from Angoulême, advised Josep Flores in his board factory in Sarrià in 1881 (Brugada and Vila 1999, p. 29). Piqués advised in 1887 and 1888 a Catalan firm owned by Ramón Romaní, who wanted to make leather board (i.e. artificial leather made from fibrous refuse materials) (MMPC, FCV, Correspondencia 1887 y 1888).

[86]The factory owned by Sociedad Española de Papelería (founded in 1912) was managed by a British engineer (WPTR, vol LVIII, no. 3, 19 July 1912, p. 134).

Table 8.4 Spanish students in foreign paper engineering/papermaking institutes, 1910–1935

Name	Institute	Period	Places of work in Spain
Francesc Batlle de Balle	Technische Universität Darmstadt	1910s	Catalan mills. His own mill from late 1920s
Joaquín Navarro Sagristá	Eidgenössische Technische Hochschule Zürich	1920s	Mills in Alcoy and Papeleras Reunidas
José Ramón Calparsoro	Engineering School of Lausanne/Ingenieurschule of Weimar	1926–1933	Pérot (wire producer)
Francesc Torras Hostench	EFPG	Early 1930s	Paulino Torras Doménech
Llorenç Miquel Serra	EFPG	Early 1930s	Miquel y Costas & Miquel
Josep M. Fornt Brunet	EFPG	Early 1930s	na
Juan González Iñurrategui	na (in Munich)	Early 1930s	San José de Belaunza

Source Collected by the author from technical journals and handbooks of the time

technical means, also advised the Spanish papermakers. Especially fruitful were the links with the École Française de Papeterie of Grenoble (EFPG).

In addition, the training of mid-level technicians improved due to the spread of general vocational schools in Spain after 1886. Another important development was the establishment of the Theoretical-Practical Vocational School of Zalla by La Papelera Española. It awarded degrees under the titles of papermaker, turner-fitter, founder, carpenter and mould-maker.

8.4.5 Technical Press: An Open Door to Update Knowledge

In the absence of media outlets that specialised in papermaking prior to the late nineteenth century, the Spanish industry was covered by the country's general technical journals. The first one that focused strictly on papermaking was *Mercado del Papel*, and it was published by Richard Gans from late 1892 to 1894. Four years later *La Industria Papelera* began publication in Tolosa, but it soon ceased because of turbulence in the papermaking industry. Over the next decade it was published only intermittently. From 1907 to 1918, La Papelera Española edited the *Boletín de la Industria y Comercio del Papel* (which reached 500 subscribers), which monitored in detail the Spanish and international industry. In addition, the forestry engineers' journals[87] and those focused on graphic arts[88] published technical articles on papermaking. Because Spain lacked its own consistently published papermaking journal, its exposure to the industry was achieved through the circulation of publications from other countries, specifically the UK (e.g., *The Paper Maker & British Paper Trade*[89]), France (e.g. *La Papeterie* and *Le Papier*) and Germany (e.g. *Papier Zeitung* and *Wochenblatt für Papierfabrikation*).

For the very first time some basic technical handbooks were also published in Spanish. The best known was *El Papel*, published in 1898 by Luis Marín, civil engineer, and it was distributed by the journal *La Industria Papelera*. The technical handbook published in German in 1875 by Carl Hofmann, *Handbuch für Papier Fabrication*, had wide diffusion in Spain in its French translation. The late 1920s saw the circulation of *Die Papierfabrikation und deren maschinen: ein lehr- und handbuch* by Friedrich Müller, who was at that time teaching at the Technical School of Darmstadt. Tomás Costa wrote in 1938 a draft of a technical handbook, which served as the basis for a book that was published some years later (Costa 1946). Its bibliography demonstrated his sound knowledge of the international papermaking industry.

8.5 Conclusions

The centrality of technology transfer in the economic growth of the developing countries is unquestionable. The mechanisms used to diffuse the new technology are varied and often overlap. Nevertheless, the transfer itself does not guarantee economic growth in the long run. It is also essential that the milieu that is receiving the new technology is conducive to doing so. The existence of the "industrial atmosphere" in the industrial districts often allows for the absorption and adoption of the new technology. All of these factors are influenced by the institutional framework of the receiving nation. For example, it is not necessary to have a high concentration in a particular sector because this deficiency can often be overcome by strong inter-industry relations. In addition, it is fundamental to have an active engineering sector and skilled labour, because technological competence allows the receiving milieu to choose the most suitable technology and to adjust it to its own needs. In these decisions lies an important part of the success of the transfer process. The failure of the most advanced technologies, even having foreign skilled labour, can be glaring.

Spain, like other peripheral countries, depended on foreign technology during the earliest days of its industrialisation process, and the papermaking sector is a good example of this phenomenon. The new technology arrived late and in a limited

[87]Revista Forestal y Económica y Agrícola, Revista de Montes, La España Forestal, Montes e Industrias.

[88]Revista Gráfica, La Gaceta de las Artes Gráficas, del Libro y de la Industria del Papel.

[89]*The Paper Maker* and its sister publication *The Paper Makers' Directory of All Nations* had representatives in Spain in 1907. See Boletín de la Industria y Comercio del Papel.

way. In addition to the low levels of paper consumption, the institutional instability caused development to be slow. Initially, French machines were hegemonic due to their characteristics, previous transfer flows and the geographical proximity of the two countries. In contrast, British machines were practically missing from the Spanish market. The foreign technicians who arrived with the machines contributed decisively to training the locals in using the new technology. The main concentrations of continuous papermaking were in the centre of Spain (with Madrid as a locus of widely dispersed factories), the Basque Country (with Tolosa as a centre) and the Catalan province of Girona. Due to the importance of the agglomeration, many of the pioneering factories, especially those located in the central regions, had short productive lives. In contrast, those located in areas with a previous papermaking tradition and/or high concentration of the new technology became major industrial districts. This was the case for Tolosa and Girona, having both a complex and rich industrial structure, within which papermaking soon achieved a dominant status. In this context, the existence of commercial networks of machinery and mechanical workshops allowed the Spanish papermakers to incorporate the technical changes more easily. Also, some areas with a high density of workshops that made paper by hand developed strategies of technological innovation that allowed them to survive for decades. In these instances, the interaction with the local workshops was decisive.

The main axes of papermaking changed beginning in the late nineteenth century, and this had a predictable impact on the industry in Spain. German and Swiss workshops emerged as dominant players at a time when the capacity of machines was growing. These machines boosted Spanish paper production from the 1890s until the outbreak of the Civil War in 1936. The main tool for transfer was the commission agent network involved in electrical supplies and turbines, because these activities were central to the most important papermaking workshops (Spain was undergoing electrification during this period). In other cases, commission agents came from the graphic arts, wood and textile sectors, and they were in charge of adjusting information to the needs of the Spanish firms. To the Basque and Catalan supremacy was added the Valencian one, which experienced a sudden process of mechanisation. These three regions came to dominate Spain's papermaking industry because of the availability of high-skilled workers and an engineering sector with a high degree of specialisation in papermaking. Both features facilitated technological change and allowed for the customizing of the machines. Some workshops, such as those from Alcoy, built continuous machines, but always of low capacity.

References

Abramovitz M (1986) Catching-up, forging ahead and falling behind. J Econ History 46:385–406

André L (1996) Machines à papier. Innovation et transformations de l´industrie papetière en France, 1798–1869, Éditions de l´École des Hautes Études en Sciences Sociales, París

Anduaga A (2011) The engineer as a 'linking agent' in international technology transfer: the case of Basque engineers trained in Liège. Eng Stud 3(1):45–70

Berg M, Bruland K (1998) Culture, Institutions and Technological Transitions. In: Berg M, Bruland K (eds) Technological revolutions in Europe: historical perspectives. Edward Elgar, Cheltenham, pp 3–16

Gutiérrez-Ravé JB, Vila A (1999) Els Flores, una nissaga de fabricants de paper. La indústria paperera a Sarriá (s. XIX). Parlem de Sarriá, 27, pp 20–32

Bruland K (1989) British technology & European industrialization. The Norwegian textile industry in the mid nineteenth century. Cambridge University Press, Cambridge

Coe DT, Helpman E (1995) International R&D spillovers. Eur. Econ Rev. 39:859–887

Clapperton RH (1967) The Paper-making machine. Its invention, evolution and development. Oxford University Press, Oxford

Cohen WM, Levinthal DA (1990) Absorptive capacity: a new perspective on learning and innovation. Adm Sci Quart 35(1):128–152

Coleman DC (1958) The British paper industry, 1495-1860: a study in industrial growth. Oxford University Press, Oxford

Costa, T (1946) Manual del fabricante de papel, Bosch Casa Editorial, Barcelona

Dasgupta P, David PA (1994) Toward a new economics of science. Res Polic 23:487–521

Díaz y Pérez N (1880) Baños de Baños: viajes por mi patria, Vda. e Hijos de JA García, Imprenta y Fundición de la viuda de Hijos de JA García, Madrid

Germán L (1994) Empresa y familia. Actividades empresariales de la sociedad "Villarroya y Castellano" en Aragón (1840–1910). Revista de Historia Ind 6:75–93

Gutiérrez i Medina ML (1997) La España Industrial 1847–1853. Un model d'innovació tecnològica, Associació d'Enginyers Industrials de Catalunya/Col·legi d'Enginyers Industrials de Catalunya, Barcelona

Inkster I, Calvo A (eds) (2010) History of technology, European Technologies in Spanish History, vol 30. The Institute of Historical Research. Continuum, London

Jeremy DJ (1991) Introduction: some of the larger issues posed by technology transfer. In: Jeremy DJ (ed) International technology transfer. Europe, Japan and the USA, 1770–1914, Edward Elgar, Aldershot, pp 2–5

Keller W (1996) Absorptive capacity: on the creation and acquisition of technology and development. J Dev Econ 49:199–227

Keller W (2004) International technology diffusion. J Econ Lit XLII 4:752–782

Memoria (1851) Memoria presentada al Exmo. Señor Ministro de Comercio, Instrucción y Obras Públicas por la Junta Calificadora de los productos de la Industria Española reunidos en la Exposición Pública de 1850, Establecimiento Tipográfico de D Santiago Sananque, Madrid

Ministerio de Industria y Comercio. Dirección General de Industria (1944) Estadísticas de la industria del papel en 31 de diciembre de 1943, Publicaciones de la Sección de Estadística Industrial, Madrid

Moreno J (1998) La industria harinera en Castilla L Vieja y León, 1778–1913, Dissertation, Universidad de Valladolid

Nadal J (1988) España durante la 1ª Revolución Tecnológica. España 200 años de tecnología, Ministerio de Industria y Energía, Madrid, pp 29–100

Nadal J (1992) Los Planas, constructores de turbinas y material eléctrico (1858–1949). Revista de Historia Ind 1:63–93

Nelson RR, Winter SG (1977) In search of a useful theory of innovation. Res Polic 6(1):36–76

Nerheim G (1991) The development and Difusión of European Water Turbines, 1870–1920. In: Bruland K (ed) Technology transfer and scandinavian industrialisation. Berg, New York, Oxford, pp 333–362

Quintanilla PM (1865) La Provincia de Gerona. Datos estadísticos, Imprenta de F Doria sucesor de J Grases, Girona

Reforma (1890) Reforma arancelaria y los tratados de comercio. Establecimiento Tipográfico Sucesores de Rivadeneyra, Madrid

Sancho A (2000) Especialización flexible y empresa familiar: la Fundición Averly de Zaragoza (1863–1930). Revista de Historia Ind 17:61–96

Scherer FM (1982) Inter-industry technology flows in the United States. Res Polic 11:227–245
Serrate JM (1888) Las grandes industrias y los grandes industriales. Estudios sobre algunas industrias españolas con datos y estadísticas relativas a su importación y producción. Establecimiento Tipográfico del Diario Mercantil, Barcelona

Part III
Local Innovations and Global Markets

Chapter 9
Technology Transfer and Local Innovation: Pulp and Paper Manufacturing in New Zealand, c.1860 to c.1960

Michael Roche

9.1 Introduction

Historical studies of pulp and paper making in New Zealand (NZ) have focussed largely on the mid-1950s and the two main companies involved, New Zealand Forest Products (NZFP) and Tasman Pulp and Paper (TPP). These few investigations have been dominated by business history (Baker 2002, 2004; Goldsmith 2009; Healy 1982; Parker 2000), biographies of corporate leaders (Parker 1994; Wallace 2001), economic history (Gould 1982; Jones 1999; Guest and Singleton 1999), and economic policy studies (McAloon 2013). This body of work has not overlooked the role played by technology in the development of the industry, but by the same token neither has it been the focus of these studies. Choices of technology and technological change in the pulp and paper industry, factors that have long been of interest in the USA, have been largely overlooked (e.g. Cohen 1984, 1987; McGaw 1985). Such work still helps pose new questions of the sector in NZ.

The indigenous forests of NZ are today mostly protected as national parks and allied reserves (25% of the land area) and commercial forestry is almost exclusively based on exotic plantations amounting to about eight percent of the land area. From European colonisation in the 1840s until the 1910s, paper was largely imported although there was some local production using rags, rope, and grasses as primary materials (mainly *Chionochloa* spp.). During this period, wood pulping techniques had been developed in Europe and North America, and thereafter NZ's initial interest was in developing a pulp and paper industry that would utilise saw mill waste from indigenous forests. Only after the creation of large-scale, state-owned plantations between 1925 and 1934 did the situation change (Appendices to the Journals of the House of Representatives [hereafter AJHR], 1935, C3, p. 2). These forests were composed of fast growing trees (maturing in approximately 25 years) and by 1934 covered

M. Roche (✉)
School of People, Environment and Planning, Massey University, Palmerston North, New Zealand
e-mail: m.m.roche@massey.ac.nz

T. Särkkä et al. (eds.), *Technological Transformation in the Global Pulp and Paper Industry 1800–2018*, World Forests 23, https://doi.org/10.1007/978-3-319-94962-8_9

406,200 acres (164,300 ha). These were established to offset an anticipated timber famine by the mid-1960s. Nevertheless, their potential to supply a pulp and paper industry was recognized in official circles. Private sector afforestation companies raising funds by selling ‘bonds’ also planted some 272,336 acres (110,257 ha) prior to 1934, with some also speculatively targeting eventual pulp and paper production.

The intersection of resources, institutions and technology in NZ gives rise to a pulp and paper story that is more complex than the local literature has heretofore recognised. This chapter keeps technology in the fore by recounting the failed schemes to pulp the indigenous forest as well as the first company efforts that eventually produced paper board in 1939 from the unheralded *Pinus radiata*. The more well-known and celebrated state and corporate wood pulping enterprises established in the mid-1950s are also discussed with an eye to the diffusion and adoption of wood pulping technology. Links to other national pulp and paper sectors are made where appropriate.

9.2 Paper Making Using Flax and Imported Wood Pulp

Ancestors of the Māori settled NZ over 700 years ago. European missionaries began arriving in 1814 and quickly set about translating and printing a bible in Māori. A large consignment of 1,000 tons of indigenous NZ flax (*Phormium tenax*) was shipped to Britain for a trial to make paper as early as 1831. Two decades after colonization paper was being made out of rags in NZ, but most of the domestic supply was imported (Smythe 1968).

Local interest in making paper out of flax was triggered in the 1870s (AJHR 1870, D14A).[1] In 1871 a government bonus of £2,500 for the production of 100 tons of printing paper went unclaimed (AJHR 1871, H7). But small papermaking ventures were independently established at Dunedin in Otago and at Matuara in Southland adjacent to the water falls; both entered production in 1876 and another located beside a tidal creek at Riverhead near Auckland followed suit in 1899 (Fig. 9.1). Commercial papermaking began in NZ on a small scale with old fashioned and worn out equipment. In Dunedin rags, waste papers, old rope, and indigenous tussock grasses (*Chionochloa* sp.) were chopped up and boiled in a digester before being washed in cold water vats. The poor quality product was suitable only for wrapping paper. At Mataura a papermaking machine imported from Australia never operated properly and there were difficulties in recruiting and retaining skilled staff. These various difficulties persuaded the three original local papermaking companies to merge in 1905 to form New Zealand Paper Mills (NZPM) a larger and more commercially

[1] The technical challenge with flax was to remove the vegetable matter without damaging its fibres. Attempts to do this continued into the twentieth century. The US Bureau of Standards, for instance, undertook papermaking tests on flax at the behest of the Department of Industrial and Scientific Research in 1928. As late as 1938 New Zealand Pulping Mills was established, but it was still unable solve the problem of mechanically stripping the flax without breaking the long fibres (NZ Flax pulping Plant 1941).

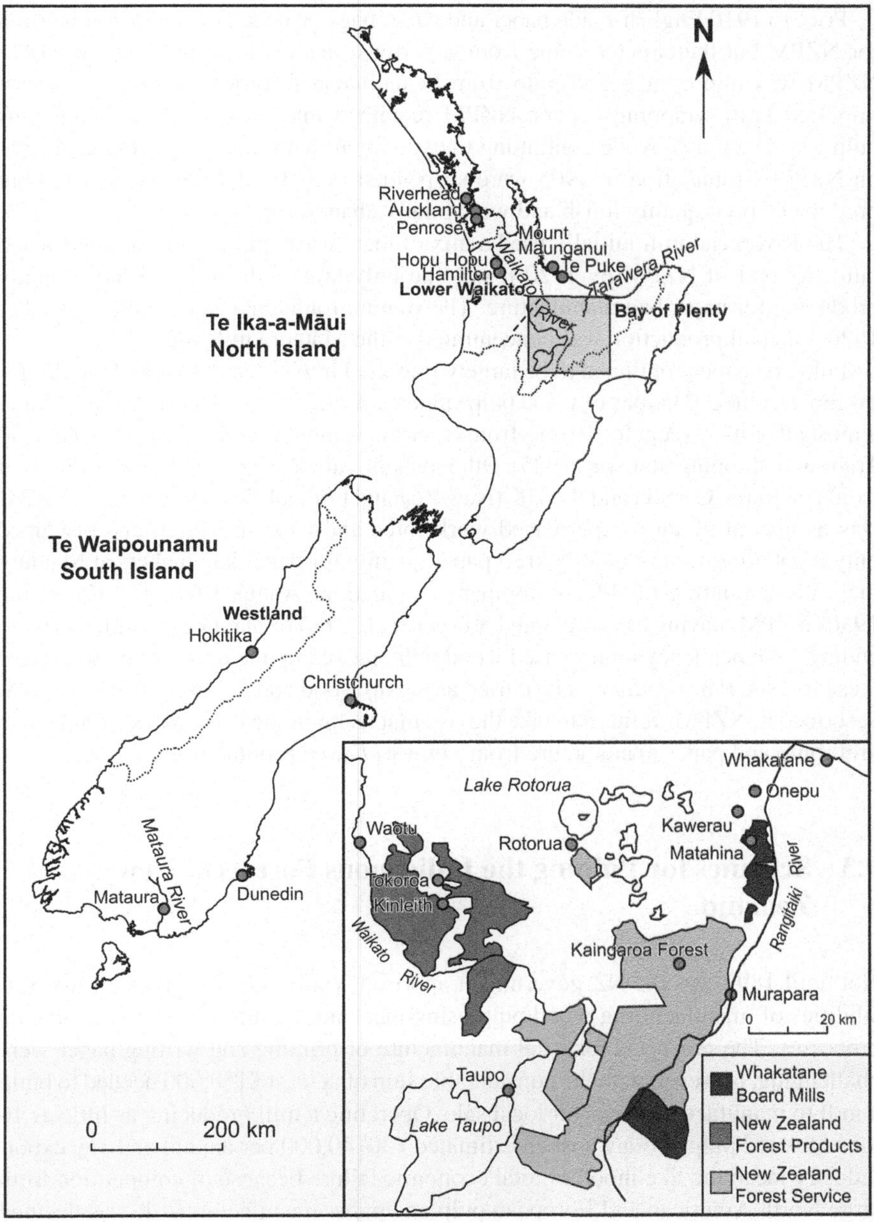

Fig. 9.1 State and company forest plantations circa 1950

viable enterprise (Angus 1976). NZPM, however, still faced strong competition from imported papers while lacking the capital to experiment with the pulping of indigenous forest species.

Prior to 1910 English-made paper and paper bags provided the main competition for NZPM but thereafter it came from superior Canadian kraft products. By 1915 NZPM was importing wood pulp from Scandinavia in order to compete against imported kraft wrapping papers. NZPM remained interested in other indigenous pulp wood ventures, while continuing to experiment with flax.[2] Experiments begun on NZPM's foundation in 1905 were suspended in 1919 and revived in 1920 but produced a poor quality finish and were finally abandoned.

The Riverhead mill initially used a mix of local waste paper and imported wood pulp. As part of NZPM by 1920 it was the only kraft mill in the country mainly producing for paper bag manufacture. The venture foundered and the mill closed in 1926 when all production was concentrated at the Mataura mill site.

Pulp and paper continued to be largely imported into NZ at this time. In 1925, for example, some 2,035 tons of wood pulp valued at £27,754 was imported into NZ, and almost all of it—except for 10 tons from Norway—came from Sweden (New Zealand Trade and Shipping Statistics 1925). Other papers, valued at £600,000, were imported from Australia, Canada and the UK (New Zealand Official Year Book 1928). NZPM was hampered by an inexperienced work force and lack of capital, and produced only 3,060 tons of poor quality kraft paper, mainly for paper bag makers at Mataura in 1930, amounting to 60% of company production (Angus 1976, p. 105). In the 1930s NZPM, having earlier planted 303 acres (122 ha) of plantations with a view to ending its dependency on imported wood pulp, passed up the opportunity to expand these forests. *Pinus radiata* was untried as a pulp wood and as Angus (1976, p. 116) described it, NZPM 'refused to take the risk that technological advances would make kraft pulp and paper manufacture from *pinus radiata* a profitable enterprise'.[3]

9.3 Schemes for Pulping the Indigenous Forests of New Zealand

Not until 1895 was the NZ government apprised, via imperial sources, of the possibilities of manufacturing wood pulp using mechanical, sulphite, sulphate or soda processes. The prospects for local manufacture of printing and writing paper were challenging, however, with the considerable sum of at least £150,000 needed to build a mill to manufacture paper for local sale. Operating a mill producing as little as 10 tons of wood pulp per day cost an estimated £30–40,000 per annum and any export industry faced the likelihood of total economic failure because of competition from large North American and European pulp and paper manufacturers. It was deemed 'a risky undertaking' and, as a warning, the failure of an Australian mill, even with subsidies, to withstand European competition was noted (AJHR 1895, H31, p. 2).

[2] NZ flax is not botanically related to the linen flax of Europe.

[3] Despite many difficulties the company survived to be acquired by New Zealand Forest Products in 1960.

The economic prospects for commercially pulping wood in NZ even for domestic consumption thus appeared bleak.

Local investors nevertheless commenced just such an undertaking in 1907 when Christchurch engineer R. C. Webb, who had connections to the Westland sawmill industry, persuaded the politicians to add a clause to an amendment to the Land Act 1907. This revision would allow 50,000 acre (20,223 ha) areas of forest land to be leased for 21 years as a resource for pulp production.[4] Two such areas totalling 50,000 acres were immediately set aside in Westland in 1908. The next year, William Butler, a prominent figure in the sawmill industry, contributed a short report on the pulp industry in Canada to the annual report of the Lands Department. Butler concentrated mainly on mechanical pulping with some brief mention of the sulphite process. He dismissed the latter as being too costly for NZ, though he appreciated that it produced higher grades of paper. He gave minimum attention to financial matters, but placed the costs of a 'first rate mill' at $Can 1,000 a day per ton of capacity (Butler 1909, p. 82). He nevertheless considered that NZ had the fibre and cheap hydro power resources necessary to enter the industry.

A more detailed first-hand report was received from another source. NZ forestry officer Richard Robinson, while en route to a forestry conference in Scotland in 1914, visited J. R. Booth's pulp and saw mills, near Ottawa, Canada; they employed 12,200 staff and produced 125 tons of pulp per day. Booth pulped only low resin spruce and balsam fir of which large quantities were available, and this gave the Canadian industry a huge advantage compared with other producers. Robinson's report accordingly alluded to the technical and financial difficulties that resinous pines posed for pulping and pointed to the large quantities of raw material required as well as the costs involved in establishing a plant (Robinson 1915, p. 46).

Concurrently the first efforts were made to assess the pulping potential of NZ's indigenous forest species (Table 9.1). Reflecting the prevailing industry view that pulping would make use of otherwise waste products from sawmilling, the first species to be tested were two of the main softwoods, rimu (*Dacrydium cupressinum*) and kahikatea (*Dacrycarpus dacrydioides*). Various hardwood species were also featured, notably the many beech species (*Nothofagus* spp.), which, though numerous in the South Island, were often small, twisted, and difficult to mill.

These tests were carried out in Britain and Europe and indicated that, while pulp made from rimu and kahikatea could be manufactured into paper, unless it was bleached its relatively dark colour would limit its use to packing or wrapping (i.e., kraft) papers. The hardwood (except the unsuitable ribbon wood) produced pulp similar in quality to European species but again it was rather dark in comparison.

The Wirtz tests of 1911 were funded by local merchants who, notwithstanding the challenges they pointed to, established New Zealand Wood Pulp and Paper Ltd. in 1916. The venture folded in 1922 with little to show for its efforts. Another company,

[4]The amendment was largely concerned with leasehold arrangements for crown settlers and the specific section concerning the pulp and paper reserves was not debated. The clause presumably at least had the support of Robert McNab, the Minister of Agriculture and Lands. Regulations were promulgated in the *New Zealand Gazette* in 1909 (Wood-Pulp Regulations under "The Land Act 1908"; NZ Gazette, 8 April 1909, pp. 970–971).

Table 9.1 Testing the pulp wood potential of indigenous NZ tree species

Year	Species	Tester	Pulping processes
1910	*rimu, kahikatea* silver beech, mountain beech, ribbon wood	Königliches Materialprüfungsamt	Soda and sulphite
1911	*rimu, kahikatea* silver beech, mountain beech, ribbon wood, kamahi	Dr. Quirin Wirtz	Soda pulping
1913	*rimu, kahikatea* mountain beech, kamahi, makomako	Landqart Paper-mills, Switzerland	Grinding test Sulphite pulping of rimu & kamahi
1921	*matai* black beech, red beech, kamahi, tawa	Imperial Institute, London	Soda process
1921?	*kahikatea*	Bertrams Ltd. Paper Machinery Manufacturers Edinburgh	Mechanical and soda
1922	*rimu, kahikatea* black beech, red beech, silver beech, tawa	Boving and Co. Pulp Machinery Manufacturers London	Sulphate process
1926	*rimu, kahikatea*	L. R. Benjamin CSIRO Australia	Soda pulping
1926	*kauri*	Imperial Institute, London	Soda pulping

Italics = softwood species
Source condensed from Entrican (1929)

Westland Wood Pulp and Paper Syndicate, was set up in 1922 inspired by a visit to NZ of representatives from British paper machinery manufacturer Boving and Co. The twofold goals of the syndicate were to secure land in Westland for pulpwood purposes under the Land Act and to develop a pulp mill drawing on slab wood and other sawmill waste. Boving and Co. tested various species for their suitability as pulpwood. In 1928 the group commissioned a report from a Canadian, Mr. A. F. Richter, of Stebbins Engineering in New York on the feasibility of the project.

Richter's report was the first external assessment of the possibilities for a pulp and paper industry in New Zealand and ran to 20 typed foolscap pages. Richter considered the available forest resources, potential mill sites, costs of production, and included some rudimentary market analysis. He proclaimed on the basis of a visual assessment alone that the 'woods of New Zealand are entirely satisfactory for the production of all grades of paper which can be made from wood fibre' (Richter 1928, p. 2). Richter also suggested that the smaller Scandinavian and continental European pulp mills would be more suitable for Westland than the larger North American plants. He identified the township of Hokitika as the best location for the pulp mill: the site was suitable for expansion, the town had rail links and a harbour, a sufficient supply of clean water was available along with inexpensive electric power and there was a considerable quantity of wood in the vicinity. He proposed building a mill producing 25–30 tons of newsprint and 12–15 tons of wrapping paper per day. Richter considered 20 tons per day the minimum profitable capacity for a sulphite pulp plant. He ventured that, because kraft papers produced by sulphate processes were little known in NZ, it would be feasible eventually to introduce a kraft paper machine into the plant.

Richter's questionable premise seemed to be that the production costs would be similar to those of pulp and paper mills on the eastern coast of the United States. Freight charges were the only unfavourable condition that he acknowledged and these were more formidable than he realized, with only a single railway running through the Southern Alps to Christchurch and shingle (i.e., gravel) bars making for difficult access to all the local West Coast ports. The Forest Service[5] contested Richter's views, disputing his observations about the mix of species suitable for newsprint and rightly dismissing his claims about the similarity of NZ's forests to those in North America, Europe and Australia. They noted dismissively that he had misidentified some species (e.g. rimu for miro [*Prumnpitys ferrunginea*]) and argued that his comments on small pulp plants badly misunderstood the local context.

On the basis of Richter's report the company cast aside all concerns about the financial and technical viability of the project and was poised to raise £350,000 capital, to obtain waste timber from millers cutting in state forests, and to secure a forest reserve for future use. The capital was to fund a mill producing 40 tons of newsprint and sulphite-based paper per day (Dunn to Prime Minister 1928). The Forest Service was concerned that the success of the enterprise would depend on the relationship between the millers with existing licenses and the company, whereby the former could effectively squeeze the profits by increasing prices for mill waste. More important, however, was the Forest Service's view that newsprint, composed predominantly of ground wood from local kamahi (*Weinmannia racemose*) hardwood and various *nothofagus* species with the remainder comprised of unbleached sulphite pulp from rimu, would, on the basis of the earlier overseas tests, lack both the strength to be run over high speed printing presses and also be too dark to make newsprint (Entrican A to Phillips Turner E 16 Jan 1929). The Chief Inspector of Forests, Arnold Hansson, also advised that it was not possible to demarcate the sought after forest reserve until the company specified what it meant by 'suitable timber' (Hansson A to Director of Forests 9 Jan 1929). The decision to grant access to the forest reserve was one for Cabinet and it rejected the company's proposal based on the Forest Service's advice in March 1929. Matters limped on until the company was wound up in 1932. In retrospect Richter can be seen to have brought to NZ a sulphite process model that was dominant in the north eastern US, whereby it depended on low resin wood and which would not have translated well to NZ's West Coast forests. The Forest Service's concerns at the time were justified. This failed venture shaped the

[5]From 1921 to 1948 this department which had national responsibilities was known as the State Forest Service and from 1949, until it was dissolved in 1987, as the New Zealand Forest Service. Forest Service has been used throughout this chapter for convenience.

subsequent attitude of the Forest Service to pulp and paper projects, particularly the amount of capital and scale of the enterprise required to be economically viable and the perils of premature attempts to establish a pulp mill.

9.4 Competing Forest Service and Company Schemes for Pulping the Exotic Forests

With the establishment of a professional Forest Service in 1920 the pulp and paper industry received renewed attention. In that year, Canadian L. M. Ellis, a Toronto forestry graduate was appointed as NZ's first Director of Forests. At only 32 years of age, Ellis was engaging, energetic, decisive, and capable of implementing bold ideas. He possessed a clear sense of what was required to introduce scientific forestry to NZ. His immediate tasks included preparing a substantial report on the country's forests and outlining a modern scientific forestry policy. Ellis faced many more pressing forestry problems, but from the onset he was convinced that 'every effort should be made to establish a pulp and paper industry in NZ' (AJHR C3, p. 15). Foremost in his mind in this regard was utilising sawmill waste from indigenous forests. Ellis also argued for the creation of a Forest Products Laboratory with two immediate tasks; investigating the pulping potential of indigenous trees and distilling wood alcohol for motor fuel (AJHR 1920, C3A). In both cases this was framed in terms of the more complete and efficient utilisation of native forests.

As a result, one of Ellis' earliest staff appointments in 1921 was an Engineer in Forest Products. In this respect Ellis was actually following the lead of Gifford Pinchot, who was the first Chief Forester of the United States Forest Service (1905–1910) (Nelson 1967). This position was filled by Alexander Entrican, who over the next few years collated existing research on the pulping of indigenous tree species. A graduate of the Auckland University College School of Engineering, Entrican was much like Ellis: a man of enormous energy, drive, and commitment. An engineer rather than a forester, he was intensely interested in issues surrounding forest utilisation.

In 1925 the Forest Service predicted a timber famine in NZ within 40 years, and embarked on an ambitious exotic afforestation programme with a target of 300,000 acres (121,405 ha) to be planted in a decade. By the time the Forest Service's efforts were curtailed in 1934–1935 some 393,998 acres (159,442 ha) had been planted, 242,600 acres (98,174 ha) being located at Kaingaroa in the central North Island. This plantation was mainly comprised of *Pinus radiata,* a native of California brought to NZ[6] in the 1850s, but also included *P. muricata, P. ponderosa, P. laricio* (AJHR 1929, C3, p. 28).

[6]Until the 1930s locally it where often described as *Pinus insignis*. Monterey Pine, Radiata Pine, or more rarely Remarkable Pine were the other popular synonyms.

9.4.1 Origins of the National Pulp and Paper Scheme

The exotic plantations were initially conceived of as meeting a future short fall of construction timbers and providing time to understand how to implement sustained yield management in the indigenous forests. Other economic possibilities were identified in 1925 when William Adamson, Technical Director of the British papermaking machinery company Charles Walsmley and Sons, saw immature exotic forest plantations planted in previous years. This was the very time that Ellis had launched the vastly increased state planting programme; Adamson announced to his tour guide Entrican that he could foresee plantations as the basis of a future pulp and paper industry. It was the amount of relatively flat land available for further planting at low cost in the central North Island that inspired Adamson. His presence in NZ was serendipitous, for he had been travelling to Japan to install a paper making machine and had detoured to see the state and company plantations. Adamson's boldness lay in believing that the resinous *Pinus radiata* would, like southern pines in the US, be able to be manufactured into pulp suitable for newsprint. Much later Entrican explained Adamson's insight in more analytical terms:

> Adamson developed a theory that as the sapwood of all pines (*Pinus* spp.) had a lower resin content than the heartwood of all spruces (*Picea* spp.), firs (*Abies* spp.), and hemlocks (*Tsuga* spp.) from which the world's supply of newsprint was made, it should be possible to develop a pine newsprint industry in New Zealand. He thought because *P. radiata* grew sapwood rapidly and in great volume, and because ample supplies of cheap hydro-electric power and water were available, production costs would be low enough to allow such an industry to compete on world markets. (Entrican 1963, p. 15)

In 1926, Ellis laid out his ideas about pulp and paper production in NZ in a memorandum to the Commissioner (i.e. Minister) of State Forests. In it Ellis highlighted the quantity of imminent first thinnings that would be available from the state plantations. Several factors would create this situation, including a lack of traditional post and firewood markets because of the distance of the plantations from centres of population and the considerable volume of mill waste from the indigenous forests. As a result, he proposed that a pulp and paper industry be established in the Rotorua area (Ellis to Commissioner of State Forests 1926). Saw millers cutting from indigenous forest wasted 60% of the log in 1923 (Entrican 1923). Ellis also observed that, at a consumption rate of 55 lb per capita, NZ was the third highest per capita user of paper products in the world behind only the US and the UK. Having depicted the imminent first thinnings as a problem, he deftly guided his minister to the view that wood pulping was the economic solution. He asserted that there was sufficient raw material and demand to justify production. In particular the best of the available pulping technologies needed to be identified and semi-commercial trials were desirable. To realize these aims he recommended that funding be provided to send Entrican to Australia, specifically to study the work that had been done there on pulping its native tree species.

Ellis had accepted Adamson's vision and converted it into a concrete proposal, but his involvement in this process would soon end. He resigned in 1928, so that the mantle of bringing a large-scale pulp and paper plant to fruition would ultimately fall to Entrican himself, something he realized on his return from overseas in 1929. Entrican remained as Engineer in Forest Products but assumed greater responsibilities as the 1930s unfolded until he became Director of Forests in 1939. Thereafter he championed a 'national pulp and paper scheme' and later the 'Murupara Scheme'. This was a task that would occupy him for the better part of 25 years. From a company perspective Entrican was regarded as seeing himself as 'guardian of the entire grand plan and [he] was deeply sensitive to any criticism of it genuine or imagined' (Parker 1994, p. 69). In the US, at the same time there were similar visionary figures, such as chemist Charles Herty, who championed the pulping of southern pines, and Ernest Kurth, founder of Southland Paper Mills in Texas (Oden 1977).

Back in 1925 Adamson had maintained that the industry should not be prematurely developed and more controversially, that the state and the companies ought to cooperate in supplying a single, large-scale plant that would be able to compete without subsidies in world markets (Entrican 1955). There were also technical challenges to be overcome relating to the commercial feasibility of pulping *Pinus radiata*. Everything depended on overseas facilities and expertise. For instance in Australia, which had its own *Pinus radiata* plantations, tests conducted by L. R. Benjamin of the Commonwealth Scientific and Industrial Research Organisation (CSIRO) demonstrated that *Pinus radiata* could make acceptable newsprint (Benjamin 1928). The NZ Forest Service then decided to conduct large-scale testing at the Forest Products Laboratory in Madison, Wisconsin. These were carried out under Dr. C. E. Curran in 1928 and were sufficiently promising for Entrican to recommend further commercial trials in the USA. The official report confidentially noted that these trials revealed it was feasible to produce newsprint pulp from a mixture of 70% mechanically pulped *Pinus radiata* and 30% sulphite *Pinus radiata* pulp, though it was noted that the latter required 'rather careful control'. In addition, kraft pulps were produced from various pines grown in NZ, and it was ascertained that bleached pulps for book paper were 'easily produced' from *Pinus radiata* using the sulphite process and from *tawa* (*Beilschmiedia tawa*) using soda and sulphite processes (AJHR 1928, C3A, p. 29). The new "semi-kraft" process did not produce paper of suitable quality or strength, although with beating it was considered potentially suitable for making insulating boards.

Entrican corresponded further with the Madison Forest Products Laboratory and Curran on wood pulping developments until at least 1939. Some of this more technical correspondence recognised that the situation was more complicated than the 1928 report had suggested. In 1932, for instance, Entrican wrote to the Director of the Forest Products Laboratory to the effect that, 'here in New Zealand where the proposal is to use pine ground wood, the problem is altogether different, the insignis pine yielding a mechanical pulp which, when mixed with even the whitest of chemical pulps is very dirty and decidedly off colour as compared with the Canadian and British newsprint in current use' (Entrican to Winslow 1932). In 1933 in a letter marked 'Private and Confidential' addressed to Herty, Entrican expressed a rare moment of doubt and anxiety about Adamson's vision. With respect to the colour problems, particularly involving groundwood pulp despite bleaching it, he explained that, 'My basic fear is […] of course, the danger of resin accumulation on the wire with continued running. The argument has been adduced that the resin content of sap pine is as low as that of spruce, but this does not necessarily mean that it is either of the same physical or chemical character or that it will behave the same as that of spruce.' (Entrican to Herty 1933) The preceding concerns aside, again in retrospect, with *Pinus radiata* being native to California and the considerable US research efforts to work out how to make wood pulp from the resinous southern pines, it is clear why the Forest Service and afforestation companies looked to North America for technological answers.

The Forest Service also gathered first hand advice for a large scale, expensive and technically challenging venture from visiting international experts. For instance in 1935, Mr. Holger Nystén, Assistant Manager of the Finnish Pulp and Paper Association, in NZ on a world tour looking for business opportunities for Finnish paper products, held discussions with NZPM but also called on Entrican. Nystén considered that the small size of the local market—NZ's total population was only 1.5 million in 1935—would make it impossible for a domestic pulp and paper mill to produce economically the range of products the country required. It would, he pointed out, require mechanical, sulphite, and sulphate processes and any overseas markets would remain dominated by the larger exporting countries (Note for File—Holger Nysten 1935). This advice reinforced the cautious approach of the Forest Service to local pulp and paper manufacture.

9.4.2 *Whakatane Board Mills*

Nystén was particularly damning of the forest utilisation plans being put forward by what became Whakatane Board Mills (WBM). The predecessor company to WBM, which dated from 1927, was the Australian registered, Timberlands New Zealand Ltd. It was one of the larger afforestation companies raising money by some questionable means through the unregulated selling of 'bonds' until the activities of the sector were criticised by a Royal Commission of Inquiry in 1934 and planting ceased (Roche 1990). Timberlands New Zealand Ltd. purchased and by 1927 had planted the 41,000 acre (16,592 ha) Pukahunui No. 2 Block in the Bay of Plenty. By 1931 through a subsidiary it planted a further 20,000 acres (8,094 ha) nearby at Matahina and in 1933 acquired another 7,000 acres (2,832 ha) of indigenous forest. The company changed its name to Timberlands Wood Pulp Ltd. in 1929 and then to Whakatane Paper Mills a few years later. After the bond selling companies were legally obliged to reorganize themselves into more conventional joint stock companies, it was renamed Whakatane Board Mills in 1935, a move that signalled a decision to concentrate on [card]board production rather than newsprint and other higher value papers.[7]

[7]In the southern US paperboard production was relatively unaffected by the cyclic fluctuations of newspaper production (Oden 1977). WBM officials had visited the Madison Forest Products Laboratory and other plants and kept themselves apprised of US developments. Even if the connection

WBM's forestry consultant Hugh Corbin, formerly Professor of Forestry at Auckland University College (1925–1934), was well placed to assist in this regard. He had inspected a pulp mill at Mannheim in 1911 while completing practical work in German forests after finishing his forestry degree at the University of Edinburgh. For a decade he was head of the School of Forestry at the University of Adelaide and consultant to the South Australian Woods and Forests Department. Corbin had collated local material on pulping *Pinus radiata* in 1915 and was familiar with later CSIRO research in Australia. In South Australia Corbin also devised a working plan for Kuitpo Forest that included a 35-year rotation for *Pinus radiata*. While neither a founder of the company nor someone who foresaw the possibilities of a local pulpwood industry, his previous experiences gave him a clearer vision of the future than some other company and state foresters. He was company technical director from 1928 and later a managing director until his death in 1950.

The Forest Service saw itself as the authoritative provider of expert advice on all aspects of forestry and had warned WBM against planting *Pinus radiata* exclusively at Pukahunui. The firm ignored this counsel and the Service was further perturbed by claims of exceptional forest growth rates in the company's promotional material. It was the beginning of a difficult relationship. In 1928 the company announced its plans for producing 16,000 tons of kraft sulphate pulp for export to the US or the UK. By 1935 this had been reduced to 12,000 tons of board products annually, three quarters for the local market and the rest for export to Australia (Draft General Memorandum Re. Whakatane Paper Mills 1935). For the Forest Service these changing plans only pointed, at best, to the firm's incompetence. In fact some of these changes in company policy stemmed from Corbin's visit to Canada and the US in 1931.

As a result of Corbin's visit to Canada and the USA in 1931, a trip that included a stop at the Forest Products Laboratory in Madison and on which he was accompanied by the managing director, Henry Horrocks, and NZ engineer, Ralph Worley, he recommended that WBM switch from making sulphite pulp to sulphate pulp instead. There were four reasons for this change of direction: *Pinus radiata* was more easily pulped by the sulphate process, recent advances in bleaching sulphate pulp made it as suitable for making white paper as that produced by the sulphite process, the small NZ market needed the same plant to produce both brown and white paper (the former could only be made by using the sulphate process), and finally sulphate pulp effluent was less polluting than sulphite effluent (Worley Evidence to Commission of Enquiry 1935). Thereafter the company organised a shipment of *Pinus radiata* logs to Curran's team at the laboratory in Madison for trials aimed at making bleached and unbleached sulphate pulp as well as groundwood pulp. These experiments did not duplicate the Forest Service trials because they focused on producing both kraft pulp and a range of fine papers using bleached pulp. Nevertheless, the trials indicated that a small amount of bleaching was essential to produce *Pinus radiata* newsprint equivalent in quality to that produced in the north eastern US using spruce. Further experimentation showed that bleached sulphate *Pinus radiata* pulp could be manufactured for writing and

was tenuous it does suggest WBM were looking to minimise the risk involved by keeping away from more price volatile paper products.

specialty papers and these were of a quality equivalent to that produced by sulphite processes. Buoyed by the results WBM pushed on.

De Guere and Worley's engineering report for Timberlands Woodpulp Ltd. in 1932 gave a new sense of scale to the enterprise. It envisaged a 26,550 ton per year sulphate plant, which was considered the minimum economic size, along with an 11,000 ton per year ground wood pulp mill. The mill would also employ new techniques. For example, salt cake and chlorine would be made from salt and sulphuric acid, both of which were cheaply available. It was consequently anticipated that 80% of the chemicals would be recovered rather than being discharged. The designated site was on the Rangitaiki River in the Bay of Plenty (De Guere and Worley 1932). What would become the contentious part of the production plan was that it annually required six million cubic feet (169,901 cu m) of thinnings with a minimum diameter of 4 inches (10 cm). This volume of fibre represented a major challenge because it was not immediately available from WBM's plantations. As a result, the firm appealed unsuccessfully for thinnings to the Forest Service, which felt it was being asked to subsidise a commercial venture from forests that had been planted for the national good to offset a future timber famine.[8]

Water was essential to industrial-scale papermaking and state support was required, particularly to permit discharges back into waterways. In 1936 the Whakatane Paper Mills Water Supply Empowerment Bill was passed into law. The legislation permitted the company to take up to 23 cubic feet per second of water from the Whakatane River for industrial purposes. Discharges into the river were subject to the 'offensive trades' conditions of the Health Act, 1920. Legislation was deemed necessary by the company and its investors because the £140,000 project needed guaranteed access to the massive volume of water that was essential to the pulp and papermaking processes. The brevity of the act belied the political scrutiny it received; it was vetted by not one but two parliamentary committees. Examination of the proposal was extended far beyond the issue of water extraction and was directed at the financial shortcomings of the earlier bond selling afforestation companies, the economic viability of the project, and the degree to which the state ought to exercise some control over the future development of the industry. On this last point there were serious differences of opinion between the members of the Labour government, elected in 1935, and the conservative National Party in opposition. Questions about the pollution of the Whakatane River were swept aside because of the broader desire to foster local economic development and break free from the impact of the Great Depression (Whatakane Paper Mills Water Supply Empowerment Bill 1936). More direct government oversight of all industrial development was to follow.

Broader economic planning developments now further complicated matters for the pioneering pulp and paper makers in NZ. The Industrial Efficiency Bill 1936 in bill form was debated at length in the country's Parliament. It was praised by supporters as 'experimental' and a means of avoiding past problems and deplored and overstated by opposition MP Sid Holland as 'the framework of a Soviet system save that private property is to remain' (Industrial Efficiency Bill 1936). The Act

[8]WBM was eventually able to source thinnings from other afforestation companies.

was intended to enable national economic planning for industry by regulating the mix, concentration, and efficiency of industrial enterprises. The Act established a Bureau of Industry, attached to the Department of Industry and Commerce, with investigatory powers and responsibility for drawing up industrial plans and licensing industries. By 1945 there were 36 licensed industries that spanned the manufacture of cigarette papers through to the taking of oysters from sea beds (AJHR 1945 H44, p. 5). Although not a target of the legislation, proposals for pulp and paper manufacturing necessarily fell under its purview, a situation that proved frustrating to companies and the Forest Service alike.

Eventually WBM was licensed to begin paperboard production. Its mill was largely designed by Swedish engineers from Karlstad Mekaniska Verkstad, who also supplied and installed the mill machinery. A Swede, E. Frastad, was brought over to be its inaugural superintendent and other key positions were filled by Swedes, Canadians, and Americans. In 1939 WBM commenced operations and produced 7,000 tonnes of paperboard. It was noted in the press that 'for the first time in history twelve-year old conifers, comparable in size to pulpwood of eighty years growth in Canada or Sweden, are being logged and converted into ground wood pulp' (Anon 1939). By 1948 its sawmill was operating. In 1949 a semi-chemical pulp mill came into production using sawmill waste and a second board machine was installed in 1955, thereby increasing production from 16,000 to 40,000 tonnes (Anon 1985, p. 24).

9.4.3 New Zealand Forest Products

The largest of the afforestation companies, New Zealand Perpetual Forests, was founded in 1923 and within three years it had planted 46,000 acres (18,623 ha), mainly in the central North Island around Tokoroa. In 1934 New Zealand Perpetual Forests became incorporated as New Zealand Forest Products (NZFP). By 1937 the firm had planted 172,307 acres (69,728 ha) (Healy 1982, p. 85). NZFP's Board of Directors had appointed as its new chair David Henry, a respected member of the Auckland business community untainted by association with the earlier bond selling era. Henry had worked as a clerk in a Scottish pulp mill in his youth. After an initial business failure in NZ, he eventually owned a successful, small plumbing manufacturing and supply business. He was later President of the New Zealand Manufacturers' Federation. As with other figures in the sector Henry was to combine vision, perseverance and determination as NZFP prepared to harvest its plantation forests. NZFP's official history describes him as persuasive but 'not noted for any great warmth or sense of humour but [he] was an intense, driving personality who kept personal relations on a formal level. He had an all-absorbing ambition to succeed in business and his manner with subordinates was often brusque and demanding' (Healy 1982, p. 77).

Some authors have argued that Entrican tripped up Henry's plans for NZFP. For example, Jones (1999, p. 201) describes how Entrican questioned the mix and distribution of NZFP's enterprises and argues that this was one of the 'short comings of industrial licensing' practices. In a small country like NZ, interpersonal relations could become fraught: Henry was described by one business biographer as 'a spiky character [… who] enjoyed the worst of relationships with Entrican' (Parker 1994, p. 75). Even though the Industrial Efficiency Act, was repealed in 1957, pulp and paper was one of three industries that remained licensed (Jones 1999).[9] This was important from an institutional perspective because the Department of Industries and Commerce from 1936 onwards under the Industrial Efficiency Act was now also involved in both licensing and also tendering Ministerial advice on forest utilisation. This meant the Forest Service, just at the crucial point when the forests were nearing maturity, was no longer the only provider of official advice to government on forest utilisation proposals. Under Henry NZFP committed itself to constructing a large pulp and associated saw mills. A pilot sulphate pulp and paper mill was purchased in New York for £25,000 with the intention of bringing it to NZ to process *Pinus radiate*. The company boasted forest assets worth £1.88 m, and it thus set about raising additional funds.[10]

With relations between the companies and the Forest Service strained, NZFP relied on the CSIRO and recruited from them senior Australian technical staff such as Stanley Clarke. Taking advice from Mr Branzell, a Swedish engineer from A.B. Karlstad who was installing machinery for WBM, NZFP decided to sell the pilot pulp and paper mill, still in storage in New York. Clarke's preference was to proceed with fibre board production immediately and delay moving into the production of other pulp and papers because, in his view, the preferred choice of chemical process for making pulp from NZ's *Pinus radiata* was still unclear (Healy 1982). Meanwhile the company established a small sawmill at Waotu in South Waikato in 1939, in order to process thinnings and understand the technicalities of sawing *Pinus radiata*. Clarke was sent to Europe and the US to investigate how insulating board and hardboard were manufactured and how saw mills were integrated into hardboard production. Subsequently Clarke developed a simple process within the mill whereby the lighter bark was separated from the heavier, chipped slab wood. As Henry later explained, 'we milled all the timber we wanted from each log and put the all the slabs through the chipper with the bark on. On being passed through a water bath the bark rose to the surface while the chips sank. Thus we secured relatively clean chips […] I would not care to estimate the importance of this floatation processes in terms of profit or economy of operation but it became abundantly clear that we could use any logs that reached the skids without reference to their sawmill properties' (Henry 1958, p. xix).

Having secured the necessary licence under the Industrial Efficiency Act, 1936, the company began production of fibre board at Penrose in Auckland in 1940. That same year NZFP, not to be distracted from its own processing plans by the possibilities of creating a larger business operation, declined to take up a stake in the now financially

[9]The other two were paua shell manufacturing - paua being a shellfish endemic to NZ, other related species are known as abalone, and pneumatic tyre and tube manufacturing.

[10]This places the actions of NZPM in not pursuing the *Pinus radiata* pulp mill option in perspective—it was beyond their financial reach, they had nothing like the asset base of the NZFP.

troubled WBM (Healy 1982).[11] Meanwhile, Industries and Commerce declined to grant NZFP a license to manufacture pulp and paper. Henry held Entrican personally responsible. Although to meet wartime needs a licence to produce multiwall bags from imported paper was granted, the decision on the pulp and paper license was delayed again in 1943. Henry appealed directly to shareholders to protest to the Ministers and in May 1943 the pulp and paper license was issued, although the earlier multiwall bag license was now revoked. NZPM and WBM as rival companies both immediately appealed the decision to grant the pulp and paper license. The appeal judge indicated that an industrial plan that provided for paper production by both the companies and the Forest Service ought to be drawn up. This decision, although it was not readily apparent at the time, would ultimately be the death knell of Entrican's cherished national pulp and paper scheme.

In 1943 NZFP immediately began planning to build a kraft pulp mill, with a view to obtaining the equipment from Australia (built to Swedish specifications), Canada, and the US, since it was impossible under wartime conditions to purchase it from Sweden. The proposal was for a 5,000 ton per year pulp plant capable of expanding to 10,000 tons annually with the goal of doubling production. An advantage of the kraft process was that it could use both young and fully mature trees without any loss of pulp whereas the sulphite process was more suited to younger trees with little heart wood. The proposed site near Tokoroa was to become known as Kinleith and in 1943 the company envisaged, optimistically as it transpired, that production would commence in 1948 (Healy 1982).

9.4.4 From a National Pulp and Paper Scheme to the Murupara Scheme

Out of Adamson's vision of a pulp and paper industry, the Forest Service developed the 'national pulp and paper scheme', one in which the afforestation companies were unwillingly ensnared from the outset as providers, along with the Forest Service, of pulp logs to a single, state-directed mill. The Forest Service's annual reports concisely trace the consolidation of the departmental appreciation of the future of the pulp and paper sector. In 1937, for example, export markets for forest products were contemplated. Discussion then turned to integrated utilisation, specifically supplying 'a complete range of forest products from firewood to pulp and paper manufactures and thus fulfilling the purpose for which the State exotic forests were originally established', or so it was claimed (AJHR 1937, C3, 4).[12] The theme of 'integrated production' was reiterated in the 1938 report, which asserted that developing new sawing and manufacturing technologies would 'necessitate state ownership' in view

[11]NZFP did ultimately acquire the company in 1960.

[12]Somewhat earlier the Director of Forests had been emphatic that, 'The view of this service is that saw logs will continue to remain for many decades the basic product of exotic forests'. (McGavock [Director of Forests] to Boas [Chief of Division of Forest Products CSIRO] 1933).

of the large volume and many grades of raw material available. Invoking the concept of scale, it was further claimed that 'anything but a State-owned enterprise would prove quite uneconomical to the State, besides involving administrative difficulties which appear insurmountable' (AJHR 1938, p. 6). This statement may be read simply as the Forest Service reflecting the views of the Labour government (which was re-elected late in 1938). At the same, it was not out of keeping with wider societal views and was also consistent with what Entrican had been advocating for some time.

The Second World War slowed company and Forest Service planning for NZ's pulp and paper industry. Entrican's wartime responsibilities as Timber Controller were arduous but did not dampen his conviction about the eventual development of the industry. In 1940, for instance, he noted advances in the US whereby newsprint was now being made from pine instead of spruce, fir and hemlock, observing that, this step 'augurs well for the establishment of an industry in this dominion (i.e., NZ)' (AJHR 1940, p. 29). Entrican was thinking specifically of the Southland Paper Company mill at Lufkin in Texas (Oden 1977; Reed 1995). A NZ pulp and paper industry, he was now convinced, would supply newsprint to the domestic market and Australia. Some fundamental work continued during the war years, for instance, into the incidence and control of sapstains (from fungal infections in the wood) and mould on exotic pulpwood (AJHR 1941, C3, p. 32).

Planning for a State led industry continued, notably with a further report entitled 'The Establishment of a local Pulp and Paper Industry on New Zealand,' authored by V. C. Rapson, a forest engineer, and Entrican. It recommended that the government establish a national pulp and paper mill in the South Waikato. The document briefly reviewed the failed attempts to pulp indigenous timbers and detailed the various companies' pulp and paper plans before outlining a 'national pulp and paper scheme'. The report endeavoured to simultaneously show that plans proposed by private enterprise were likely to be uneconomic while a state led initiative drawing on both the company and state plantation forests could be the basis of a successful national pulp and paper industry. While criticising the past actions of the afforestation companies, the report did offer them room to collaborate with the government, albeit under the latter's control. Companies could contribute in proportion to their planted forest area and, by supplying pulpwood and capital they could participate in profit sharing. Rapson and Entrican further presented the 'national pulp and paper scheme' as a necessary extension of sound forest management practices. They declared that it was 'now clear that the time has arrived when all forests irrespective of ownership must be managed in accordance with acceptable technical standards which ensure not only their safety against fires and disease etc. but their orderly management under proper forest working plans' (Rapson and Entrican 1943, paragraph 11).

To achieve this management goal Rapson and Entrican recommended that the government pass legislation requiring afforestation companies to employ qualified foresters and prepare working plans to Forest Service standards as well as appointing foresters or forest engineers to their boards of directors. This degree of intervention was politically unacceptable even to the Labour government and it is doubtful that there were sufficient qualified foresters available. In any case nothing was really possible while the war continued.

Building on the report by Rapson and Entrican the Forest Service's annual report in 1943 declared that, 'a combined pulp and paper mill with an annual production of almost 40,000 tons is advocated in the Lower Waikato as the most economical unit to establish meantime' (AJHR 1943, C3, p. 3).[13] This sentence was most likely drafted by Entrican and set out government policy on the necessary scale and preferred location of the mill.

Another layer of bureaucracy now became involved in official discussions about the industry in the form of the Organization for National Development (OND). Set up in May 1944, the OND was intended to convert the economy from a war-time to peace-time footing. Maintaining full employment was one of its responsibilities. It was disbanded in 1946 after displaying a combination of paralysis by committee and meddling in operational matters (Baker 1965). Nevertheless, in June 1944 cabinet asked the OND to investigate the national pulp and paper scheme. A small sub-committee comprised of representatives from the OND, Treasury, Works Department, Forest Service, and Industries and Commerce was proposed (Entrican to Director 1944). Entrican did not appear to resist scrutiny of the national pulp and paper scheme by another government committee. In any case the cabinet directive was a step forward, but perhaps he also sensed that the more important contest still lay ahead with Treasury head Bernard Ashwin, over the cost of building a pulp and paper mill (Baker 2004; Easton 2001).

At its first meeting the committee discussed the question of the impact of the national pulp and paper scheme on existing licenses held by NZFP and WBM. The OND representative considered revoking their existing licenses, a move that reflected the centralised command powers of war-time planning, but even Rapson, the Forest Service representative on the committee, argued for a more cooperative approach, citing the example of the Public Utility Corporation in the United Kingdom (Rapson 1944). Endeavouring to salvage the situation Entrican quickly proposed that WBM and NZPM be encouraged to concentrate on their highest quality products after which, unlikely as it seemed he believed they would voluntarily surrender their licences to produce chemical pulp and paper.

Adamson travelled from Australia to appear before the OND committee and answered many technical and economic questions that were of concern. In considering the current and anticipated demand for paper, the OND noted that *Pinus radiata* alone could now provide a pulp suitable for making paper and newsprint. In addition, the OND favoured an integrated pulp and paper plant so that pulp in slush form could be used for papermaking instead of the more costly process of reconstituting dried pulp. Moreover, hydro power options existed on the Waikato River adjacent to coalfields. River pollution was acknowledged as being a problem resulting from the sulphate process, but it was noted that the volume of toxic discharge had been greatly reduced by recent technical developments while the by-products of the sulphite process could now largely be converted into saleable chemicals. Hopu Hopu on the Waikato River 10 km south of Hamilton was the proposed mill site. This

[13] A 40,000 ton plant was double the size of the company proposals.

choice was the outcome of estimating the cost of transporting state and company logs from forests to mill sites and paper from the mill to Auckland City and its port.

The OND committee's minutes and subsequent correspondence also expressed some clear views about a new structure for the industry. Entrican wrote to the Chair of the OND early in 1945 defending his existing proposals. 'As regards the powers the government to enter into such an undertaking [building the pulp and paper mill]', Entrican declared, 'I explained that this was already covered by the Forests Act 1921–22, but admitted that special legislation would probably be required in order to give effect to the almost unanimous opinion that the organisation should be built up on company rather than departmental lines' (Entrican to OND 1945).

While the presumption was that all the capital would be provided by the State, it was 'generally agreed that the industry should not operate as a government Department, even though financed by the state. A broad supervision of state interests in connection with the under-taking would be exercised through the State appointee(s) on the board' (OND Executive Committee Minutes 1945). The OND recommendations to Cabinet thus moved the national pulp and paper scheme forward but away from the original Forest Service conception of it as a State funded and State run operation. Further changes came in the form of a significant decision on the part of the Forest Service now to locate the pulp mill not adjacent to the Waikato River but instead beside the Tarawera River. This decision marked the transition from a 'national pulp and paper scheme' to the 'Murupara Scheme' because the identification of the new location marked final acceptance by the Forest Service that NZFP would be pursuing its own utilisation trajectory. More importantly, it also foreshadowed that the Murupara Scheme would not be a departmental undertaking. The Labour government's Commissioner of State Forests was also in favour of company control of the processing plant. A change of government late in 1949 saw Labour voted out after 14 years and a centre right National Government elected under Sid Holland, who had in 1936 been so very critical of the Industrial Efficiency Act. Holland was supportive of a pulp and paper utilisation project but neither of the original 'national pulp and paper scheme' nor of processing under foreign ownership.

Adapting to new political circumstances after a change of government, Entrican in his Forest Service annual report for 1950, under the heading of the Murupara project, returned to the initiative's technical challenges. He stressed how an integrated sawmill and pulp mill would combine a groundwood mill for the logs without heartwood and a sulphate plant to handle the crocked logs and those that could not be milled into lumber. Significantly he also referred to commercial trials undertaken for the Forest Service at Southland Paper Mills in Lufkin, Texas, in 1949, which had been organised before the change of government but not undertaken until National was in power. These tests had aimed to address the technical issue related to treating the resin in *Pinus radiata* ground wood pulp so that it did not clog the papermaking machines. Southland had participated on the basis that the testing would be kept confidential, a pledge that was important from Southland's perspective. As Kurth explained to Entrican, the former anticipated other pulp and paper mills to be established in the southern US in the next 15 years and 'I am really of the opinion that we would not care to go in for further publicity at this time' (Kurht to Entrican 1949).[14] The tests demonstrated that the *Pinus radiata* newsprint was stronger than its Canadian equivalent, thereby making it possible to run the papermaking machinery at 1,440 feet per minute for a 28 lb sheet or several hundred feet faster than normal (AJHR 1950, C3, p. 13). Entrican considered this manufacturing production advantage would compensate for any inferiority in the NZ newsprint's colour as compared to that produced in Canada or Scandinavia.

9.4.5 The Kaingaroa Tender and Tasman Pulp and Paper

The main difficulty for developing pulp and paper production as Entrican presented it now lay in the 'inherent psychological difficulties hindering public acceptance of the scheme' (AJHR 1950, C3, p. 13). NZFP obviously would have dissented from this viewpoint. Entrican continued to argue that NZ was 'a country of individualists' (AJHR 1950, C3, p. 13), but he pointed to large scale developments in the co-operatively organised dairy industry as a sign that such enterprises could succeed. In a somewhat risky move intended to push the scheme forward, he also commented that, in the near future, 'the forests will have grown too much heart wood and there will be insufficient sapwood for newsprint production' (AJHR 1950, C3, p. 13). But there were also practical considerations; freed from any obligation to take company wood at this late stage the site for the plant was again revisited. The availability of geothermal energy for generating electricity was decisive and a site near Onepu on the Tarawera River was now selected. This site was named Kawerau and was 14.5 miles (23 km) inland.

The NZ government's Kaingaroa forest tender for 23,000,000 cubic feet of logs per year was let in 1951 with the purchaser obligated to establish an integrated sawmill and pulp and paper mill on the site at Kawerau. The log supply was for 25 years with the right to two additional 25-year terms. Sawmill and wood waste would supply the mill furnace and mill effluent was to be bound by the requirements of the Health Act, 1920 (Corbett 1951). The expected production was to be 70,000,000 board feet of sawn timber, 50,000 long tons of newsprint, 10,000 long tons of printing and writing papers, and 25,000 long tons of sulphate pulp.

Although the tender documents were distributed internationally, expressions of interest from British company Bowater and the Florida-based St. Joe Paper Company were not firm offers. Bowater, which was once thought likely to tender, was preoccupied with another pulp mill project and did not submit a bid. The lack of international interest was a disappointment for the National Government (Baker 2002) and probably indicated that foreign companies regarded the project as too risky, involving as it did large scale commercial pulping of an untried species in a small domestic market and distant from overseas markets. There was only one complete application under

[14]Kurth was correct in his assessment—in 1950 the Coosa River Newsprint Company opened in Alabama and in 1954 Bowater Paper Corporation commenced operations in Tennessee (Oden 1977).

the name of TPP, and local construction company Fletcher Holdings was behind it.[15] Managing Director Jim Fletcher had driven the proposal forward, but not before having to persuade his father, company founder Sir James Fletcher, of the potential of the project. Jim Fletcher, Sir James' second son, had risen to head the business and was described as a 'risk-taker'; the family firm had shifted away from large-scale, state housing construction projects as the programme was wound down in the 1950s into industrial construction, and initially Fletchers were only intending to construct the mill (Parker 1994, p. 76). Fletcher only informed the board of Fletcher Holdings at the eleventh hour that the company was not only contracting to build the plant but also to operate the mill. Fletcher made use of earlier connections with American firms Raymond Concrete Pile Company of Delaware and Meritt Chapman and Scott of New York to design the mill. The plan was based on an earlier Meritt Chapman design for the Southland Paper Mill in Texas (Baker 2002).[16]

From a Fletcher Holdings perspective the TPP story has some different touchstones to that of the Forest Service (cf. Entrican 1955; Parker 1994). Ultimately, however, TPP would have a share capital of £1,000,000 from its Tasman sponsors (largely Fletchers) and an equal sum from the NZ government, a public share issue of £2,000,000 and £2,000,000 from Australia and NZ newspaper groups as well as US and UK pulp and paper interests. A loan of £7,000,000 was also secured from the US Export-Import bank. This ownership model was considerably different from that envisaged under the national pulp and paper scheme. The NZ government further assisted the project by constructing infrastructure to support it. This was facilitated by legislation to begin work on a connecting railway and upgraded port facilities at Mount Maunganui. Legislation was also enacted in 1954 to permit the company to draw up to 50 cubic feet per second of water from and discharge mill wastes into the Tarawera River.

Even though the financial arrangements were now in place the project still faced some demanding technical challenges. When it went into operation in 1955 the No.1 paper machine at Tasman, nicknamed 'jumbo', was the third biggest in the world (Parker 1994, p. 101) From a technical viewpoint the Forest Service happily noted that the 'elimination of pitch troubles' from pine-based newsprint had not caused the anticipated difficulties (AJHR 1956, C3, p. 18). By 1956 the mill was operating at 66% capacity. The company's version of the same story was more fraught. Unlike in the Lufkin trials, which used only carefully selected logs, Tasman had to cope, on a commercial basis, with logs of various dimensions plus *Pinus radiata* being 'a sappy wood: about 2% of it by weight is composed of resinous substances. The problem was that resins were not removed in the normal ground wood pulping process, and thus the challenge was to find ways to bind the material into the pulp fibres. Anything

[15]The 'Tasman' name is derived from that of Dutchman Abel Tasman was the first European to sight and map part of the NZ coast line in 1642.

[16]Southland Paper Mill executives did not share Fletcher's confidence and expressed the view to Entrican that 'we cannot doubt Merritt-Chapman and Scott's ability as constructors and designers of Kraft mills; however we have a considerable doubt as to their ability to design properly a newspaper mill' (Wortham to Entrican 1952). This knowledge probably added to the later difficulties between Entrican and Fletcher's over the management of TPP.

that does not bind, even in small amounts, turns into the newsprint worker's enemy, a rogue by product called pitch [...] it took some thirty years to reduce the problem of pitch to a mere nuisance. But in the early days it was nightmare' (Parker 1994, p. 105).

The government had three directors on TPP's Board, Ashwin (Secretary for Treasury), Entrican (Director of Forests) and McKillop (Works Department). Sir James Fletcher, patriarch of the construction company, served as its first chair. Fletcher, Ashwin and McKillop were well known to each other. Fletcher had successfully tendered for the contract to construct state houses and was Commissioner of Defence Construction during the Second World War. Even so differences of opinion amongst the government directors and between the government and business directors defined the company's early years (Easton 2001; Goldsmith 2009).

Constructed at a time when environmental concerns were embryonic, and with a license to pollute in the form of the Tasman Pulp and Paper Enabling Act, 1954, the mill breached even these limits. Data from the year 1960 revealed that the mill had exceeded its suspended solids limit every year since 1955, with the exception of 1957 (O'Halloran to Minister of Marine 1962). The original standards for toxic substances, temperature and suspended sediments imposed in 1955 were considered inadequate by 1962. The Pollution Advisory Council also imposed new environmental restrictions on TPP in 1961, including limiting discharges of solids into the river to a maximum of 12 tons during two out of every five days and ensuring waste was held in settling ponds for two hours before being discharged (Tasman Pulp and Paper Company Disposal of Trade Wastes into Tarawera River 1961). In contrast NZFP used sedimentation, land irrigation and soakage systems at Kinleith to deal with its effluent, approaches that the Marine Department described as 'excellent and effective' (O'Halloran to Minister of Marine 1962).

In the late 1960s the Kawerau Borough Council complained to TPP about air quality and dust residues, including the hydrogen sulphide smell, even though the town was located upwind from the mill. National legislation, the Clean Air Act, 1972, now affected the company by imposing tighter limits on its operations, including dust and sulphur emissions from the recovery furnace (e.g. 40 kg per hour for No. 1 Furnace). This included a limit of 200 milligrams per normal cubic metre from vents discharging chlorine (Medical Officer of Health 1973). Establishment of the Tasman and NZFP plants marked the first wave of largescale pulp and paper making in NZ. A second round occurred in the 1970s with the Oji CHH joint venture and Karioi mill employing different technologies in 1971 and 1977 respectively. The original two companies were bought out or disappeared in the 1990s (Parker 1994; Roche 1993; Wright 1999). Fletcher Holdings, TPP and Challenge Corporation merged in 1981 to form Fletcher Challenge, which successfully pursued its goal of becoming an international pulp and paper concern. In 1992 it was the sixth largest forestry company in the *Fortune 500* world rankings. Nevertheless, after underperforming for too long, it was broken up in 1999 and Tasman was purchased by Norske Skog in 2000 (Roche 1993; Wallace 2001).

9.5 Conclusion

The historical literature on pulp and paper making in NZ has a business, economic history, and political orientation in which technology tends to play a limited role. In this chapter, in contrast, pulp and paper making technology has been given a more prominent place alongside the forest resource base and institutions. Political and economic factors were not set aside to do so but rather the focus was on highlighting how existing technologies, experimentation and new technologies and their diffusion were integral to the development of a wood-based pulp and paper industry in NZ.

It amounted largely to the diffusion and adoption of industrial processes and organisational practices from both North America and Scandinavia. In terms of timing, the pulp and paper industry in NZ emerged against a backdrop of an international transition from sulphite to sulphate processes. In planning for the actual pulp and paper making plants the Forest Service, and Entrican in particular, were influenced in the long run by Scandinavian models of integrated sawmills and pulp and paper plants rather than the large-scale, North American industrial pulp and paper mill model. Nevertheless, indirectly they recognized the importance of creating an industry on a scale that would be able to compete in export markets, and thus American engineering expertise was called upon to construct the Tasman plant. Similarly, WBM and NZFP recruited Scandinavian technical experts to get their plants up and running as well as train the local labour force.

To focus only on the roles of the state and companies, as most accounts do, is to overlook the importance of North American pulp and paper making research capacity. The main technological challenge in NZ was that of producing wood pulp from the untried *Pinus radiata*. There was minimal capacity for undertaking these trials locally and so they were conducted initially in Australian and later American laboratories and mills. The success the Americans enjoyed in discovering how to pulp southern pine provided a beacon of hope. NZ thus borrowed American technology that had been developed to overcome resin accumulation when manufacturing newsprint. This goal was finally realized in semi-commercial trials conducted in Lufkin, Texas. One unexpected outcome from these experiments was the discovery of the surprising strength of newsprint made from *Pinus radiata*, which meant earlier plans for complicated mixes of pine and tawa pulp could be dispensed with and which enabled papermaking machines to be run faster than normal. It is still unclear whether the *Pinus radiata* tests in Madison in 1930 provided any encouragement to the American researchers that southern pine could be pulped. The debarking experiments undertaken by NZ Forests Products stand out, however, as a story of local innovation, specifically to facilitate the separation of logs from bark prior to grinding.

The wood fibre sources for the NZ industry were also distinct for another reason. In addition to being plantations that were both state and company owned, these forests grew at a remarkable rate. They were ready to harvest in 25 years. This contrasts with national industries based on slower growing natural forests. Forest age was a factor of increasing importance in other ways for both state and company forest utilisation plans in the early 1950s as the trees reached optimal harvesting. In NZ, a few visionary individuals from both the public and private sectors were involved across the entire spectrum—from planting to processing the trees—as the industry was being founded. It meant that NZ's various pulp and paper projects became deeply personal ones, such that modifications of their plans or delays in realizing these enterprises were deeply felt and sometimes accompanied by recriminations. The roles of Entrican, Henry, and Jim Fletcher have been previously considered but the existing historical work also tends to overlook the failed efforts to pulp indigenous forest species and the pioneering role of WBM in favour of the NZFP/Forest Service dynamic. Entrepreneurship resulted in failure well as success.

Individuals aside, the broader institutional setting was still important. The nineteenth century NZ lacked a financially well-endowed private sector, and so the state, well into the twentieth century, became a major actor in infrastructure and natural resource development. The Forest Service for many years nurtured a scheme for the pulp and paper industry and believed that this agenda fell within its mandate under the State Forests Act, 1921–22. However, the Industrial Efficiency Act, 1936 gave a new role to the Department of Industries and Commerce, and provided a legitimate space for companies to develop their own utilisation plans with the result that different parts of the state apparatus had different visions for pulp and paper making in NZ. The overarching political context also plays a part in the story. In 1928 when the first *Pinus radiata* wood pulping trials were underway and up until 1935, NZ had a conservative government, which was replaced by a Labour government (1935–1949) whose goal was a planned economy with full employment. After 1949, however, a more free enterprise National Government developed the Murupara Scheme in its final form with state wood and a private processing plant.

It is appropriate to give the final word on the relationship between forest resources and technology to Entrican:

> For any natural resource to be successfully developed, its development must have inherent economic and technical advantages. Nothing illustrates this better than the production of newsprint. New Zealand raw material grows many times faster than most of the raw material grown abroad for newsprint production. This means a lower growing cost. It also means that the raw material for any plant of economic size can be grown in an area so much smaller than elsewhere that transport costs are also low. Add to these facts that the raw material makes the strongest standard newspaper in the world; that eventually geothermal power and steam for newsprint production will be generated at or below world parity; and that ample and cheap water supplies are available. New Zealand obviously has here a group of inherent technical and economic advantages which insure financial success once the operating bugs inseparable from any pioneering effort are eliminated (Entrican 1958, p. 34).

This statement most aptly summed up the situation prior and up to the 1950s. That said Entrican's assessment so confidently expressed and forward looking, obscured the natural resource, organisational, technical and financial challenges that faced the state and companies, and the tensions between Forest Service and companies. These challenges have been explored in more detail in this chapter and emphasise that the successful establishment of a pulp and paper industry in New Zealand had to overcome a number of obstacles and was by no means a foregone conclusion.

References

Angus J (1976) Papermaking pioneers, a history of New Zealand Paper Mills Limited and its Predecessors. New Zealand Paper Mills, Mataura

Anon (1939) A great industry takes shape. NZ Financial Times Supplement, 10 Jun 1939

Anon (1985) Whakatane. New Zealand Forest Products News, vol 21, pp 21–24

Appendices to the Journals of the House of Representatives (AJHR) C3 [various years] State Forest; C3A, 1920, Forest Conditions in New Zealand and Proposals for a Forest Policy; C3A, 1928, Pulp and Paper Making; D14A, 1870, Report from New Zealand Commissioners relative to the manufacture of New Zealand Flax; H7, 1871, Report of the joint Committee on Colonial Industries; H44, 1945, Industries and Commerce; H31, 1895, Cellulose Industry

Baker A (2002) Private and public enterprise: Fletcher Construction and the building of New Zealand. Essays Econ Bus Hist 20:39–55

Baker A (2004) Governments, firms, and national wealth: a new pulp and paper industry in New Zealand. Enterpr Soc 5(4):669–690

Baker JVT (1965) War economy, Official History of New Zealand in the Second World War. Government Printer, Wellington

Benjamin LR (1928) Kraft pulp and paper from Pinus insignis. Council for Scientific and Industrial Research, Melbourne

Butler J (1909) The wood-pulp industry. Append J House Represent C4:81–83

Cohen A (1984) Technological change as historical process: the case of the US pulp and paper industry. J Econ Hist 44:775–799

Cohen A (1987) Factor substitution and induced innovation in North American kraft pulping: 1914–1940. Explor Econ Hist 24:197–217

Corbett EB (1951) Proposals for the Sale of Logs from Kaingaroa State Forest New Zealand. Government Printer, Wellington

De Guere, Worley RG (1932) Report by De Guere and Worley on NZ pulp and paper project. Papers of Commission of Inquiry into Whakatane Paper Mills—general. AADY W3564 828 29/5/102 [part 2] Archives New Zealand, Wellington

Draft general Memorandum Re Whakatane Paper Mills (1935) [written by Entrican], Papers of Commission of enquiry into Whakatane Paper Mills—General. AADY W3564 828 29/5/102 [part 2], AADY W3564 828 29/5/102 pt 1 Archives New Zealand, Wellington

Dunn to Prime Minister (18 Sept 1928) C366 527 F1 661 5.20 Pulp and Papermaking—general 1922–34. Archives New Zealand, Wellington

Easton B (2001) The Nationbuilders. Auckland University Press, Auckland

Ellis to Commissioner of State Forests (29 Nov 1926) T1 354 52/539, Paper Manufacture Wood Pulp 1926–1940. Archives New Zealand, Wellington

Entrican A (1929) Paper pulp from New Zealand grown woods. N Z J Sci Technol 11:65–80

Entrican A (1923) Our forest resources. N Z Build Prog 18:175–180

Entrican A (1955) Romance of Murupara: dream of 30 years ago becomes reality this month. N Z Timber J 11:36–39

Entrican A (1958) The state and the development of natural resources. N Z J Public Adm 21:29–36

Entrican A (1963) Influence of forestry on forest policy and forest products trade in Australia and New Zealand. HR Macmillan lectureship address delivered at the University of British Columbia, Vancouver

Entrican A to Director (29 Jun 1944) Policy and administration—Organisation for National Development—national pulp and paper project examination. F1 W3129 19 1/29/5. Archives New Zealand, Wellington

Entrican to Herty C (25 Oct 1933) Miscellaneous correspondence (H) 1920–1937 FW607 11c. Archives New Zealand, Wellington

Entrican A to Organisation for National Development (22 Feb 1945) Policy and administration—organisation for national development—national pulp and paper project examination. F1 W3129 19 1/29/5. Archives New Zealand. Wellington

Entrican A to Phillips Turner E (16 Jan 1929) Wood Pulp Westland Wood Pulp and Paper Syndicate. F1 570 47/6/2. Archives New Zealand, Wellington
Entrican A to Winslow CP Director, Forest Products Laboratory Madison (21 Jul 1932) Miscellaneous correspondence (Madison) 1929–1939, FW607 3 3d. Archives New Zealand, Wellington
Goldsmith P (2009) Fletchers, a centennial history of Fletcher Building. David Ling, Auckland
Gould JD (1982) The Rake's progress? The New Zealand economy since 1945. Hodder and Stoughton, Auckland
Guest M, Singelton J (1999) The Murupara Project and industrial development in New Zealand 1945–1965. Aust Econ History Rev 39:52–71
Hansson A to Director of Forests (9 Jan 1929) Wood Pulp Westland Wood Pulp and Paper Syndicate, F1 570 47/6/2. Part 2, Archives New Zealand, Wellington
Healy B (1982) A hundred million pine trees. Hodder and Stoughton, Auckland
Henry D (1958) Integrated utilization of the exotic forests of New Zealand. APPITA 11:xviii–xxi
Industrial Efficiency Bill (1936) New Zealand Parliamentary Debates, vol 247. Government Printer, Wellington, 2R, Hof R, 523–567
Jones SRH (1999) Government policy and industrial structure in New Zealand, 1900–1970. Aust Econ History Rev 39:191–212
Kurth E, Entrican A (4 Nov 1949) The Murupara Scheme—techniques general correspondence with Southland 1950. F1 W3129 344 de 80.761. Archives New Zealand, Wellington
McAloon J (2013) Judgements of all kinds, economic policy making in New Zealand 1945–1984. Victoria University Press, Wellington
McGavock AD, Boas IF (18 Oct 1933) Entrican AR Correspondence Re 1928 American and Australia Trip W2479 FW2479 2/ 0/23/0/1. Archives New Zealand, Wellington
McGaw J (1985) Accounting for innovation: technological change and business practice in the Berkshire County paper industry. Technol Cult 26:703–725
Medical Officer Ministry of Health 1973 Tasman Pulp and Paper. Air pollution—Work—Auckland Tasman Pulp and Paper 1967–1974. AAFB W4349 181 340-381 part 2. Archives New Zealand, Wellington
Nelson CA (1967) Born and raised in Madison: the Forest Products Library. For Hist Newsl 11:6–14
NZ Flax Pulping Plant (14 Mar 1941) Timber utilisation Wood Pulp and paper investigations—New Zealand Pulping Mills. F 1 W3129 47/6/7 Archives New Zealand, Wellington
New Zealand Official Year Book (1928) Government Printer, Wellington
New Zealand Trade and Shipping Statistics (1925) Government Printer, Wellington
Note for File—Holger Nysten (6 Aug 1935) Timber utilisation—wood pulp and paper investigations—New Zealand Pulping Mills. F1 W3129 47/6/7. Archives New Zealand, Wellington
O'Halloran G to Minister of Marine (11 Jun 1962) Water classification—Tarawera River 1969–1983. ABQU 632 W4452 631 126/2/1. Archives New Zealand, Wellington
Oden J (1977) Charles Herty and the birth of the Southern newsprint paper industry, 1927–1940. J For Hist 21:76–89
Organisation for National Development, Executive Committee Minutes (22 Feb 1945). Policy and Administration—Organisation for National Development—National Pulp and Paper Project Examination. F1 W3129 19 1/29/5. Archives New Zealand, Wellington
Parker S (1994) Made in New Zealand. The Story of Jim Fletcher. Hodder and Stoughton, Auckland
Parker S (2000) Cutting Edge. 100 years of innovation leadership and sheer grit. The Carter Holt Harvey story. Penguin, Auckland
Rapson VC, Entrican A (1943) The establishment of a local pulp and paper industry in New Zealand. PR [Permanent Records Collection] 347. [Author's Collection—Ex-NZFS Library]
Rapson VC (1944) Director of Forests Note 16 Aug 1944. Policy and administration—organisation for national development—national pulp and paper project examination. F1 W3129 19 1/29/5. Archives New Zealand. Wellington
Reed M (1995) Realization of a dream: Charles Herty and the South's first newspaper mill. For Conserv Hist 39:4–16

Richter AF (1928) Report covering investigations relating to application of New Zealand timber for pulp and paper. Timber utilization—wood pulp and paper industries—economic pulp wood survey Westland. F1 W3129 259 47/6/2/part 3. Archives New Zealand, Wellington

Robinson RG (1915) Report on forestry conference in Scotland and through America and Europe. Append J House of Represent C1B:34–53

Roche M (1990) The New Zealand timber economy 1840–1935. J Hist Geogr 16:295–313

Roche M (1993) Geography, internationalization and the condition of forestry in New Zealand. N Z Geogr 49:23–31

Smythe R (1968) The pulp and paper industry in New Zealand and its future. APPITA 22:9–15

Tasman Pulp and Paper Company Disposal of Trade Wastes into Tarawera River (1961) Paper manufacture—Wood Pulp 1949–1961. T1 512 52/539. Archives New Zealand, Wellington

Wallace B (2001) Battle of the Titans: Sir Ronald Trotter, Hugh Fletcher and the rise and fall of Fletcher. Challenge, Penguin Books, Auckland

Whakatane Paper Mills Water Supply Empowerment Bill (1936) New Zealand Parliamentary Debates, vol 246. Government Printer, Wellington, pp 715–728

Wright M (1999) Working together—the history of Carter Oji Kokusaku Pan Pacific Ltd 1971–1993. Pan Pac Forest Industries, Napier

Wood-Pulp Regulations under "The Land Act 1908" (1909) N Z Gaz 29:970–971, 8 Apr 1909

Worley RP (1935) Evidence to Commission of Inquiry, 1935. Papers of Commission of Inquiry into Whakatane Paper Mills—General. AADY W3564 828 29/5/102 [part 1], Archives New Zealand, Wellington

Wortham RW (1952) (Executive Vice President of Southland Paper Mills) to Entrican A 15 Jan 1952. The Murupara Scheme—techniques general correspondence with Southland 1950, F1 W3129 344 de 80.761. Archives New Zealand, Wellington

Chapter 10
Making Paper in Australia: Developing the Technology to Create a National Industry, 1818–1928

Gordon Dadswell

10.1 Introduction

This chapter analyses the factors that led to the establishment of the papermaking industry in Australia. It started out as a cottage industry that initially used rags, waste paper and straw as its raw materials. With the application of science, however, it moved towards the use of indigenous Australian hardwoods, specifically the eucalypts. Central to the growth of this industry, and to the story told in this chapter, is the concept of industrial efficiency, a term that emerged in the United Kingdom in the late nineteenth century in the context of the failure of the British army in its war in South Africa between 1899 and 1902. The upshot saw Britain recognize that the national focus on personal liberty needed to shift towards a German model. This prototype included a model army, social insurance and a highly structured education system that was connected with science-based industries. The discourse within Britain during the war was that any advantage achieved by the United Kingdom would be lost if the German model was not followed (Searle 1971).

What followed was a concept that would bring together industry and science throughout Britain and Australia. The focus was on improving the processes used by industry and thereby strengthening the national economy. The efficiency movement arrived in Australia in the early twentieth century (Hagelthorn 1915). Of specific importance to this chapter was the shift in thought by researchers towards using science in three Australian national forest products laboratories in the production first of pulp and subsequently paper. This chapter tells this story.

It is framed by four periods of Australian papermaking. First, the years 1818–1917 witnessed a variety of materials being used as raw materials, including imported and locally sourced waste products and imported pulp and paper. In the second period,

G. Dadswell (✉)
School of History and Philosophy of Science, University of Melbourne, Melbourne, VIC, Australia
e-mail: gdadswell@dcsi.net.au

T. Särkkä et al. (eds.), *Technological Transformation in the Global Pulp and Paper Industry 1800–2018*, World Forests 23, https://doi.org/10.1007/978-3-319-94962-8_10

1918–1922, the first trials to manufacture paper from eucalypts were conducted in a very primitive laboratory, and this story is told in some detail. The third section, 1922–1923, saw a shift to conducting research that involved the use of commercial mills, and the fourth section covers the years 1924–1928 and discusses the beginning of a truly national paper industry.

10.2 Colonisation, Federation, Tariffs and Early Papermaking in Australia, 1788–1917

Australia was colonised by the British beginning in 1788. The first settlement was established at what became known as Sydney Cove. A total of six colonies were created in the early nineteenth century along with the Northern Territory that was part of the colony of South Australia. Around the year 1818 paper mills were established through the eastern and southern colonies, with the first erected in Sydney; it closed in 1834. However in the late 1860s new mills were erected in Liverpool, New South Wales and South Melbourne in the colony of Victoria. Additional mills were built in Victoria around the 1890s, and one of these, the Fyansford mill located in the rural city of Geelong, was significant and will be discussed further (Donath 1957).

On 1 January 1901 these colonies and the Northern Territory became the Commonwealth of Australia and the colonies were renamed States. The Australian Capital Territory was created out of the state of New South Wales (Macintyre 2001). One of the first pieces of legislation placed before the inaugural national parliament was a Bill relating to customs duties; it was assented to on 16 September 1902 (Customs Tariff Act 1902). The act established a schedule of fees for import duties that included a section relating to paper and stationery. The schedule listed various types of paper that would be taxed, but it initially exempted pulp and paper shavings and recycled paper used for papermaking. The Act was amended in 1908 (Customs Tariff Act 1908) when the schedule was expanded to include the previously exempt products of paper pulp, paper shavings and waste paper for papermaking. This decision was rescinded later that year. However, while the general tariff required that a duty be paid on these items, those manufactured in the United Kingdom were exempt from the tariff. The 1921 Act confirmed this situation (Customs Tariff Act 1921). The result of not having a tariff on paper pulp meant that there was a relatively weak incentive to establish a paper industry in Australia.

Additional mills were established in the early 1900s in New South Wales and Queensland. The latter was established at Yarraman in 1913 and aimed to produce pulp from two of Australia's native conifers, *Araucaria cunninghami* and *Araucaria bidwillii*, rather than using the materials mentioned above (Pulp and paper 1988). The pulp from the mill was subsequently bought by the Sydney Paper Mills in New South Wales. The Queensland mill also tested other species of timber including *Aleurites moluccana*, *Eucalyptus paniculata*, *Flindersia Australis* and *Flindersia oxyleyana*. The overall results, however, were not promising. Factors such as the distance from the mill, problems with transporting the logs to the mill as well as the unavailability of labour effectively militated against a sustainable wood pulp industry being founded in Queensland (Lightfoot 1919).

The State Government of Tasmania also sought to establish a papermaking industry. In 1915 it engaged the services of American Henry Surface to investigate the potential of various species indigent to that State for making paper pulp. Surface was a consulting engineer at the Forest Products Laboratory in Madison, Wisconsin, in the United States. The native species he used included *cunninghamii*, and *eucalypti globulus, obliqua* and *regnans*. Surface's report described, in considerable detail, the various processes used to make pulp, in particular the soda technique, as well as providing a useful reference to one of his earlier publications, which was even more specific about methodology (Surface 1914). The report clearly identified what was required to make pulp, namely the requisite volume, concentration, causticity of chemical solution, and the pressure of the cooking process. The tests were undertaken in the Technical School at Hobart, the capital city of Tasmania, using equipment from the school as well that manufactured in Hobart. Surface conducted four tests on each of the species, and concluded that the yield of unbleached pulp produced was as little as 30%, which would limit the commercial viability of a paper industry in Tasmania (Surface 1915).

Surface's studies concerned more than merely making paper using Tasmanian timber. His mandate also required him to consider forest conditions, logging methods, transportation, potential markets, labour and other matters pertaining to the creation of a new industry. He concluded that the manufacture of wood pulp for papermaking was unlikely to be profitable, primarily due to the hardness of the woods tested, their short fibres and the lack of infrastructure in the logging industry (Surface 1915). The results of the investigations were sent to the Advisory Council and widely reported in the Tasmanian newspapers, but not on the mainland (National Archives of Australian (hereafter NAA), A10095, minutes 1916).

A year later, the South Australian government sought to identify potential raw materials that were suitable for papermaking. The underlying reason for this interest was the price of imported paper, which had doubled with the onset of the First World War. The state had no mills and was dependent on imports from other states in Australia and from overseas. The reports of these investigations can now be considered pivotal to understanding the attempts to manufacture paper in Australia in the early twentieth century. The author of the report, William Hargreaves, traced the history of papermaking in South Australia back to 1848. He also discussed in considerable detail various possible raw materials, including sawdust, wood shavings, straw, grasses, rags and waste paper, as well as the current processes that were being used at the time to break down the fibres. He acknowledged the research undertaken by Surface and concurred that, while Surface's tests had resulted in the production of good pulp, economically the costs of production would be uncompetitive when compared with imports (Hargreaves 1916).

Hargreaves' report also included a detailed economic analysis of the practicality of establishing a paper industry in South Australia. His emphasis was of course on the cost of production, however, he found that while producing wood pulp was not an option, making paper from straw would be possible. As there were no mills in South Australia he also identified the capital required to erect a mill and concluded that overall, there was potential for papermaking in South Australia (Hargreaves 1916).

The developments in the State of Victoria took a slightly different direction to those in the other States. In 1917 the Institute of Victorian Industries had sent samples of *Eucalyptus regnans* to Norway for testing using the mechanical grinding technique. The results were poor because the fibres were mashed and therefore useless except for "filling" in a papermaking process that was based on conifers. The secretary of the newly created Australian Advisory Council for Science and Industry would comment in 1918 that the soda process might have produced better results (Lightfoot 1919).

Also in Victoria, as mentioned earlier, a paper mill had been established at Fyansford. The mill, now renamed the Australian Paper Mills Co. Ltd., was interested in manufacturing pulp because it could use the pulp produced in Queensland and add it to the imported material, which would result in the reduction of the amount of imported pulp. The results of this activity were mixed, however, but the overwhelming conclusion was that the cost of sourcing and delivering the products was uneconomical (Benjamin 1923). The mill also tested *Eucalyptus regnans*. While this timber apparently produced good pulp and paper, the firm was unable to access enough raw material to meet the demand for paper and it returned to using imported pulp. Other materials, such as grasses, rush, sedge, prickly pear and palms, were also tested gratuitously by the mill at Fyansford on behalf of the Advisory Council (Lightfoot 1919).

10.3 Australian Papermaking in Perth, Western Australia 1918–1922

The second stage of paper making in Australia was the result of a report from Alphonse Mathey, the Conservator of Forests in Dijon, France. He was visiting Western Australia to attend a Forestry Conference in 1918 and his report described how good pulp could be made from *Eucalyptus globulus* and furthermore, that in tests it had produced excellent paper. The research had been undertaken at the Grenoble School of Paper Making and used eucalypt wood grown in Spain. The report from the laboratory contained two critical pieces of information about pulping eucalypts. The first was to use immature trees which were not more than 25 years of age and that would produce a greater yield of pulp than mature trees and second, that the critical feature for papermaking using hardwoods was not the length of wood fibre but rather the ratio between length and width, which had to be less than 0.02. The report concluded by recommending the sulphite method for cooking the wood (NAA, A8510 34/15).

The need for establishing an Australian paper industry became an imperative following the end of the First World War. The armistice resulted in a serious shortage of raw material, primarily wood pulp, with the result that Australia had to import paper for domestic consumption (Benjamin 1923). What ensued was a concerted approach to establish a paper industry using Australian eucalypts that took place in Perth, Western Australia, beginning in 1918. With regard to this region, Mathey had commented to the West Australian Conservator of Forests about the timber wastage that was occurring in the local eucalypt forests. The State had licensed these woodlands to Millars Karri and Jarrah Forests Ltd., in 1902 and during the harvesting of timber, primarily for export, smaller logs had been left on the forest floor. The Conservator approached the Perth Technical School and two of its staff to carry out tests to determine if the smaller logs could be used for papermaking. He was seeking a more efficient use for them instead of having them decay or get burned in a fire.

The researchers who carried out these investigations were employed by the School. In order to use its laboratory, they had to work when the School was closed. It would appear that the initial tests started in May and were completed by October 1918. The primitive nature of their investigations can be appreciated when it is recognised that they used chemicals and apparatus supplied by the school and wood, immature *Eucalyptus diversicolor*, which was eight years of age, that was supplied by the Conservator. Their aim was ascertain the possibility of producing wood pulp that could be used for papermaking from this eucalypt. The process involved cutting the timber into ½″ sections and then placing the chips in a water oven at 100 °F to obtain the moisture content of the wood before cooking the chips. Before adding additional liquids, they needed to establish the water content of the wood as it varied from tree to tree (NAA, A8510 34/20).

Initially, Isaac Boas and his co-researcher Louis Benjamin, had been completely ignorant of the process required to make paper pulp. Both men were chemistry lecturers at the school, however they were unfamiliar with how to reduce timber into pulp. They certainly sourced a considerable amount of information from international journals as well as reading the reports from Mathey and Surface, which had recommended using the soda process as opposed to the sulphite and grinding processes. One critical piece of information was provided by Mathey, the ratio between length and width (i.e., less than 0.02) (NAA A8510 34/15; Advisory Council of Science and Industry Executive Committee. Paper pulp. Eucalypts and Casuarina. Experiments by M. Mathey Benjamin 1959).

The cooking process used by Boas and Benjamin was critical. Using the soda process, but with an eye to the commercial potential of their investigations, they identified the lowest concentration of soda solution possible in an effort to minimize the loss of weight in the resultant pulp. A loss of weight in pulp meant less pulp, and therefore less paper. They also used the minimum amount of chemicals to achieve that goal, an approach that would result in lower costs for manufacturers (Benjamin 1923). Both researchers were cognisant of the commercial imperative of these requirements.

They carried out a second set of tests in September 1918. The objective was to establish whether older *Eucalyptus diversicolor* would produce an equivalent amount of pulp as immature wood of the same species. The rationale for this test was to compare the results of using different ages of trees with the goal being to identify pulp that would be commercially viable for making paper. The same process described above was replicated, but the trees were fifteen years of age as opposed to eight. The results showed that the percentage of pulp produced was less than that achieved using younger trees, 44% as opposed to 56%, and also that the cooking process consumed a higher amount of soda and bleach, and was therefore more expensive. They recognised that, if they used wood that was eight to 10 years of age, it was viable to pulp *Eucalyptus diversicolor* (Benjamin 1923; NAA A8510 34/20, 1918). A seminal moment in the development of a paper industry in Australia occurred when Boas made some paper from the pulps using a hand frame. He sent samples of this paper to Western Australian and Federal parliamentarians.

The initial success achieved by the researchers was immediately published in various newspapers around Australia. Their results demonstrated that science could be applied to identify the most appropriate method for producing pulp from at least one indigenous hardwood. While the publishing of their findings in a variety of newspapers (Journal 1919; Telegraph 1919; West Australian 1919) was clearly aimed at the paper industry, it was also aimed at announcing to the general public that papermaking using Australian hardwoods was feasible. Just who submitted these almost identical articles is unclear, however it could be assumed that it was Gerald Lightfoot, Secretary to the Advisory Council, and someone who was very interested in promoting the use of Australian woods for paper making. The results were also noted in *Science and Industry* (Activities of the Institute 1920), a publication of the Advisory Council.

Further investigations into other Australian hardwoods was now dependent on the establishment of a forest products laboratory. The formation of such an organisation was crucial because the laboratory in the Perth Technical School was no longer available. The most active impetus for promoting the idea of a laboratory was the interstate forestry conferences. At the first conference, a resolution was passed to the effect that there was a need to investigate ways of identifying methods of converting timber into commercial products in order to reduce the waste of wood (Report 1911). In other words, the goal was to use science for industrial purposes. A following conference also addressed the matter of a national forest products laboratory whose primary role should be to investigate the suitability of Australian timber for making wood pulp for paper (Report 1912). The conferences went into abeyance for several years, however the idea of a laboratory surfaced in the press, particularly in Tasmania (Examiner 1913; Examiner 1914; Mercury 1914). Through 1913 and 1914 these articles canvassed both the leading work carried out in the US by the Forest Products Laboratory in Madison as well as the need to conserve timbers and wood pulp.

The resumption of the conferences in 1916 saw its participants assume a much more assertive posture regarding the need for a national forest products laboratory. Conference papers recorded the need to pursue various investigations in connection with forest products and recommended that special research work in that direction be undertaken. A Conference on Australian timbers also supported the establishment of a laboratory. These events occurred amidst a growing recognition of the need to link science specifically to the timber industry to conduct research into the properties of woods (Carron 1985). The pressure continued until 1917 when it was resolved that an institution for education and forest products should be established (Report of the proceedings 1917).

Also during 1916 and 1917 pressure was exerted on the recently created Advisory Committee, established to co-ordinate the Australian national scientific research. The Committee was contacted by many interested groups including those who sought to establish a laboratory to conduct research on indigenous wood species for papermaking. Many of the enquiries also came from businesses that used timber to make products other than paper. Another pressure group came from within the Council and by members of the public enquiring about the need for research into Australian woods (NAA A10095 1916; NAA A10095 1917). While the dominant issue was still the use of Australian woods for paper pulp, the need for a laboratory was also stressed. The decision to establish a laboratory was made in January 1919 but it was not implemented. To ensure the laboratory was established, the federal government was requested to provide £5,000 that would be matched by the Western Australian government. The proposed arrangement failed, principally because the state government declined to commit such a large amount of money. Nonetheless, the pressure on the Advisory Council and the Federal government from Western Australia had been, and would continue to be, unrelenting throughout 1919 (NAA, A10095 1919), since the initial papermaking investigations carried out in that state had shown their potential to be successful.

Meanwhile the researchers returned to their lecturing duties. With the likelihood of a laboratory still some time away Boas sought, and was granted, leave of absence for six months from his position at the Perth technical school to travel overseas to look at forest products laboratories in the US, the UK, France, Norway and India. He returned in late 1919 and provided a very detailed report to the Executive Committee of the Advisory Committee. While he commented on all aspects of the laboratories he visited, he also identified what research was being undertaken in papermaking. He found that, not only in the US but also in Europe, the idea of using hardwoods for papermaking was deemed to be effectively impossible and of value only as a 'filler' (NAA, A8510 32/39). His report also identified the equipment needed to conduct future investigations in Australia. Boas' investigation and his findings were extensively covered in *Science and Industry* as well as Australian newspapers (Argus 1920; Daily Commercial News and Shipping List 1920; Daily Telegraph 1920; Robinson, Our forest products, Science and Industry 1920).

The motivation for Boas to visit forest products laboratories was clearly the publication of Lightfoot's, *Paper-Pulp: Possibilities of its manufacture in Australia*, in early 1919. This *Bulletin* had been foreshadowed by the Advisory Council and endorsed in May (NAA, A10095, 1917). The *Bulletin* presented a clear message about the need to conduct research into the viability of establishing a paper-pulp industry in Australia, and the utilization of Australian hardwoods. Lightfoot also recognized that this research needed to be conducted in a specialised laboratory, in conjunction with the State forest commissions and the papermaking industry.

In early 1920 a laboratory was finally established in Perth following the creation of the Institute of Science and Industry that had been approved by the Federal Parliament in September 1920. The laboratory was an outbuilding located behind the Perth

Technical School and was named "Forest Products Investigations" and Boas was nominally appointed as Officer-in-Charge. He was responsible for the pulp work being conducted as well as monitoring a chemist who was working on wood tannins at the University of Western Australia. However, having returned from his overseas trip, he was offered an opportunity to work for a private firm rather than head up the nascent laboratory (NAA, AA1964/52 6; NAA A8510 32/40). He agreed to remain in-post for twelve months and his role became more that of an administrator than a researcher.

During that twelve months Boas' central tasks were to obtain money to pay for salaries, acquire suitable equipment and obtain timber from the east coast states. He made contact with the forestry commissions in the states of New South Wales, Queensland, Tasmania and Victoria, all of which contributed money and wood to his cause. As for equipment, Boas sought the assistance of the Australian Paper Manufacturers, mentioned earlier, which supplied autoclaves and a beater. However, the movement of materials and equipment from the east coast to the west was bedevilled by the tyranny of distance. The shipping process took up to three months when it involved transporting goods from Brisbane, Sydney, Melbourne and Hobart to Perth.

The final piece of equipment that was required was a papermaking machine. During his overseas trip Boas had been offered a machine but was refused permission to purchase it. The machine he had seen in England was a model Fourdrinier manufactured by T. J. Marshall & Co. in London. In order to purchase this machine he approached the Perth newspaper companies to enlist their assistance. He was successful in this endeavour, and four companies donated £500 each and the machine was duly purchased (NAA, A8510 34/40; State Record Office Western Australia, 4 November 1914). The reason for the companies' willingness to donate this substantial amount was clear. They had accepted Boas' ultimate research agenda, specifically that Australian woods could be used to manufacture newsprint.

In 1919 with equipment, financing and wood finally lined up, Benjamin established the specific research objectives. Six projects were identified that would use eucalypt species from Western Australia and the east coast of Australia. The projects included investigating soda, sulphate and sulphite processes using small and large autoclaves followed by blending tests and fibre studies (NAA, A8510 34/53). The timbers used were *eucalypti diversicolor*, *maculata*, *marginata*, *pilularis*, *regnans*, *rubida* and a non-eucalypt, *Grevillea robusta*.

The results of the cooking experiments were divided into three sections. First, those using mature *Eucalyptus diversicolor* established that good yields could be achieved and they required small volumes of bleach. The tests on mature *eucalypti diversicolor* and *marginata* identified that mature *eucalyptus marginata* that was older than 20 years could easily be bleached, but if it were not bleached it had limited use due to the wood's reddish colour. The second investigations identified that using very young wood should be excluded as a papermaking material because of its low yield, shortness of fibre and cost of milling. The results from the third investigation, those sent by the New South Wales, Queensland and Victorian forestry commissions, revealed that the best pulps and bleaching was achieved using immature wood. The best pulpwood came from *Eucalyptus regnans*. The others, *eucalypti maculata*, *pilularis* and *rubida*, were found to be difficult to cook and bleach. The tests made on *Grevillia robusta* revealed that it was much easier to cook than the other specimens, and the resultant pulp was easy to bleach. The results of the papermaking investigations followed immediately (NAA, A8510 34/53). Earlier tests had confirmed that the felting characteristics of six of the species would allow them all to produce paper, although their strength was in doubt. *Eucalyptus marginata* was excluded from the papermaking process for reasons of its colour.

Initially, the researchers were very focussed on the goal of producing fine and writing grade papers. They recognised that it was essential to utilise science in producing newsprint and wrapping papers but also that there was an immediate need for high grade papers. To this end they set out with the goal of avoiding using a filler pulp, usually an imported one, and instead addressed the process of beating the pure pulp, which revealed that the process of beating was pivotal to the production of quality paper (NAA, A8510 34/53).

Research which followed also identified that *Eucalyptus regnans* not only produced the best pulpwood, but also the best paper. Their results identified that the pure hardwood sheets were actually stronger, even when not calendared, than a variety of good office envelope and bond writing papers taken apparently randomly from the stationery supply in the laboratory. Benjamin also identified that under different cooking conditions *eucalypti rubida* and *diversicolor* could also produce a sheet equally strong to that of *Eucalyptus regnans*. Further tests were conducted on the strength of the paper. Also, the results were compared against imported papers that were made using the same techniques: sulphite, sulphite and soda (NAA, A8510 34/73). The findings showed that paper from *Eucalyptus regnans* was stronger than the best of the imported papers. Paper from *Grevillea robusta* also excelled.

10.4 Australian Papermaking in a Semi-commercial Mill, 1922–1923

Following the investigations undertaken in the laboratory in Perth, the final stage of paper making in Australia shifted from the laboratory to a commercial mill. The focus of this trial was to ascertain whether results achieved in a laboratory setting could be replicated in a semi-commercial situation and, more importantly, it would bring together science and industry (NAA, A8510 34/52). While it was always intended to test the commercial value of the woods from the three States participating in the project, this shift in thinking moved the investigations directly into the commercial environment, one that used commercial scale digesters, beaters, bleaching containers and a full sized paper machine. The woods used for pulping were *eucalypti delegatensis*, *diversicolor*, *pilularis*, and *sieberi*.

The investigations into the pulping and paper making at Geelong was the responsibility of the Director of the Institute for Science and Industry. The most important requirement was the availability of the commercial mill at Geelong, and the second, to obtain financial support from the Federal government. The request to utilize the mill included seeking permission to erect the pulping plant that was being manufactured in Perth, using steam in the digester, and providing water and directions from the mill as to the process required by the mill for sending the pulp from the digester to the paper machine. The Director also briefed the mill owners about the woods that would be used in the experiments: eucalypti *diversicolor*, *globulus*, *pilularis* and *regnans*. He finalized the arrangements by making explicit that the purpose of this research was to determine the potential of establishing a commercially viable paper industry. The company responded positively to the requests (NAA, A8510 34/57).

The second task, to find the money to pay for the semi-commercial research, was more complex. Benjamin had set out the specific conditions for the research that included the need to follow up the pulping tests by making paper, using the results to inform future pulping procedures, the necessity for papermaking to be directly supervised by the person who oversaw the pulping process, the necessity of engaging in close discussions with the papermaker at the mill and to ensure that previous research carried out in Perth could be validated (NAA, A8510 34/57).

In response the Institute Director contacted the Federal Minister of Trade and Customs, outlining the history of the pulp and paper research. In his submission he set out the details of the financial contributions from the States, the woods that would be tested as well as the reasons for moving the trial to Victoria. The Minister acknowledged the recommendation and then dropped what appeared to be an unintentional bombshell. The Forest Products Laboratory in Perth was to be transferred to Melbourne. Regardless of this move the Federal government agreed to fund the semi-commercial research (NAA, A8510 34/57).

Before it could be undertaken the equipment had to be tested. Benjamin had returned to Perth to expedite the delivery and shipping of the plant being manufactured by the State implement works. However, his report on the readiness of the equipment made clear that the government factory in Perth that was manufacturing some of its pieces had not been completed due to confusion about the specifications Benjamin had prepared (NAA, A8510 34/53). He decided instead to erect the equipment in Perth, and test its functionality before sending it to Geelong. The tests identified that no alterations were required, and the plant was shipped at the beginning of April. With the plant in transit by sea, Benjamin and a laboratory assistant left Perth for Melbourne by rail, a trip that would take four days. As has been mentioned earlier, the tyranny of distance was ever prevalent in early 1920 Australia (NAA, A8510 34/57).

On their arrival in Victoria at the Fyansford mill, Benjamin and his assistant concentrated on erecting additional equipment in addition to that used by the mill. The researchers erected frames for the equipment that was being manufactured in Perth, and also installed steam, water and electrical systems. The paper mill was a working factory that was underutilized and the company had agreed to erect the equipment manufactured in Western Australia so long as it did not interfere with its current operations. Benjamin was able to install the equipment because a lack of orders for fine paper from the federal government caused the mill to run only intermittently. The periodic closure of the mill was to be a continual challenge for the research program (Argus 1922; Geelong Advertiser 1922; NAA, A8510 34/52).

The major pulping occurred between June and November 1922 although the reports did not indicate which woods were used. What is known is that the cooking of *eucalypti delegatensis*, *diversicolor*, *pilularis* and *sieberi* and something named 'mixed Tasmanian Eucalypts' generally proceeded smoothly. However the closures of the mill, a result of a lack of orders from the Federal government for paper products, effectively prevented the use of any equipment such as the paper making machine (Register 1922; Maitland Daily Mercury 1922; Daily Telegraph 1922). Nonetheless the research confirmed the results that had been produced earlier in the Perth laboratory.

It is unclear what Benjamin did when faced with the mill closure. However there were other problems that he had to manage. A critical one occurred when pulp was found to contain dirt from an unknown source, although it had probably come from logs that had not been properly washed. The issue of dirt in the pulp was raised by the management of the mill, who insisted that the pulp must be perfectly clean. Benjamin managed this situation by developing a vibrating screen and improving the washing process. There was also an issue that involved the pulp on the drainers having been left for five days, resulting in it having to be dumped; to prevent future desiccation more attention was paid to the other activities that were taking place in the mill. The researchers also found that bleaching below a temperature of 100 °F was ineffectual, a situation that was easily managed by increasing the temperature. There is little information about the attitude of the mill staff to working alongside researchers, although there is evidence that the staff were uneasy. There was one significant area of tension between the mill's personnel and the researchers, and it concerned the manner in which paper pulp was beaten. Benjamin retrained the "beatermen" to use blunted knives and bedplate and to lightly brush the pulp with a pressure not exceeding 10 lbs per square inch (the Fourdrinier process beats pulp using blades. Because eucalypt pulp's fibres are much shorter than other species, if they are beaten in the usual manner the result is mush. Hence Benjamin's request to blunt the knives). The more usual pressure was 18 lbs per square inch used for longer wood fibres. As expected, Benjamin's methodology resulted in pulp that showed no evidence of the fibres having been cut.

A further challenge had little to do with the direct application of the apparatus. Benjamin's approach for documenting research plans and reports can be viewed as having been random (NAA A8510 34/57). In late September 1922, he recommended that the staff remaining in Perth should move east to assist in the work at Geelong. His rationale for making this request was that the research was behind schedule, and the work that had been undertaken required analysis and publication. Benjamin had not managed these requirements because of his hands-on involvement with the cooks and papermaking. In anticipation of the conclusion of the experiments conducted in Geelong, he contacted the Director about the necessity of approaching the Forestry Commissions to supply information about their forest reserves and thereby establish the cost of production of pulp using their resources. This request was ignored by the Director, who recommended that the work in Geelong should be completed before any attempt was made to publish the results and to identify the potential for a full-scale papermaking industry using indigenous woods (NAA A8510 34/52).

The semi-commercial experiments demonstrated to both scientists and paper mill staff how eucalypt pulp reacted in an industrial context. The researchers learned how a mill operated and some of the difficulties it typically encountered in doing so, such as the lack of orders and the time frame for production. The mill staff also learned about the brushing technique and moisture requirements when using eucalypts.

The paper produced from the commercial process was sent to a local printer as well as being used in three government annual reports. Paper samples were sent to McCarron Bird, printers in Melbourne, who commented that they were equal to those that were currently being imported. The annual reports of the Institute of Science and Industry, the Forestry Commission of New South Wales and Forestry Commission of Victoria were also printed on this paper (NAA A8510 34/47; NAA A8510 34/53; NAA A8510 34/62).

Following these successful projects the focus shifted to the economic potential for Australian papermaking. The state forestry commissions were requested to supply information about the possibilities for establishing paper mills in the country. They were asked to comment on their projected costs (i.e., for chemicals, power, harvesting and delivering logs, fuel), access to fresh water and the potential of their forests to support an industry (NAA A8510 34/50; NAA A8510 34/56). The responses from the states were published as a *Bulletin* in May 1923. It provided information about the possibility of establishing a pulp and paper industry in Australia, as well as including samples of paper made in Geelong.

The *Bulletin* provides an excellent summary of papermaking activities at the time in Australia. The report addressed experimental work, the making of pulp and paper in a semi-commercial context and provided an economic analysis of the potential for manufacturing paper in Australia using Australian hardwoods. The *Bulletin* explicitly addressed the demand for various types of paper and board based on information garnered from the official record, as well as details of import duties applying to paper and stationery. Based upon this information the document concluded that, although Australia's mills were capable of supplying the country with paper board, only 40% of wrapping paper was made locally, and therefore there was potential for its manufacture in Australia. Referring to fine paper that was manufactured using rags, the conclusion was that there was little potential for a local industry. However, Benjamin turned his attention to printing papers, for which he found that the need was substantial and the demand could be supplied from local mills, assuming that the pulp could be economically produced (Benjamin 1923).

The results of the semi-commercial investigations had revealed several characteristics of papermaking in Australia. The increase in demand for paper could be met from the various native species that had been investigated. However the characteristics of eucalypt pulp, which resulted in less shrinkage and subsequently delivered a larger sheet for use, had responded differently to the non-eucalypt pulps (*Aleurites moluccana, Angophora subvelutina, Callitris glauca, Casuarina torulosa, Fagus cunninghamii, Grevillea robusta, Schizomeria ovata, Sterculia discolor, Tarrietia argyrodendron, Tristania conferta*). Also, the research established the viability for the commercial potential of pulp and paper companies in Australia. Finally, the document underscored the beneficial engagement that occurred between a commercial paper mill and a national forest products laboratory and the value of science to improve the efficiency of manufacturing paper in Australia by reducing the need to import paper and pulp.

The *Bulletin* also laid out a plan for establishing a pulp and paper industry in Australia (Benjamin 1923). It included details of forest conditions, manufacturing conditions and cost of production. The final section considered the commercial possibilities that emerged from the research, which revealed an abundance of wood, primarily in Victoria and Tasmania. It also identified the need to reduce the elimination of cellulose when using Australian woods as well as recognising the potential of using domestic pulp to make a variety of papers. Furthermore, the report also reiterated the need for a mill to produce fine papers and newsprint in Australia entirely from chemical pulp.

10.5 Australian Papermaking in a Commercial Mill, 1924–1928

With the publication of the *Bulletin*, the staff moved from Perth to a new laboratory in Brunswick that was owned by the State Government of Victoria (NAA A458 P346/1). The entire pulp and paper section, whose staff had been boosted from three researchers to six, moved to the new laboratory. The foci for the researchers were threefold. First to ascertain whether it was possible to produce eucalypts-based pulp for making newsprint; second, to investigate the potential of producing eucalypts-based sulphite pulp for papermaking; and third, to determine the possibility of using *Pinus insignis* to make kraft pulp and paper.

The focus on newsprint using eucalypt pulp was triggered by a significant increase in demand for it in Australia (Benjamin 1927). The country imported all the newsprint it required and the demand for it had increased substantially from 1921 to 1926, rising from 60,000 tons to 110,000 tons during that period. The researchers used two processes for making newsprint. The first processed sulphite pulp and the second used groundwood pulp. The species used were *eucalypti delegatensis*, *globulus*, *regnans and sieberiana* (Benjamin 1927). It is not proposed to discuss the details of the grinding or cooking process because grinding is self-explanatory and the cooking process for eucalypts has been noted previously. As well as publishing the results in a *Bulletin* the results were also published in Australian newspapers (Examiner 1927; Sydney Morning Herald 1927; Telegraph et al. 1927; West Australian 1927).

The report produced by the researchers on the production of pulp for newsprint was radically different from the earlier one (i.e., in 1923) discussed above. Following the production of the sulphite pulp, Benjamin reported on the feasibility of using Australian pulp as opposed to that supplied from the United Kingdom or Canada (Benjamin 1927). His opening comment established that it would be viable to produce newsprint using 70% sulphite and 30% groundwood from eucalypts. He also identified that the species used in the United Kingdom and Canada were predominantly spruce and balsam fir, which produced a yield of 1,000 lbs of pulp per cord of wood with a delivery price into Australia of 50/- shillings.[1] The results from using *Eucalyptus obliqua* yielded 1,800–1,900 lbs per cord and from *Eucalyptus globulus*, 2,000–2,200 lbs with a delivered price of 25/- shillings. Benjamin compared the cost of producing one ton of sulphite pulp from Australian timber with the cost of using imported sulphite spruce pulp. The latter was £5 per ton for production, but with additional costs including loading, unloading, agent's fees and shipping costs the landed price in Australia was about £10 per ton. Benjamin also identified that a similar figure for sulphite pulp imported from the United Kingdom was £14 per ton (Benjamin 1927). He also observed that, in terms of wood supply, eucalypts grow faster than spruce or fir. With these broad brush reckonings, Benjamin established that it was economically feasible to manufacture newsprint in Australia using eucalypts.

Having provided a brief commentary on the economic differential between countries he detailed the actual costs incurred in Australia. His calculations were based on the production of 100 tons of both sulphite and groundwood pulp. They encompassed three key cost areas: stock that included wood, sulphur, limestone and lime; conversion which identified manufacturing, repairs, wages, coal, wires, screen plates, belting, supplies and replacements; and establishment and overheads including depreciation, supervision, rent, taxes, insurance and miscellaneous items. The total cost for sulphite pulp was £5/17/6 and for groundwood, £3/7/6, for an average of £4/12/6 (Benjamin 1927).

Benjamin then identified the issues that needed to be considered if a newsprint industry were going to be established in Australia. He recognised that one ton of pulp from the United Kingdom, which consisted of 75% groundwood, 25% sulphite and 10% clay, would cost £8/15/0 and for Canadian pulp consisting of 75% groundwood and 25% sulphite, the cost would be £5/17/0. His conclusions from these figures were that the United Kingdom, which sourced its pulp from mills in Norway, was less competitive than Canada. He also observed that while the Canadian product was cheaper than that imported from the United Kingdom, Australian publishers preferred the English product primarily because of the colour and texture of the Canadian pulp which was harder, creamier, and tended to be red in colour and over-blued (Benjamin 1927). Newsprint was traditionally dyed blue and local publishers were accustomed to printing on a paper that had a high degree of sulphite, was clay-filled and was white or slightly blue.

Benjamin also identified the effect of the duty that was applied to newsprint. The product from the United Kingdom was not subject to a tariff while the Canadian pulp initially had a tariff of £3 per ton but this was removed in 1925. As a result, imports of pulp from Canada increased from 2.5% in 1923 to 23% in 1926. However, the key cost issue was the c.i.f (i.e. cost, insurance and freight that was the responsibility

[1] The discussion on economics uses British imperial measurements: lbs = pound; ton; and currency = pounds, shillings and pence—£/0/0.

of the seller). Shipping agreement that resulted in an overall cost of one ton of pulp brought into Australia of £18/5/0. Commenting on this situation, Benjamin believed that it was highly unlikely that the Canadian pulp would be able to sell below £18 per ton. He concluded that if production was based in Australia, and even including c.i.f. for shipping from one Australian port to another, the cost per ton would be £17. Finally he identified that, if pulp were manufactured in Australia, the total margin between the imported product and the local would be £150,000 per annum (Benjamin 1927).

At the beginning of this discussion, it was identified that the detail provided by Benjamin ran counter to previous papermaking reports. What had changed was the nature of the national science organisation from the Institute of Science and Industry to the Council for Scientific and Industrial Research after a Commonwealth Act was assented to in June 1926. With it came a new executive committee, rather than a single director, which resulted in an organisation that was specifically required to engage with both primary and secondary industries in terms of scientific research (Commonwealth of Australia. Science and Industry Research. An Act to amend the Institute of Science and Industry Act 1920, 21 June 1926; Currie and Graham 1966). With a new committee that had as its mandate to engage with industry, research staff were required to work directly with industry, which of course Benjamin was already doing. He was using science to improve the efficiency of a specific industry.

The investigations into grinding and the sulphite process for the production of newsprint were important because of the effect they had on commercial interests. As a result of the research, Benjamin made contact with a private financial organisation known as the Collins House Group. It was effectively a conglomerate of various companies which were instrumental in developing Australia's natural resources, including a newsprint interest from Australian timbers (Cochrane 1980). Benjamin met with the organisation's technical adviser, and travelled with him to the United Kingdom and the Netherlands to test the potential of *Eucalyptus regnans* as a suitable timber for making newsprint pulp (Radford 1979).

His visits overseas were financed by a company supported by Collins House, namely Amalgamated Zinc. It had been approached to consider the potential of using Australian timber for making paper. The arrangement, whereby a public scientist visited another country, paid for by an Australian commercial company, was reflective of just how far papermaking had come in Australia. It was clear that there was an overarching industrial relationship between the Collins House group and the laboratory that recognised the potential of applying the latest technology to make paper using indigenous timbers.

The team returned to work at the Brunswick laboratory to perfect the techniques it had learned overseas. Examples of the paper made in the United Kingdom and the Netherlands, as well as samples made at Brunswick, were sent to senior staff at the *Herald Newspaper* organization in Melbourne. As expected, the company could not distinguish between the papers made overseas and the local product made in the Brunswick laboratory. The researchers were also engaged in establishing a kraft mill with Australian Paper Manufacturers Ltd., which was built in Fairfield, a suburb of Melbourne. The researchers were responsible for erecting the plant and then overseeing the production of the paper using *Pinus insignis* (Benjamin 1959).

To further the research and development of a paper mill using indigenous timbers, adjustments were required to the tariff. During the period 1926–1927 pressure was applied on the Federal government by Amalgamated Zinc to increase the tariff and to create a bounty on the domestic production of paper-grade pulp. The major issue for prospective papermakers who would be using Australian timber for producing pulp was to seek a duty of 40% on pulp, paper shavings and waste paper, which were used for making paper, as well as an increase on newsprint and printing paper that were imported in rolls. They also sought to encourage the government to pay a bounty of £5 per ton for pulp manufactured in Australia (Weekly Times 1925; Examiner 1925). Neither the tariff (i.e. a financial penalty imposed on imported products) nor the bounty (i.e. a financial grant for Australian-made products) were approved by the Australian federal parliament following strong opposition from Australian newspaper companies that argued that the tariff would increase their cost of production (Customs Tariff Act 1926).

With no support from the Federal government, Amalgamated Zinc went ahead and built a test project mill for the purpose of confirming the pulping of eucalypts using the sulphite process (Argus 1927; Telegraph 1927). Research using this process had been carried out earlier using *Eucalyptus delegatensis* in the new Division of Forest Products laboratory created under the Council for Scientific and Industrial Research that was formed in mid-1926 and which superseded the Institute. The results of the research confirmed the investigations undertaken in the Netherlands (Benjamin 1959).

The paper researchers from the new laboratory moved to Tasmania to work at the Kermandie mill. They were still employed by the Australian government but resigned in 1928 when they were formally employed by the Tasmanian Paper Pty. Ltd. However before they left they produced two further *Bulletins* that summarized their work on kraft pulp and paper derived from eucalypts using the sulphite process (Benjamin et al. 1928; Benjamin and Somerville 1928). With their departure the papermaking research went into abeyance in the national laboratory. The lack of research in the Melbourne laboratory discouraged individual state agencies from undertaking papermaking research. It had moved away from a national laboratory into a commercial paper mill.

10.6 Conclusion

This chapter has identified four stages of paper making research in Australia from 1818 to 1928. Underpinning the chapter was the concept of national efficiency which, with the application of science by industry, would produce more products more efficiently. In the first period of papermaking (1818–1917) we saw many forms of fibre being used, ranging from rags to re-cycled newsprint. Regardless, research was carried out by several states, most notably Queensland, New South Wales, Tasmania and South Australia. The work carried out by Surface in Tasmania and Hargreaves in South Australia has been discussed. Surface's research was identified as pivotal to future research because of his attention, testing the papermaking potential of various species and his identification of its financial implications. This period also saw the Victoria State government send wood samples to Norway and the establishment of a paper mill in Victoria that used pulp from a Queensland mill.

The second period discussed ran from 1918 to 1922 when research was carried out in Western Australia. The methodology used in Perth had been strongly influenced by the work conducted by Mathey in France. The chapter has shown that the research conducted in Australia would continue not only because of Mathey's work, but also through the demand for paper following the cessation of the First World War when supplies from Europe were limited. The research in Western Australia has been discussed in detail and included the methodologies used and the age of the various species. The success of this research was evident when samples of handmade paper were sent around Australia. This period also saw the creation of a formal forest products laboratory and the decision to send Boas, the primary researcher in Perth, overseas to assess foreign laboratories. The model he sought for Australia was that of the Forest Products Laboratory, Madison, Wisconsin, US. The chapter has also identified the problems encountered in outfitting a laboratory in Australia.

The third section of the chapter covered the years 1922–1923. It addressed the transfer of the lab from Perth and the use of a commercial mill in Victoria. The move required the development of new equipment and shipping from west to east across the Australian continent. A further challenge has been identified, namely the frequent closure of the mill due to a lack of orders. Once work was re-commenced it has been shown that Benjamin, who was in charge of the operation in Victoria, was required to train the mill staff in how to manage eucalypt pulp.

The final period ran from 1924 to 1928. It witnessed the establishment of a commercial mill in Tasmania for the sole purpose of producing newsprint. In addition, following trials Benjamin formally published the results of his work and established the economic advantages for manufacturing newsprint in Australia from indigenous species as opposed to importing paper and pulp. Also noted were the attempts by the company and Benjamin to realize a tariff on imported pulp and the finished paper product.

This chapter has demonstrated that the application of science could establish a specific industry. The work of Boas and Benjamin, and their co-researchers, was to use science to produce paper. The principle of efficiency is to create a product using the latest scientific knowledge. The chapter has identified that the initial research was random, however that which was commenced in 1918 was more structured. Various technologies have been identified that were used to produce wood pulp. Also discussed was the move to a purpose built national laboratory; identifying specific species of tree, using fewer chemicals, modifying equipment and through national legislation.

References

Activities of the Institute Paper Pulp (1920) Sci Ind 2(1):39–40

Argus [Melbourne], Our timber resources. Investigations abroad. Lessons for Australia, 15 January 1920, 7

Argus, Closing of paper mill, 17 March 1922, 4

Argus, Paper pulp industry. Amalgamated Zinc plan, 4 March 1927, 8

Benjamin L (1959) The challenge of the eucalypts. APPITA Newsl 13(3):90–103

Benjamin L (1923) The manufacture of pulp and paper from Australian woods. Government Printer, Melbourne

Benjamin L (1927) Newsprint. Preliminary experiments on the grinding of immature eucalypts for mechanical pulp and possibilities of manufacturing newsprint in Australia. Counc Sci Ind Res Bull 31. Government Printer, Melbourne

Benjamin L, Somerville J (1928) Paper-pulp and cellulose from the eucalypts by the sulphite process. Counc Sci Ind Res Bull 37. Government Printer, Melbourne

Benjamin L, Somerville J, Jeffreys R, Cohen W (1928) Kraft pulp and paper from Pinus insignis. Counc Sci Ind Res Bull 35. Government Printer, Melbourne

Carron L (1985) A history of forestry in Australia. Australian National University Press, Canberra

Cochrane P (1980) Industrialization and independence. Australia's road to economic development. University of Queensland Press, Brisbane

Commonwealth of Australia (1926) An act to amend the Institute of Science and Industry Act 1920. Sci Ind Res

Currie G, Graham J (1966) The origins of CSIRO. Science and the Commonwealth Government 1901–1926, Commonwealth Scientific and Industrial Research Organization, Melbourne

Customs Tariff Act 1902 (Cwlth) Schedule—Division xiii (Austl.)

Customs Tariff Act 1908 (Cwlth) Schedule—Division xiii (Austl.)

Customs Tariff Act 1921 (Cwlth) Schedule—Division xiii (Austl.)

Customs Tariff Act 1926 (Cwlth) Schedule—Division xiii (Austl.)

Daily Commercial News and Shipping List [Sydney], Forest products. Scientific research, 17 January 1920, p 4

Daily Telegraph [Launceston], Timber resources. Investigations abroad. Lessons for Australia, 21 January 1920, p 3

Daily Telegraph, Paper mills closing, 12 July 1922, p 4

Donath EJ (1957) The Australian paper industry—past, present and future. Aust. Q 29(3):61–68

Examiner [Launceston], Forestry in America. Conservation of timber, 11 October 1913, p 3

Examiner, Wood pulp. A suggested experiment, 28 April 1914, p 6

Examiner, The paper bounty, 3 October 1925, p 4

Examiner, Manufacture of newsprint, 18 August 1927, p 4

Geelong Advertiser [Geelong], Paper mills resume, 2 May 1922, p 4

Hagelthorn F (1915) Introduction. National efficiency. Public Works Department, Melbourne, pp 1–2

Hargreaves W (1916) An investigation into the prospects of establishing a paper-making industry in South Australia, Department of Chemistry Bulletin 1, Adelaide, Government Printer, pp 1–42

Journal [Adelaide], Paper from eucalyptus trees. Important discoveries, 12 March 1919, p 1

Lightfoot G (1919) Paper pulp. Possibilities of its manufacture in Australia. Advisory Council of Science and Industry, Bulletin 11. Government Printer, Melbourne

Macintyre S (2001) Settlement. In: Davidson G, Hirst J, Macintyre S (eds) The Oxford companion to Australian history, rev. ed. Oxford University Press, Melbourne, pp 585–586

Maitland Daily Mercury, Paper mills closing. Imported article cheaper, 11 July 1922, p 7
Mercury [Hobart] Forest products. Important experiments in America, 6 March 1914, p 4
National Archives of Australia (NAA): AA1964/52 6, Summary of proceedings at meeting of the Commonwealth Advisory Council, held in Melbourne, with adjournments, 9–11, July 1917
NAA, A458 P346/1, letter 11 June 1923
NAA, A8510 32/39, report 1920
NAA, A8510 32/40, letter 6 February 1920
NAA, A8510 34/15, report 27 August 1918
NAA, A8510 34/20; report 15 August 1918
NAA, A8510 34/20, report 26 October 1918
NAA, A8510 34/40, letter 31 May 1920
NAA, A8510 34/47, letter 9 November 1922
NAA, A8510 34/50, letter 23 September 1922
NAA, A8510 34/52, report June 1922
NAA, A8510 34/52, report 30 April 1922
NAA, A8510 34/52, report October 1922
NAA, A8510 34/53, summary 15 January 1921
NAA, A8510 34/53, diagrams 12 January 1922
NAA, A8510 34/56, letter 23 May 1921
NAA, A8510 34/57, letter, 31 January 1922
NAA, A8510 34/57, telegram 10 February 1922
NAA, A8510 34/57, memorandum 14 February 1922
NAA, A8510 34/57, memorandum 6 March 1922
NAA, A8510 34/57, memorandum 23 September 1922
NAA, A8510 34/62, letter 12 July 1922
NAA, A8510 34/73, letter 14 January 1921
NAA, A10095 1916, minutes 13 May 1916
NAA, A10095 1916, minutes 19 September 1916
NAA, A10095 1917, minutes 20 March 1917
NAA, A10095 1917, minutes 1 May 1917
NAA, A10096 1919, minutes 28 January 1919
NAA, A10095 1919, minutes 24 February 1919
NAA, A10095 1919, minutes 10 March 1919
NAA, A10095 1919, minutes 11 August 1919
Pulp and paper (1988) In: Technology in Australia 1788–1988: a condensed history of Australian technological innovation and adaptation during the first two hundred years, 1788–1988. Australian Academy of Technological Sciences and Engineering, Melbourne, p 245
Radford J (1979) David Avery. In: Nairn B, Serle G (eds) Australian dictionary of biography, 1891–1939, vol 7. Melbourne University Press, Melbourne, p 126
Register [Adelaide] Paper factory to close, 11 July 1922, p 6
Report (1911) of the resolutions, proceedings, and debates of the interstate conference on forestry, Sydney, November, pp 2, 3
Report (1912) on the resolutions, proceedings, and debates of the interstate conferences on forestry, paper presented in Melbourne, November, 1912, p 23
Robinson E (1920) Our forest products. Sci Ind 2(2):65–67
Searle G (1971) The quest for national efficiency. A study in British politics and political thought, 1899–1914. The Ashfield Press, London
Surface H (1914) Effects of varying certain cooking conditions in producing soda pulp from Aspen. Bulletin 80, US Department of Agriculture, Washington
Surface H (1915) Feasibility of manufacturing paper pulp from Tasmanian timbers. Parliament of Tasmania, Hobart
Sydney Morning Herald [Sydney] Newsprint. Commercial production. Promising results, 21 June 1927, p 11

Telegraph [Brisbane], Paper pulp. Important Australian discoveries, 18 March 1919, p 3
Telegraph, Making newsprint, 8 September 1927, p 3
Weekly Times [Melbourne], Australian paper. Protection needed, 16 May 1925, p 9
West Australian [Perth], Paper pulp. From Australian material. Important discoveries, 21 March 1919, p 7
West Australian, Newsprint, 7 September 1927, p 14

Chapter 11
The Quest for Raw Materials in the British Paper Trade: The Development of the Bamboo Pulp and Paper Industry in British India up to 1939

Timo Särkkä

11.1 Introduction

Paper is manufactured from fibrous plants, which can be planted or grow naturally under favourable conditions of climate and soil. Today, the typical plants used for paper manufacture are coniferous trees, such as pine, spruce, fir and hemlock, and deciduous trees, such as eucalyptus, with the rest of coming from various non-wood feedstocks. Natural growing reeds and grasses form an important source of fibre in developing countries, where about 60% of fibres originate from plants such as esparto, sabai and bamboo, (Bajpai 2012, p. 7) which are discussed in this chapter.

The properties of paper depend on the colour, length, diameter, flexibility and strength of the fibres used, which is why paper manufacturers tend to devote serious attention to the selection of the best available raw material for the finished paper. While the selection of a papermaking material is based on research done in the field of cellulose chemistry, this chapter highlights that it also involves a function of changes in product market demand, investments, knowledge, technology, the surrounding institutional environment and organisational solutions.

The chapter deals with the relationships between technology transfer, technology leadership, and product variety in the context of the British paper trade. By the paper trade we refer to the manufacture of printing, writing and wrapping paper for both the export trade and domestic consumption. The research focuses on the roles played by the availability of technology, knowledge, capital, and raw materials on the one hand and demand characteristics on the other, within British domestic, colonial and organisational frameworks, and considers specifically the development of the bamboo pulp and paper industry in British India.

The investigation takes its inspiration from the changes in the British paper industry's structure since the beginning of the twentieth century and looks back at its per-

T. Särkkä (✉)
Department of History and Ethnology, University of Jyväskylä, Jyväskylä, Finland
e-mail: timo.sarkka@jyu.fi

T. Särkkä et al. (eds.), *Technological Transformation in the Global Pulp and Paper Industry 1800–2018*, World Forests 23, https://doi.org/10.1007/978-3-319-94962-8_11

formance while it was a leader in world markets in contrast to times when its leading role was challenged by various competitors. Since the birth of the mechanised paper industry in early nineteenth century Britain, the outlook of the global paper industry has changed radically. While Britain was the industry leader during the earlier part of the nineteenth century, toward the end of the century its dominance was contested by producers in Canada, Finland and Sweden. These countries were endowed with abundant supplies of hydro-electric energy, wood as a raw material and efficient transport systems of lakes, rivers and ports, and they thus became the major paper-making players. In the Nordic countries (i.e., Sweden, Finland and Norway), pulp and paper manufacture adopted the character of an export industry, which was built to supply the growing British and global demand for wood pulp and finished paper products.

The lack of indigenous raw materials was an enormous impediment to the British paper industry, whose existence was threatened by increased overseas paper imports, low investments in new plants and machinery as well as low profit margins (Särkkä 2012). By the mid-nineteenth century, increased domestic demand for paper combined with the scarcity in the supply of rags compelled British papermakers to look for new raw materials in the form of tropical and semi-tropical grasses. The requirements of the trade were first met in the shape of semi-tropical esparto grass from Spain, and later also from North Africa. The success of esparto encouraged enterprises in India beginning in the late 1870s, a location that offered great natural advantages for manufacturing paper because of abundant supplies of indigenous tropical grasses such as sabai and bamboo. Although theoretically known since the mid-1870s, only during the 1920s did pulping bamboo started on a commercial scale in India, the first country to use bamboo as the primary raw material anywhere in the world. As it transpired, bamboo raw material provided the basis for the long-term development of the pulp and paper industry in India.

This country is the setting for this investigation, which will follow the business historical tradition of in-depth and rich historical descriptions based on primary sources and secondary literature. In his seminal study *Industry and Empire* (1968), Eric Hobsbawn claimed famously that Britain's relative decline as a leading technological innovator (he was witnessing this process in the 1960s) was due to its early and sustained involvement as a leading international industrial power. To understand why the industrial revolution and many of the technological innovations attendant upon it took place in Britain and not another country, Hobsbawn claimed that we must focus on the 'world' economy of which Britain was a part. In other words, he was referring to the 'advanced' areas of (mainly) Western Europe and their relations with the colonial or semi-colonial dependent economies (Hobsbawn 1999). More recently, an increasingly global perspective in business history has emerged. It advocates the study of the behaviour of firms over extended periods of time and an understanding of the global framework, composed of markets, institutions and organisations, in which this behaviour occurs (e.g. McNeill 1990; O'Brien 2006; Pomeranz 2000). With this case method, it is hoped to provide information that will add to our understanding of the British paper trade's history and possibly extend or clarify the history of papermaking as a whole.

11.2 Quest for a New Raw Material

Britain was the first country in the world to possess the capital, the enterprise and the skill necessary to develop its industrial capacity in the field of mechanical papermaking. This process began when experiments were undertaken with papermaking technology at Frogmore Mill, on the River Gade, near Hemel Hempstead in Hertfordshire, and the adjoining Two Waters Mill, where the first Fourdrinier papermaking machines were installed in 1803 and 1804. The stimulus given by this early mechanisation revolutionised the whole process of paper manufacture, and by the 1860s mechanisation had become widespread in the paper industry (Coleman 1958, pp. 179–183).

The production of handmade papers had been established to satisfy the local demand for paper, but improvements in papermaking technology and building of a network of canals and railroads made possible the production of greater quantities for wider markets. An increasing population contributed directly to the rising domestic demand for paper. The census of 1801, shows that the population of the British Isles was 15 million, whereas sixty years later it was nearly 29 million (Census records for England and Wales; Scotland; Ireland 1861). New patterns of social interactions, increased literacy and heightened social consciousness also contributed to the increasing demand for paper. The extension of education and literature, the need for cheap newspapers and serial publications, the increased demand for writing paper for writing and manufacturing and commercial purposes generally, greatly stimulated domestic consumption. In addition, the growth of the industries of the Second Industrial Revolution increased the demand for papers of all kinds. The mechanisation of industry indirectly gave people and institutions more reason to need paper. The early growth of mass communication through new forms of cheap publications was made possible by mechanical printing and papermaking. The single factor that most contributed to the increased demand for paper was the demand generated by the London publishing industry. Penny and halfpenny newspapers, journals, magazines, reviews and cheap editions of books came within the reach of the very poorest members of society. Without paper manufacture, it would not have been possible to bring many of these cheap publications into existence (Coleman 1958, pp. 210–211).

To meet the increased demand, within less than fifteen years from the abolition of the Customs and Excise duties in 1861, the annual production of paper had more than doubled. The improved productivity, wider markets and the increased demand for papers meant that a sufficient supply of raw materials became essential to the industry. Before the middle of the nineteenth century, the refuse material from cotton, hemp, flax and jute mills formed the principal and almost sole source of fibre. Besides the waste from textile mills, other refuse materials such as thread, string, ropes, burlap, gunny bags, cotton linters and waste paper were pulped. Raw fibres were but little used in paper manufacture, and only small quantities of cereal straws and cotton stalks were used in some grades of paper. Recycled fibres were preferred over raw fibres because the former had undergone the process of semi-manufacture, and thus this material could very easily be reduced by simple mechanical means to a mass of fine fibres, which interlaced and formed a continuous, even-textured web. Old garments were especially valued in paper manufacture because, having been exposed to sun and weather and repeatedly washed, all but the most highly resistant material had already been removed from them. Being essentially pure cellulose, it was possible to treat old garments in a very weak alkaline solution by digesting them for a relatively short period of time with little or no pressure, which greatly reduced the need for chemicals and fuel and therefore kept the costs of treatment low (Herring 1855, pp. 49–55; Podder 1979, p. 92; Routledge 1875).

One of the most pressing problems faced by the British paper manufacturers was failure to supply enough rags to keep up with the increasing demand for paper, and this problem recurred at various times prior to 1861. There were two developments that helped to alleviate the pulp famine. The first was the continued expansion of the cotton textile industry in nineteenth century and the second was an early nineteenth century invention, the introduction of chlorine bleaching, which made coloured rags suitable for papermaking. Nevertheless, Britain continued to need to import 20% of its total rag needs (15,000 tons p.a.) to meet the requirements of its paper industry. By the middle of the nineteenth century, rags were imported to Britain from various ports all over the world, especially from Hamburg. This was not a cost-efficient way to obtain papermaking materials, not only because of the high transportation costs it entailed but also because many nations increased their existing export duties on rags. Owing to increased demand for paper and papermaking materials, the price of rags doubled between 1848 and 1855. In the British paper trade, over half of all running costs were accounted for by raw materials. It was thus regarded essential for the successful introduction of a substitute for rags that the substance should be available in abundance to guarantee the continuity of supply at a constant price (Coleman 1958, pp. 214, 338, Appendix IV; Hills 1988, pp. 128, 131; Dykes Spicer 1907; diagram III, p. 32; Magee 1997; Shorter 1971, pp. 113–115, 139).

In an elementary way, several British papermakers tried out modern processes for the isolation of cellulose from raw fibres. The first and one of the most important attempts was made by Matthias Koops, who started to experiment with wood, straw and vegetable fibres as early as 1800, but only cereal straws proved to have any commercial value in Britain. Being an easily pulped material, the main advantage of straw was that pulping required little energy and thus the need for fuel was low. In addition, since the cereals are rich in carbohydrates with a low lignin content, they provided a good yield with little chemical consumption. In terms of paper quality, the main disadvantage of straw as a papermaking material was that its fibres are short, giving the paper it produced a low tear strength (Herring 1855, pp. 55–56; Podder 1979, p. 60).

In Britain, a sustained supply of straw proved difficult to assure, however, and therefore the use of straw as a papermaking material remained marginal and temporary. Being agricultural residues (i.e. by-products of agricultural operations), the availability of cereal straws was conditioned by seasonal variations. In addition, cereal straws had a great liability in terms of their tendency to deteriorate while in storage. Owing to the consumption of straws for agricultural purposes, the continuity of supply at a constant price was difficult to assure. To economise the use of straw, it was often pulped with rags. Cereal straws and straw pulp were imported from Holland and Belgium into Britain up to about 1860, when the commercial manufacture of paper from straw ceased.

The outbreak of the American Civil War, immediately following the repeal of the Customs and Excise duties, threatened to cause a cotton famine in the British paper mills. It seemed evident that, unless a new raw material suitable for papermaking was speedily introduced, the British paper trade would have been seriously crippled. To address this problem, in 1861 the House of Commons ordered a Select Committee to inquire into the situation. In its report, the committee directed its special attention to the possibility of identifying a new raw material, which could be utilised directly, without having to pass through the process of semi-manufacture (Routledge 1875).

As it transpired, it was esparto grass that first offered a solution to the raw material shortage in Britain. Esparto constitutes two perennial grasses, *Stipa tenacissima* (Spanish grass) and *Lygeum spartum* (African grass), that are both endemic to the Western Mediterranean (growing in Portugal, Spain, Morocco, Algeria, Tunisia and Libya). It grows in clumps grouped in relatively dense formations in dry or semiarid Mediterranean areas (rainfall between 200–400 mm/year), often in poorly developed soils on limestone, and at elevations from 0 to 1,000 m. above sea level (Fajardo et al. 2015). The first papermaker in England, who devoted his attention to the use of esparto in the manufacture of paper, was Thomas Routledge (1819–1887), who begun to experiment with various vegetable fibrous plants at Eynsham Mill, on the River Evenlode, Oxfordshire in the 1850s. Having visited Spain, Routledge entered into contracts with grass harvesters for their produce. After being dried like hay, sorted into different qualities and baled, the grass was ready to be transported to the British markets (Esparto Paper 1956; Hills 1988).

There were two developments in the British paper and chemical industries that alleviated Routledge's experiments with esparto. The first was the introduction of the so called "soda process" by Charles Burgess and Hugh Watt in 1851 at Frogmore Mill. Burgess and Watt used "caustic soda" (sodium hydroxide) for pulping the wood fibres, but in the absence of a readily available wood raw material, there was little financial support in Britain for the development of a pulping process based on wood. In 1854, Burgess took out a patent in the United States, where the first mill, which utilised the soda process, began its operations near Philadelphia in 1855 (The Paper Trail at Frogmore Mill 2013). What Routledge found at the Eynsham Mill was that the soda process was ideal for esparto, which gave easy bleaching pulp in good yield when cooked with caustic soda. Another significant development took place in the 1860s, when the Weldon process of bleaching powder was introduced. It greatly increased the power of production and diminished the cost of manufacturing bleaching powder (CNJIS, 24 December 1875, p. 299). After the introduction of these two processes, the comparative expense of chemically reducing raw fibres was no longer an obstacle to progress. Except for the use of more chemicals, there was little cost difference in processing paper from esparto or from rags.

Up to 1860 Routledge himself was the only paper manufacturer who used esparto, but it did not take long before other British papermakers availed themselves of the new raw material. By 1865, the esparto imports into Britain rose to 50,800 tons and by 1871 to 146,300 tons. The import of esparto reached its peak in 1888, with 249,000 tons compared to 41,000 tons of rags (Routledge 1875; Dykes Spicer 1907; Diagram V, Export of Esparto, pp. 13–17, 35, 89–90; Magee 1997, pp. 118–127; Shorter 1971, pp. 141–142). Under these circumstances, the demand for esparto was not being met, and this led to a rapid extinction of the Spanish grass. Being a wild grass, or, botanically speaking, a sedge, reproduction from a seed was a laborious, slow and costly process. In the 1870s, African esparto (known to the paper trade as alfa) was substituted for the Spanish imports. Most alfa imports came from Algeria, where the French Government had offered monetary incentives and concessions to induce railway communication with the interior district, where the plant abounded on mountainous plateaus. 61,000 tons of alfa were imported into Britain from Algeria in 1874, but the difficulty of procuring labour and the cost of railway carriage added considerable transportation costs to British mills. Furthermore, being a French colony, British producers had little hope of obtaining a reduction in prices. Owing to the increased demand and transport costs the price of esparto almost doubled between the early 1860s and 1875. The real cost of the finished paper was even higher because the value of alfa paper was proportionately lower in the market compared to paper made from Spanish esparto (Routledge 1875).

11.3 Introduction of Wood Pulp for the British Paper Trade

Before the mid-1870s British paper makers had been using a wide variety of raw materials, consisting mainly of refuse from the cotton, flax, linen and jute industries together with amounts of raw fibres, which were mainly obtained from cereal straws and esparto. Wood emerged as the predominant source of fibre in the late 1870s, when international breakthroughs in the production of chemical wood pulp had made the product available for the British manufacturers. Wood pulp began its steady rise to prominence in Britain initially at the expense of esparto but then later at the expense of both esparto and rags (Coleman 1958, pp. 342–343; Hills 1988, pp. 150–153; Shorter 1971, p. 114).

In the absence of readily available domestic wood resources, British manufacturers quickly availed themselves of Scandinavian timber, and there were several reasons for this development. These included its close proximity, the suitability of the product for newsprint and finally the low cost of the product compared to other available materials. Much of Scandinavia was covered in coniferous forests, and primarily two species—Norway spruce (*Picea abies*) and Scots pine (*Pinus sylvestris L.*)—were being harvested. The uniformity of this raw material permitted pulp- and papermakers to obtain a regular product, and because cutting was occurring on a large scale, standard products were available in international markets at very low rates. The process of wood pulping at the time was fairly straightforward. The logs were first debarked and cut in small pieces, and then pounded into fibre by a grinding machine. Next water was added to the fibre and the resulting pulp was cleaned by pressing it through sieves. Then the pulp was graded into equal lengths of fibre, thickened and fed into a machine which spread it into an even layer. The excessive water was removed by means of a high pressure press. Finally, the layer of pulp was cut into sheets, which were pressed into bales with a hydraulic baling press and dried in a machine (The Finnish Sawmilling, Pulp and Paper Industries 1936).

In terms of quality, the product—dry, mechanical pulp—was unsuited to the manufacture of better qualities of paper, as sunlight caused it to fade and change its colour. The product was, however, perfect for the manufacture of newsprint by being mixed with about 20–25% chemical pulp. It was also used in manufacturing board of various kinds as well as cheaper types of wrapping, writing and printing papers. Since these papers have a short life, the paper on which they are printed need not possess either purity or permanence. They must, however, be as cheap as possible.

By the early twentieth century, the British paper industry had grown increasingly dependent on wood pulp imports to maintain its rate of growth. In 1913 the share of mechanical and chemical wood pulp was 79.6% (993,000 tons) of the total imports of raw materials into the UK (1,249,000 tons), while the import of esparto and other fibrous vegetable materials constituted 18% (225,400 tons) and rags a mere 2.4% (30,000 tons). The First World War caused a violent disruption to the British paper trade, but it recovered rapidly during the interwar period. In 1919, the total imports of raw materials were about one million tons but doubled by 1929. The share of wood pulp imports of total imports of raw materials was constantly over 80% (Table 11.1). The Nordic countries were collectively the dominant suppliers of wood pulp to Britain; for instance, they delivered 1,300,000 tons in 1929. In the same year British pulp producers, which sourced their raw material from Norway, stood at 209,000 tons, which represented 13.9% of the total wood pulp use (Statistical Year-book of the League of Nations (LN) 1938–1939).

By 1959, the British papermaking industries imported over two million tons of raw materials. The share of wood pulp had risen to nearly 100% of the total imports of raw materials (2,289,800 tons) (Table 11.1). For instance, within the Bowater organisation, the sources of supply were in Scandinavia for mechanical pulp and in Newfoundland for pulpwood and sulphite pulp (Reader 1981, pp. 209, 245). The early esparto and wood pulp preparation plants established in Britain during the interwar period did not survive the post-war period but were closed down by the end of the 1950s as they could not operate economically (Hills 1988, p. 142). Thereafter, wood pulp was delivered to British mills exclusively from the countries in which the wood was grown,[1] and the paper mills in Britain depended for survival entirely on imports of wood pulp.

[1]The first integrated paper and pulp mill in the UK that tried to utilise domestic wood resources in any significant scale was Fort William of Wiggins, Teape & Co., Ltd. in the Scottish Highlands that went into production in 1966. The venture was economically unviable, and the pulp mill remained operational only for the first fifteen years (until 1980) Owen (2000).

Table 11.1 Raw material imports into the UK, 1913–1959 (1000 tons and %)[a]

Year	Rags	Wood pulp	Esparto	Other fibrous vegetable material	Total
1913	30.0	993.6	208.2	17.2	1,249.0
1919	6.6	952.9	71.8	1.3	1,032.6
1929	20.0	1,664.5	317.7	9.7	2,011.9
1939	15.5	1,147.0	225.0	15.7	1,404.1
1949	20.7	1,326.0	366.4	20.6	1,733.7
1959	na	2,263.6	na	26.2	2,289.8
Total 1913–1959	92.8	8,347.6	1,189.1	90.7	9,720.2
% of total	1.0	85.9	12.2	0.9	100.0

Source Accounts Relating to Trade and Navigation of the United Kingdom (1913–1959)
[a]The figures are calculated from imperial tons. One imperial ton equals 2,240 lb avoirdupois or 20 cwt (about 1,016 kg)

11.4 Bamboo Considered as a Papermaking Material

Bamboo is an old papermaking material that was subject to a new processing treatment at the beginning of the twentieth century. In East Asia, where the plant is indigenous, bamboo had long been used to produce good quality pulp in the production of handmade papers (Herring 1855, pp. x, 22–23, 31–33). In Britain, the increased dependence on wood pulp, the likelihood of a pulp famine, and the consequent increase in the price of imported wood pulp were the factors that raised the possibility of making large volumes of good quality pulp from bamboo. The most significant factor that contributed to undertaking experiments to use bamboo for pulp was the simple fact that the plant flourished in many parts of British India. For the early twentieth century British paper trade, which was nearly crippled by the lack of a readily available indigenous papermaking material, bamboo groves in India provided a seemingly inexhaustible supply of raw material. From the very beginning, it was obvious that it would not be economical to import bamboo in large quantities into Britain owing to the bulkiness of the material and its relatively low value. Furthermore, the plant's tendency to be damaged from fermentation made importing raw bamboo difficult. Even if it were dried, its high cost of carriage made bamboo imports commercially unviable. It was evident, therefore, that to ensure a continuous supply of bamboo, it would have to be reduced into fibrous stock where grown. This development induced export capital into India, where industrial interests had been shunned as inferior to plantation and mining undertakings (Bagchi 1972; see also Cain and Hopkins 1994).

Bamboo is a grass, which can be propagated by seed and transplanting. The life cycle for bamboo varies with different varieties and different climates and soil. In India, *Bambusa arundinacea* has a 12-year reproduction period and a 34-year life cycle. The plant is gregarious and frequently exterminates other vegetation: a clump throws out 15 to 20% new shoots every year. Thus old plants were cut every fifth year (Podder 1979, p. 53). The main benefit of bamboo as a papermaking material is that, under favourable conditions of climate and soil, the plant produces a heavy crop with minimal care and cost. In addition, the plant grows thick in clusters or clumps and not in patches, which makes its collection easy (Watt 1885; Sindall 1909; Raitt 1928, pp. 155–165).

Bamboo's suitability for paper manufacture had been under consideration since 1875, when Thomas Routledge initiated the first serious attempt to utilise it as a regular source of supply for pulp at the Ford Paper Mills at South Hylton, on the River Wear, Sunderland. Routledge published the results of his experimental investigations in two pamphlets, *Bamboo considered as a paper making material* (1875) and *Bamboo and its treatment* (1879). They outlined how he had confirmed that bamboo as a papermaking material had definite benefits over other fibrous plants. The modest experiments preceding Routledge's had failed both industrially and commercially by virtue of bamboo having been collected and treated without regard to its age. Owing to the presence of a large quantity of silica, and the extreme hardness of the stem, the only possible means of converting bamboo into pulp was to subject it to a prolonged digesting process in very strong solutions of caustic alkali at an elevated temperature. By this means pulp had been produced but at a great cost and under perilous conditions (caused by digesting the material under high pressure). The chief breakthrough Routledge made was to reveal that cutting bamboo stems at an early stage of their growth (before the cellulose and lignin have become indurated and silica deposited) would allow it to be processed using only a mild chemical treatment (Routledge 1875, 1879).

Nevertheless, it took twenty-five years before Routledge's experimental work led to the first serious attempts to utilise bamboo for mechanical papermaking in India. In 1900, the Government of India decided to experiment on a commercial basis with using bamboo for papermaking, and requested Robert Walter Sindall, a consulting chemist to the wood pulp and paper trades, to conduct experiments with it in Burma. Sindall's experiments, carried out under the auspices of the government between 1900 and 1904, demonstrated the technical possibility to produce soda pulp bamboo. The results of Sindall's work were published in a pamphlet titled *Bamboo for papermaking* (1909), and the general impression it presented was that bamboo was difficult to treat. The main challenge in digesting bamboo lay in the removal of its starchy carbohydrates, which required a large amount of alkali. Sindall also noted that digestion turned the pulp dark brown, which made it difficult and expensive to bleach. A further challenge was related to making the nodes reducible (i.e., the bamboo stem is a hollow culm with solid joints called nodes). Cutting the nodes out by means of a circular saw was experimented with but it proved to be very time-consuming and costly. Despite these difficulties, in 1908 some eight to nine tons of selected *Bambusa polymorpha* were shipped to Britain and this was converted into paper by Messrs. Thomas & Green Ltd. at Soho Mills on the River Wye, in Wooburn Green, Buckinghamshire. The experiment was most satisfactory. The yield was over 50%, and the use of 8 to 10% bleach gave the paper a considerable brightness.

The manufacturer reported that the raw material worked exceedingly well on the paper machine and produced a very good sheet of paper, capable of withstanding considerable wear and tear, and therefore it was specially suitable for book printing. Some of the paper manufactured at Soho Mills was later sent to the North of Ireland Paper Mill Company for lithographic printing (Sindall 1909).

Encouraged by these positive results, further steps for developing a pulping process suitable for bamboo were taken at the cellulose and paper branch of the Forest Research Institute and College in Dehra Dun (some 150 miles north of Delhi). The history of pulp and paper development work at the Forest Research Institute can be traced back to 1909, when Sir Ralph Pearson, a forest economist, commenced an investigation of several bamboo areas in India and Burma with a view to ascertaining whether bamboo existed in commercially exploitable quantities. Pearson's report established the definite commercial possibilities of manufacturing pulp from bamboo (Pearson 1913). In the following year, W. Raitt, the mill manager of the Bengal Paper Mills Co. at Raniganj, was employed to conduct laboratory experiments into developing a pulping process suitable for bamboo (The Work of the Forest Department in India 1920).

It began with transporting the raw material to the factory. The stems were then passed through heavy crushing rolls in order to split and flatten them, and at the same time crush the nodes. Then the material was tightly packed in the digester where it was cooked for a period of six to seven hours with caustic soda of suitable strength. After the boiling process, the material was put through the washing and breaking engine, which reduced it to pulp. Finally, the thoroughly washed fibre was dried and baled up in hessian for storage or export (in a similar manner to cotton or jute in textile industry). Using this method, Pearson and Raitt succeeded in manufacturing bamboo pulp by the soda process on a limited scale but the process did not work satisfactorily enough for commercial production (London Metropolitan Archives [LMA] 1919), CLC/B/022, BPC, MS28965:1, pp. 241, 281; Sindall 1909; Clapperton and Henderson 1947).

The technology of paper manufacture from bamboo was perfected over the next few years, and this resulted in an improvement in crushing technology and the economic use of chemicals in treating the pulp. A decisive step in the commercial utilisation of bamboo as a papermaking material was taken on 27 July 1911, when Samuel Milne patented a new type of crushing machine that was able to break the nodes without destroying the fibre (Espacenet. British Patent 20560). This meant that bamboo stems could be treated as a whole. The obvious advantage was that it saved the expense of cutting the nodes out, and there was also a saving in raw material. Furthermore, crushing the nodes before digestion allowed for easier chemical treatment, and the period of digestion was shortened owing to the greater speed with which the cooking liquors penetrated the pulp. It also permitted the use of a weaker solution of cooking chemicals.[2]

[2]According to Podder (1979) it was James C. Lowe, who designed the apparatus for node crushing (59).

The next step in the development work was taken on 16 March 1916, when two British papermaking and cellulose experts, James Lockhart Jardine and Thomas Nelson, patented a process for the production of an acid magnesium sulphite solution for the extraction of cellulose from bamboo. The patentees found that the acid sulphite process was cheaper than the alkali sulphate or the caustic soda processes. They also found that the sulphite process produced paper with more enduring whiteness. The acid process was similar to that employed on coniferous wood but instead of calcium bisulphite it used magnesia bisulphite after calcium was found to be too acidic for bamboo (Espacenet. British patent 2509).

Although Routledge, Sindall, Pearson and Raitt had all demonstrated that using a properly adapted treatment made it possible to obtain good quality pulp from bamboo, they received no encouragement from financiers for commercial production. Competing in markets against wood pulp or finished wood pulp paper imports was impossible unless financiers were assured that bulk supplies could be delivered to a mill at rates that would permit the finished paper to be sold at a profit. Raitt's experiments were, however, said to have impressed Lord Minto, the Viceroy and Governor-General of India, 1905–1911. As a result, in 1914 the Forest Department started negotiations with Thomas Nelson, one of the original patentees, to build a mill in Beypore but it appears that no agreement was made. Further investigations into the cost of cutting and transporting bamboo were made in 1917, but these, too, proved to be in vain (Marsden 1922, p. 39). In 1919 the Government of India followed up the initial success of the Forest Research Institute in the production of bamboo pulp. Raitt's laboratory experiments justified setting up a pilot plant, which went into operation using the soda process in 1924 (Bagchi 1972, p. 395; Podder 1979, p. 36).

11.5 Experiments with Bamboo on a Commercial and Organised Basis in India

Mechanical papermaking in British India was introduced in the late 1870s when Britain's increasing demand for paper combined with the scarce supply of rags compelled papermakers to look for new raw materials, specifically tropical and semi-tropical grasses. The success of using esparto in Britain encouraged business interests to try it in India, where indigenous sabai grass (*Eulaliapsis binata*) was the primary raw material of the trade. It is a perennial plant belonging to the family *Poaceae*, and is grown in wide areas in India, East Asia and South-East Asia. The main advantage of sabai as a papermaking material was that its pulping required little chemical treatment to give good quality pulp. Like esparto, sabai yielded easy bleaching pulp in good volumes when cooked with caustic soda. Furthermore, the plant's thin and long leaves possess high-quality fibre, which were ideal for papermaking purposes (Khandual 2016). The extraction of sabai was difficult and cumbersome, however. The grass grew in widely scattered patches, which made it hard to keep collection costs low (Podder 1979, pp. 51–52). By the early 1900s, the known sources of sabai in

Table 11.2 Paper and board imports into British India, 1899–1910 (tons and %[a])

Year	Printing paper	Other kinds of paper	Board (all kinds)	Total
1899–1900	4,854.1	4,353.3	763.9	9,971.3
1900–1901	5,472.0	3,325.0	1,005.8	9,802.8
1901–1902	5,981.5	4,747.1	1,012.0	11,740.6
1902–1903	6,464.8	4,737.2	1,321.7	12,523.7
1903–1904	5,576.0	4,821.5	1,491.3	11,888.8
1904–1905	7,883.3	11,226.3	1,567.8	20,677.4
1905–1906	7,715.1	15,879.6	1,916.7	25,511.4
1906–1907	11,032.0	16,055.1	1,866.3	28,953.4
1907–1908	14,392.3	18,136.5	2,223.4	34,752.2
1908–1909	12,014.3	19,930.8	2,406.0	34,351.1
1909–1910	12,924.5	22,354.2	3,101.5	38,380.2
Total 1899–1910	94,309.9	125,566.6	18,676.4	238,552.9
% from total	39.5	52.7	7.8	100.0

Source Annual Statement of the Sea-borne Trade and Navigation of British India with the British Empire and Foreign Countries 1899–1910

[a]The figures are calculated from imperial hundredweights (cwt). One imperial cwt equals 112 lb avoirdupois (about 50.8 kg)

the sub-Himalayan tracts were already tapped by the existing Indian paper mills. The biggest Indian paper producers, the Titaghur Paper Mills Co., Ltd. (established 1882) and the Bengal Paper Mills Co., Ltd. (established 1887), were forced to transport sabai over long distances inland (from Nepal, the Punjab, United Provinces, Bihar, Orissa, and Central Provinces) to their mills in Bengal, which greatly increased their transportation costs (Bagchi 1972, pp. 395–396, 398).

The gradual expansion of the wood pulp industry in Europe and North America put a stop to further progress of the paper industry based on natural growing reeds and grasses and shifted the focus of interest from tropical regions to Scandinavia and North America. Favoured by seemingly inexhaustible supplies of cheap timber in these regions, not only was the demand for wood pulp met but it was exceeded, thereby leading to the sale of the excess at "dump prices" in the world markets. In 1912–1913, wood pulp imports into British India totalled 13,250 tons (The Work of the Forest Department in India 1920) and comprised mechanical pulp from Canada and sulphite pulp from Sweden. During the period 1899–1910, paper and board imports into British India peaked at 38,380.2 tons (Table 11.2). British imports represented between 35 and 67.4% of the annual total paper and board imports into British India, with the rest arriving from European paper mills on the Continent, particularly from Germany, Austria, Sweden and Norway (Annual Statement of the Sea-borne Trade and Navigation of British India with the British Empire and Foreign Countries (TN) 1899–1910).

As paper manufacturing is a highly capital intensive industry and it is characterised by rather inflexible combinations of inputs of investments, knowledge and technology, Indian paper mills could not respond to the threat of imports either by producing wood pulp or newsprint cheaply in India or by the extensive substitution of labour for capital (or chemicals) in the manufacture of paper. Besides, Indian timber resources were poor from the point of view of developing a low cost wood pulp industry. The growth of coniferous woods was limited to the Himalayas, and their comparative inaccessibility and difficult terrain rendered their exploitation unviable (Podder 1979, p. 46). A factor which favoured the use of imported wood pulp over indigenous materials was that it took a much larger quantity of sabai to produce the same quantity of paper (it took between two and a half and three tons of sabai to make a ton of sabai pulp). Furthermore, the cost of imported wood pulp went down in relation to that of indigenous raw materials (Bagchi 1972, pp. 395–396, 398; ITB 1931, p. 68).

The increased price for sabai pulp as well as increased imports were the factors that drew attention to the possibility of making good quality and large volumes of pulp from bamboo. But the factor that most increased interest towards the use of bamboo as a papermaking material was the outbreak of the First World War. During the conflict a shortage of shipping caused trade between India and the rest of the world to collapse. The difficulty of obtaining wood pulp and paper during the war forced manufacturers in India to switch to indigenous raw materials instead (Marsden 1922, p. v). It also revealed India's vulnerability in terms of supply of overseas imports. The decreased imports and the consequent increase in prices meant that Indian mills were not in a position to meet the demand for paper. The effects of the war, then, gave the necessary economic incentive for the papermakers in India to utilise bamboo for large-scale paper manufacture.

11.5.1 Capital and Entrepreneurship

Before 1914, most of the capital employed in the Indian paper industry was controlled by two British managing agency houses, F. W. Heilgers & Co. and Balmer Lawrie & Co. The former controlled the largest Indian producer, The Titaghur Paper Mills Co., and the latter the second largest, the Bengal Paper Mill Co. Such a large degree of concentration of capital facilitated the smooth working of Indian paper markets through various price and market-sharing arrangements (Bagchi 1972, pp. 134, 158–159, 175–176, 178, 391–392). At the beginning of the twentieth century, there was no sharp distinction between the commercial and industrial interests of the managing agents. Typically, managing agents acquired a diverse group of subsidiary companies and built up a portfolio of investments primarily in British companies. Whether a subsidiary company was registered in the UK or India depended primarily on the convenience of the managing agents (Bagchi 1972, pp. 162, 200; see also Jones 1992).

The Bamboo Paper Company was the first London-based company that was established to carry on the business of manufacturing and dealing in pulp from bamboo in India. On 27 January 1919 the incorporated company was 'freestanding' in the sense that it had limited managerial resources, and it operated with the support of the financial and marketing services provided by commercial and financial City firms (Wilkins 1988, p. 264). The Bamboo Paper Co. was a private company under the meaning of the Companies Acts of 1908–1917. Its investments were made through channels other than the Stock Exchange, and no invitation was made to the public to subscribe for any shares, debentures or debenture stock of the company. The private capital invested in the company constituted reinvested profits and capital raised from their partners of the managing agency firm in India, Andrew Yule & Co., which had its registered offices in Calcutta (LMA, CLC/B/022, BPC, Memorandum and Letters of Association. MS28961).

After its establishment, the Bamboo Paper Co. left technical and production matters for the India Paper Pulp Co., which was incorporated on 4 April 1918 by Andrew Yule & Co. According to an indenture, which was made on 4 November 1919 between the Bamboo Paper Co. and the India Paper Pulp Co., the latter company got exclusive rights in the use of inventions relating to improvements in making bamboo pulp and authority to manufacture and sell it. In return, the India Paper Pulp Co. allotted and issued 3,800 fully paid up shares of its capital to Bamboo Paper Co (i.e., 20% of the equities). (LMA, CLC/B/022, BPC, Indenture, MS28963).

The Bamboo Paper Co. did not invent the process for extracting cellulose from bamboo using acid sulphite but had acquired the patent rights by assignment from the original patentees, Thomas Nelson, his brother Ian Nelson and J. L. Jardine, who acted as the company's paper and cellulose experts. The patents covered two principal areas: the actual pulping process by which it was possible to manufacture high-grade pulp on a commercial basis and the machinery for crushing bamboo (LMA, CLC/B/022, BPC, Letter Book, MS28964:1, pp. 126, 241). Thomas Nelson & Sons, Ltd., which had its registered premises at Parkside Works, Dalkeith Road, Edinburgh, acted as the patent agents for the Bamboo Paper Co. The former tried to eliminate potential competitors in the field by securing foreign and colonial patents for the sulphite pulping process and the crusher.

According to the requirements of the Patent Office, working on a commercial scale should have been carried out within four years of the issue of the patent (Espacenet. British Patent 14421, 16 June 1914) but this provision was suspended during the war and the period for complying with it was extended until 1922. A certain amount of crushing, pulping and paper manufacturing was initially done at the works of James Brown & Co., Ltd., the Esk Mills, Penicuik, Midlothian (LMA, CLC/B/022, BPC, Letter Book, MS28964:1, p. 335; MS28964:2, p. 178). For the commercial development of the process, the India Paper Pulp Co. required a large plant, which it started to construct in 1919 on the bank of the River Hooghly, at Hazinagar, near Naihati, Eastern Bengal Railway, 30 miles from Calcutta. The experimental and development work was carried out over the next four years at the Naihati mill, and this resulted in improvements in the mechanical treatment as well as the economic use of chemicals in treating bamboo. The mill started to pulp bamboo by using the acid sulphite process in 1922 (ITB 1924, p. 312; Podder 1979, pp. 8–9, 55–56).

11.5.2 *The Essential Conditions for a Successful Bamboo Pulp and Paper Industry*

Taking for granted the commercial value of the bamboo in its application to paper-making, the following points needed to be considered in the erection of the Naihati mill. First, to reduce transportation costs, it was imperative that the mill be located close to the source of raw materials. The amount of raw material needed was calculated by estimating the quantity of bamboo required for the manufacture of a stated annual amount of finished, air-dry pulp, and the area of land that was estimated to be necessary for the production of that quantity. Air-dried bamboo typically yielded 42 to 50% pulp depending on which species was used. On the basis of a 45% yield, the manufacture of one ton of pulp required the treatment of two and a quarter tons of bamboo (Sindall 1909).

Initially, the India Paper Pulp Co. adopted a business model by which the supply of raw material was ensured from the point of view of the system of cropping and cultivating the Kasalong Bamboo Reserve, which the company leased and was located in Chittagong Hill Tracts. By gradually harvesting the older growth for the manufacture of certain qualities of pulp and introducing a system of cropping, it was possible to control the growth of bamboo and thus ensure a permanent supply of raw material. The bamboo was procured by first cutting it and then floating it in rafts down to the company's crushing and baling plant in Chittagong. The capacity of this crushing plant was approximately 24 tons of raw bamboo per day. The crushed material was then transported by steamer to the Chittagong Jetties, and finally delivered by rail to the Naihati mill (the total distance from the bamboo reserve to the mill was 470 miles). Working the concession directly proved to be too costly, however, and before long the company started to obtain its raw material directly from contractors in Assam as well as Chittagong (ITB 1924, pp. 312, 362; ITB 1931, pp. 40, 45). Although this approach was cost-effective, its obvious disadvantage was that it was difficult to ensure the uniformity of the raw material.

Second, the location of the mill needed to be considered. Up to the late 1930s, Indian paper mills were almost entirely dependent on imports from abroad for their requirements of machinery, machine tools, various chemicals and other materials (Bagchi 1972, p. 434). To ensure the delivery of the supplies of raw material, coal for energy and various chemicals and loading materials and the transfer of the finished paper, a railway siding and a port contiguous to the mill had to be built. In wood pulping, hydro-electric energy resources were essential in choosing a location for the mill, but this consideration did not apply to the Naihati mill, which obtained electric power from a steam driven power station situated in the mill. Coal was brought by rail from the Bengal coalfields (a distance of 150 miles) (LMA, CLC/B/022, BPC, Letter Book, MS28964:2, p. 197).

The technology in connection with the crushing process was perfected in Edinburgh at Messrs. James Bertram & Son, Ltd., which supplied the crushing plant. The same company delivered a 98-inch wide paper machine to the mill together with various fittings. Besides the actual machinery needed for the production of pulp and paper from bamboo fibre, the equipment for the Naihati mill varied widely and included, among other items, corrugated roofing materials from British Everite & Asbestilite Works, Ltd. of Manchester, a Turbo-Generator from Messrs. Bruce, Peebles & Co. of Edinburgh, an engine from Belliss & Morcom, Ltd. of Ladywood, Birmingham, and a steam turbine from Ljungström Steam Turbine Co. (Aktiebolaget Ljungströms Ångturbin) of Sweden (LMA, CLC/B/022, BPC, Letter Book, MS28964:2, p. 197). The supplies of heavy chemicals and minerals included, for instance, 50 tons of refined rock sulphur per month from Sicily, together with volumes of beaching powder, China clay, dyes and magnesite. In terms of the latter, which was imperative for the extraction of cellulose from bamboo, the company relied on supplies from the Salem Mines, located in Kancheepuram, near Chennai (LMA, CLC/B/022, BPC, Letter Book, MS28965:1, pp. 246, 273, 325). It also considered acquiring its own magnesite deposits in India (ITB 1924, p. 315).

Finally, the total operating costs, which would determine whether the finished paper could be sold at a profit, needed to be considered. The enormous demands for all kinds of machinery, equipment and chemicals on the one hand and the existing industrial problems both in England and India, and the serious shortage of building materials and the shortage of shipping and communication in general (in 1921 cables from London to India frequently took from seven to eight days) on the other meant that reliable cost estimates were difficult to make (LMA, CLC/B/022, BPC, Letter Book, MS28965:1, p. 317).

In particular, fluctuations in the exchange value of the rupee created difficulties for the India Paper Pulp Co. Between 1898–1899 and 1919–1920 the value of the rupee was maintained at the fixed rate of 1 s. 4d. in sterling. The breakdown of the pre-war monetary mechanisms and the inability of the Government of India to introduce a paper currency to replace the old silver rupee meant that the fixed value of the rupee became linked to the soaring price of silver (Bagchi 1972, pp. 64–65). Fluctuations in the rupee's exchange value combined with intense competition from abroad made the potential development of the bamboo pulp and paper industry look extremely gloomy. In 1919, it was estimated that a total expenditure on the Naihati mill was Rs. 750,000 (£50,000 when the exchange value of rupee was 1 s. 4d in sterling) (LMA, CLC/B/022, BPC, Letter Book, MS28965:1, pp. 275–277). This was significantly higher than Sindall's estimate some ten years earlier. In 1909, the cost of a dry-pulp plant, which had a weekly output of 200 tons pulp, was in total about £36,000 (Sindall 1909).

The estimated production cost at the Naihati mill was Rs. 80 per ton of wet pulp and Rs. 140 per ton of paper (LMA, CLC/B/022, BPC, Letter Book, MS28965:1, p. 273). These estimates were exclusive of labour costs, which amounted to 38% of the total production costs in the UK, but were significantly lower in Bengal, where cheap and abundant Indian labour was one of the main assets of the pulp and paper industry (ITB 1924, p. 311; Bagchi 1972, pp. 152–153). In general, from the mid-1920s onwards there was abundant labour in most of India's major industrial centres. Since working conditions in mills were much better than in mines and plantations, and wages were higher, particularly for skilled workers, it was unnecessary for the managing agents to make any special effort to recruit labour (Bagchi 1972, pp. 135, 138). In 1924, it was estimated that the Naihati mill employed some 800 to 1,000 labourers (inclusive of those workers who were extracting and collecting raw material) (ITB 1924, p. 315). The wages of Indian workers varied from Rs. 190 per month at the high end of the scale (i.e., the Head Mechanic) to Rs. 14 per month at the low end (i.e., a 'Boy') (ITB 1924, p. 318). These wages were significantly higher than average monthly wages of Indian workers in the Calcutta jute mill, where workers earned between Rs. 12 and Rs. 20 a month (Bagchi 1972, pp. 126–127). The management at the Naihati mill was European, earning approximately 50% more than the average wage rate that would have been drawn in England (ITB 1924, p. 318).

11.5.3 The Paper and Pulp Trades and Indian Tariff Policy

Up to 1914, the Government of India pursued the policy of free trade and non-intervention in industry, and there was virtually free trade as far as imports into India from other countries were concerned. This pattern of trade primarily served the British imperial system; Britain had a large market, in which Lancaster and Manchester piece goods (i.e., fabrics woven in standard lengths for sale) and other industrial products in demand in India such as machinery and hardware, entered duty-free at the time when other markets in the world were closing against them. On the other hand, Britain was the largest purchaser of raw materials and manufactured jute from India. The system's smooth functioning was facilitated by the fact that the Indian money market was directly linked to the City, the world's chief money market before 1914, and also the fact that there was little industry in comparison with trade (Bagchi 1972, pp. 5, 48–50, 58–59).

An early attempt at import substitution in the Indian paper industry was choked off before the First World War by the rapid progress of the wood pulping process in Europe and North America. The war caused a violent disruption to the trade owing to the lack of shipping capacity. By 1918–1919, the amount of wood pulp imported into India had recovered to merely 2,090 tons, causing soaring prices. At this time, the Raneegunge Mill of the Bengal Paper Mill Co. was buying imported sulphite pulp at the very high price of £20 a ton. Its average cost of manufacture from all materials was £14 a ton (LMA, CLC/B/022, BPC, Letter Book, MS28965:1, pp. 198–199). By the end of 1922, competition from European paper makers had started in earnest, and thus prices fell rapidly. In this regard, competition was most intense from the UK, Norway, Sweden, Finland, Germany and Holland. In terms of wood pulp, Sweden, Finland and Norway were the main suppliers to the Indian markets (ITB 1924, p. 321). The average price of imported Scandinavian pulp was in the neighbourhood of £10–15 a ton. By way of comparison, the price of Scandinavian wood pulp at an English mill was reported to be approximately £9 a ton (ITB 1931, pp. 19, 90).

The first step in erecting a tariff on imported pulp and paper occurred in June 1923 when the Indian Paper Makers' Association submitted its claim for protection to the Government of India. The question was referred to the Indian Tariff Board (hereafter ITB) one year later. The ITB considered it inadvisable for the government to commit itself firmly to the protection of the bamboo pulp and paper industry until the bamboo pulping process had really proved itself commercially. Instead, the ITB suggested that the government support the India Paper Pulp Co. by providing capital either in the form of loans or by guaranteeing a public issue of debentures for an extension its capacity. Secondly, it recommended that, in place of the existing 15% ad valorem duties on printing and writing paper, a specific duty of one anna per pound should be imposed on all writing paper and on all writing paper other than newsprint containing 65% or more of mechanical pulp. Newsprint was exempted from the protective duty on the grounds that mechanical pulp had never been made from either sabai grass or bamboo and that the existence of cheap newsprint on the Indian markets depended on imports.[3] The government rejected the ITB's recommendation to provide financial assistance to the India Paper Pulp Co. for the three reasons. First, the India Paper Pulp Co. was a private company. Second, the acid sulphite process was covered by patent rights held by Jardine and the Nelsons, who were members of the company. And finally and most importantly, financial assistance to the Indian paper industry should assist equally all competitors within the industry and should not benefit the India Paper Pulp Co. alone and thereby give it an undue advantage over its rivals (ITB 1931, p. 3; Bagchi 1972, p. 397).

Ultimately, the ITB ended up endorsing protection instead of financial assistance to the India Paper Pulp Co. because it considered that bamboo could serve as the raw material for the long-term development of the Indian paper industry. The two major producers of paper in India, the Titaghur and the Bengal Paper Mills, did not have promising prospects, on the one hand, because the supplies of sabai grass from known sources were rather limited, and on the other, because the both companies' mills situated far from the known sources of sabai. The Bamboo Paper Industry Protection Act of 1925 marked the end of the free trade era. It imposed an import duty of one anna per pound on printing and writing paper, and on all writing paper other than newsprint containing 65% or more mechanical pulp. Papers made from bamboo were protected for seven years (up to 31 March 1932) (ITB 1931, pp. 3–5; Bagchi 1972, pp. 5, 396–397; Podder 1979, p. 56).

The protective duty for the bamboo pulp and paper industry in India had numerous effects. The bamboo pulping process, although theoretically known since the mid-1870s, only entered the period of commercial production in 1922 and developed in an economically viable manner only after tariff protection had become effective. The technology was first developed at the Naihati mill, but was later adopted by the other

[3] In general, the underdeveloped countries like India imported most of their newsprint from the outside world before 1939. The first Indian newsprint mill, National Newsprint and Paper Mills, Ltd. started production in 1955–1956 (Podder 1979, pp. 177, 196).

Indian mills in response to the increased cost of sabai pulp (Bagchi 1972, p. 419). In 1923, the Titaghur Paper Mills Co. and the Bengal Paper Mill Co. had twelve paper machines altogether (an aggregate capacity of 31,000 tons). The India Paper Pulp Co. had initially only one paper machine (its capacity was 6,000 tons and it installed a second machine in 1927), with which it had managed to produce (by using entirely their own bamboo pulp) 4,228 tons of paper by the end of 1924 (ITB 1924, p. 311; ITB 1931, p. 10). Although it had started production only in 1922, it already showed lower production costs than its competitors in Bengal. Including labour, its costs were Rs. 485.27 a ton of paper for 1923–1924 (ITB 1924, p. 335). In 1930–1931 the costs were approximately one third lower (Rs. 330.65 a ton) (ITB 1931, p. 54). The fall in costs is attributable to the growing familiarity with bamboo as a papermaking material, the employment of contractors as opposed to using men who were on the company's payroll to work the concession and improved efficiency in production, which reflected positively on the demand for coal (ITB 1931, pp. 63, 65–66).

The tariff protection was not much of a success in terms of increasing production of bamboo paper in relation to India's total output of paper. The rate of growth of the bamboo paper industry under tariff protection was initially very modest. Production of 2,500 and 5,000 tons were recorded in 1925 and 1931 respectively, while the total output (from all materials) of the Indian mills increased from 25,000 to 40,000 tons. In 1925, the protected bamboo paper industry represented 14.7% of the total Indian paper production and dropped to a mere 8% in 1931 (ITB 1931, pp. 18, 25). Protection was primarily effective in restricting the consumption of certain grades of paper, but a substitution of unprotected for protected varieties took place. The total Indian consumption of all grades of paper increased from 112,000 tons in 1924–1925 to 154,300 in 1930–1931, and the consumption of tariff protected paper increased from 43,300 tons in 1924–1925 to 49,000 in 1930–1931 (ITB 1931, p. 105).

One of the unintended effects of protecting the bamboo pulp and paper industry was a rise in the proportion of imported wood pulp used in the manufacture of paper. This was a natural outcome because the bamboo pulping process developed very slowly. At the time of the 1931 enquiry by the ITB, besides the India Paper Pulp Co., only the Titaghur Paper Mills Co. seemed to have succeeded in developing a truly effective bamboo pulping process. Titaghur relied first on the soda process but later applied the alkali process together with other Indian mills. Only the India Paper Pulp Co. continued to rely on the acid sulphite process it had developed (ITB 1931, pp. 47, 106). Between 1924–1925 and 1930–1931 the use of both indigenous and imported materials increased, but there was a drastic decline in the ratio of indigenous compared to imported materials used. All the major companies were using a higher percentage of wood pulp in 1930–1931 than they had done in 1924–1925. In the case of the India Paper Pulp Co., for example, imported wood pulp accounted for 21.02% by weight of the finished paper it produced in 1924–1925. In 1930–1931 wood pulp accounted already for 63.04% by weight of the finished paper produced and bamboo some 30%, with the rest coming from recycled fibres. The total quantity of wood pulp imported into India in 1930–1931 was 22,715 tons, which was equivalent to nearly 20,000 tons of finished paper or about half the total Indian paper production (ITB 1931, pp. 55, 87, 93; Bagchi 1972, p. 402).

The Bamboo Paper Industry (Protection) Act of 1932 extended protection by another seven years since the withdrawal of the protective duty would have meant the disappearance of bamboo as a raw material for the manufacture of paper. The ITB estimated that without tariff protection it would be far too expensive for the mills to use bamboo pulp. This would have wasted all the development work, left indigenous raw material resources undeveloped and kept the Indian paper industry dependent on imported wood pulp. During the First World War, the scarcity of wood pulp had nearly crippled the paper industry and highlighted the importance of having a domestic pulp industry (ITB 1931, pp. 82–84; Bagchi 1972, p. 45). During the period from 1932 to 1937 India's total production increased every year. This was due to the better utilisation of existing production capacity under the stimulus of higher duties. The protective duty on paper was raised in 1932 from 18.75% to 30% ad valorem, with a preferential duty for grades manufactured in Britain (Bagchi 1972, p. 402). This was a result of the imperial preference system, created at an Imperial Economic Conference in Ottawa in July-August 1932, which aimed at expanding trade among the members of the British Commonwealth in a world of shrinking commerce and rising trade barriers (Pollard 1963).

A period of seven years (1932–1939) afforded the Indian paper industry an opportunity to consolidate its position and encouraged it to undertake further expansion. It also made considerable progress in substituting bamboo pulp for wood pulp. This was primarily the result of the imposition in 1932 of a specific duty of Rs. 45 per ton on imported pulp by the Bamboo Paper Industry Act of 1932. Since all protective duties were subject to the revenue surcharge imposed in November 1931, the effective rate of specific duty on imported pulp came to Rs. 56.25 per ton. The replacement of imported pulp by domestically produced pulp was also facilitated by resolving the problems involved in the mechanical treatment of bamboos (ITB 1931, pp. 95–96; Bagchi 1972, p. 399).

The monetary incentive given by the Bamboo Paper Industry Protection Act of 1932 was sufficient to introduce bamboo into the paper industry. As it transpired, bamboo raw material provided the basis for the long-term development of the Indian paper industry. Between 1936 and 1939, a number of new mills appeared, including Star Paper Mills in 1936, Mysore Paper Mills in 1937, Orient Paper Mills in 1938, Sirpur Paper Mills in 1938 and Rohtas Industries in 1939. In the late 1970s, bamboo formed 67% of the country's pulp raw material (Podder 1979, pp. 66, 177). In the more recent decades, the scarcity of bamboo, softwoods, and recycled fibres has led to the development of eucalyptus for plantation purposes and pulping hardwoods.

11.6 Conclusions

The British paper trade history was defined since the mid-1850s by a quest for a new raw material to replace rags. Increasing demand for paper combined with the scarcity in the supply of rags induced British papermakers to look for new raw materials in the shape of tropical and semi-tropical grasses. The requirements of the paper trade were first met by Routledge's discovery that esparto grass from Spain, and later from North

Africa, could be utilised. Since the late 1870s, the success of esparto encouraged mill developments in British India, where the primary raw material was sabai grass. The gradual expansion of the wood pulp industry in Europe and North America put a stop to further development work based on naturally growing grasses and shifted the focus of British papermakers from tropical regions to those of Scandinavia and North America. Favoured by seemingly inexhaustible supplies of cheap timber in these countries, and assisted by their cheap hydro-electric power and efficient transport systems, imported wood pulp assumed a pre-eminent position as a papermaking material in Britain, whose large paper industry offered a convenient market for wood pulp and finished papers produced in Scandinavia and North America. Not only was the demand for wood pulp met but it was exceeded, which inevitably led to "dumping" of the excess in world markets.

A decade before the outbreak of the First World War, competition from European paper and wood pulp had started in earnest in India. Although theoretically known since the mid-1870s, bamboo began being treated only after the First World War, when the Government of India offered financial incentives and concessions for the exploitation of forest areas to induce the creation of pulp and paper industry based on utilising bamboo as the raw material. The increased dependence on wood pulp, the likelihood of a pulp famine, and the consequent increase in price for imported wood pulp were the means for drawing attention to the possibility of making commercial volumes of good quality bamboo pulp. The bamboo pulping process entered the period of commercial production in 1922 but developed in an economically viable manner only after tariff protection had become effective in 1925. The technology was developed under British auspices, but was later adopted by Indian paper producers in response to the rising costs of imported wood pulp.

In India, coniferous species were limited to the Himalayas, making exploitation of softwood resources economically unviable. In their absence, papermakers in India adopted technologies and organisational solutions that distinguish them from the other British Empire and Commonwealth countries. These developments took place within British national, colonial and organisational frameworks, and reflected the availability of technology, knowledge, investments, and raw materials on the one hand and demand characteristics on the other.

References

Accounts Relating to Trade and Navigation of the United Kingdom (1913–1959). His Majesty's Stationery Office, London; Andrew Yule & Co. Ltd., 1863–1963 (1963)

Annual Statement of the Sea-borne Trade and Navigation of British India with the British Empire and Foreign Countries (TN) (1899–1910) Office of the superintendent of government printing, Calcutta

Bagchi AK (1972) Private investment in India 1900–1939. Cambridge University Press, Cambridge

Bajpai P (2012) Biotechnology for Pulp and Paper Processing. Springer-Verlag, New York

Cain PJ, Hopkins AG (1994) British imperialism: innovation and expansion, 1688–1914. Longman, London/New York

Census records for England and Wales; Scotland; Ireland (1861) The online historical population records (Histpop). http://www.histpop.org/ohpr/servlet/. Accessed 24 Feb 2018
Chemical News and the Journal of Industrial Science (CNJIS), 24 December 1875, 32(839):299
Clapperton RH, Henderson W (1947) Modern paper-making, 3rd edn. Basil Blackwell, Oxford
Coleman DC (1958) The British Paper Industry, 1495–1860. A Study in Industrial Growth. Clarendon Press, Oxford
Dykes Spicer A (1907) The paper trade. A descriptive and historical survey of the paper trade from the commencement of the nineteenth century. Methuen & Co., London
Espacenet. The Online Patent Search, European Patent Office. https://worldwide.espacenet.com/. Accessed 24 Feb 2018. British Patent 20560, S. Milne, Improved Method of and Means for Treating Bamboo Preparatory to Making it into Pulp, issued 27 July 1911; British Patent 14421, issued 16 June 1914; British patent 2509, JL Jardine and TA Nelson, Improved Process for the Production of an Acid Magnesium Sulphite Solution and the Application of Same for the Extraction of Cellulose from Fibrous Vegetable Materials, issued 16 March 1916
Esparto Paper (1956) The Association of Makers of Esparto Papers, London
Fajardo J, Verde A, Rivera D, Obón C, Leopold S (2015) Traditional Craft Techniques of Esparto Grass (*Stipa tenacissima L.*) in Spain. Economic Botany XX(X): 1–7
Herring R (1855) Paper & Paper Making, Ancient and Modern. Longman, Brown, Green, and Longmans, London
Hills RL (1988) Papermaking in Britain 1488–1988. A Short History. Athlone Press, London and Atlantic Highlands, NJ
Hobsbawn E (1999) Industry and Empire. From 1750 to the Present Day, Penguin, London
Indian Tariff Board (ITB) (1924) Enquiry regarding the grant of protection to the paper and pulp industries. Evidence tendered by applicants for protection. The Government of India, Calcutta
Indian Tariff Board (ITB) (1931) Report of the Indian Tariff Board on the grant of protection to the paper and pulp industries. The Government of India, Calcutta
Jones S (1992) Merchants of the Raj. British managing agency houses in Calcutta in yesterday and today. Macmillan, Basingstoke and London
Khandual A (2016) Sabai grass: possibility of becoming a potential textile. Springer, Singapore
London Metropolitan Archives (LMA) (1919) CLC/B/022. Bamboo Paper Company Limited (BPC). MS 28961, Memorandum and Letters of Association; MS 28963, Indenture, 4 November 1919; MS 28964/1 Letter Book, letters 1–500; MS 28964/2, Letter Book, letters 1-241; MS 28965/1, Letter Book, letters 1–503
Magee G (1997) Productivity and performance in the paper industry. In: Labour, capital, and technology in Britain and America, 1860–1914. Cambridge Studies in Modern Economic History, vol 4. Cambridge University Press, Cambridge
Marsden F (1922) Paper and paper pulp production in the Madras Presidency. Bulletins of Indian Industries & Labour 24. Superintendent Government Printing, Calcutta
McNeill W (1990) The rise of the west after twenty-five years. J World Hist 1:1–21
O'Brien P (2006) Historiographical traditions and modern imperatives for the restoration of global history. J Glob Hist 1:3–39
Owen G (2000) From empire to Europe. The decline and revival of British industry since the Second World War. HarperCollins, London
Pearson RS (1913) Note on the utilization of the bamboo for the manufacture of paper-pulp. Indian Forest Records, vol IV, part V. Superintendent of Government Printing, Calcutta
Podder V (1979) Paper industry in India. A study. Oxford & IBH Publishing Co, New Delhi, Bombay and Calcutta
Pollard S (1963) The development of the British economy 1914–1950. Edward Arnold, London
Pomeranz K (2000) The great divergence: China, Europe, and the making of the modern world economy. Princeton, NJ, Princeton University Press
Raitt W (1928) Bamboo Hope. Progress Report of Forest Research Work in India. For the Year 1926–7. Government of India, Calcutta, pp 155–165
Reader WJ (1981) Bowater. A history. Cambridge University Press, Cambridge

Routledge T (1875) Bamboo considered as a paper-making material with remarks upon its cultivation and treatment. E. & F. N Spon, London
Routledge T (1879) Bamboo and its treatment: cultivation and cropping. Sunderland
Särkkä T (2012) The British Paper Industry, 1800–2000. In: Lamberg J-A, Ojala J, Peltoniemi M, Särkkä T (eds) The Evolution of the Global Paper Industry 1800–2050, A Comparative Analysis. Springer, Dordrech, pp 167–190
Shorter AH (1971) Paper making in the British Isles. A historical and geographical study. David & Charles, Newton Abbot
Sindall RW (1909) Bamboo for papermaking. Marchant Singer & Co, London
Statistical Year-book of the League of Nations (LN) (1938–1939). League of Nations, Geneva
The Finnish Sawmilling, Pulp and Paper Industries (1936) Central Association of Finnish Woodworking Industries, Helsinki
The Paper Trail (2013) Frogmore Mill
The Work of the Forest Department in India (1920) His Majesty's Stationery Office, London
Watt G (1885) A dictionary of the economic products of India. Superintendent of Government Printing, Calcutta
Wilkins M (1988) The free-standing company, 1870–1914: an important type of British foreign direct investment. Econ Hist Rev 41(2):259–282

Chapter 12
Creating Global Markets: Seaborne Trade in Pulp and Paper Products Over the Last 400 Years

Jari Ojala and Stig Tenold

12.1 Introduction

Companies in the pulp and paper industry have traditionally been nationally located although their markets have been international, or rather intracontinental. In other words, globally operating multinational companies are a rather recent phenomenon (Ojala et al. 2006; Sajasalo 2003; Siitonen 2003). Today though, the major companies operate globally and have built global production chains to take advantage of regional comparative advantages in terms of factor endowments. Thus, where to locate production has always been the crucial question for companies in this business. The availability of raw materials and markets might explain the emergence of the paper industry in different regions, but also technology, labour, access to energy, and institutional structures supporting the industry are important drivers for both the success and failure of the industry. In addition, transport solutions are important throughout the production line: from forests to factories and from factories to the customers. Thus, the technological solutions for transport and their associated costs are crucial for a high-volume, bulk product business such as the pulp and paper industry.

The declining cost of transport made it possible for a bulk industry such as pulp and paper to serve markets far from the source of the raw materials. The decline in transport costs was not only caused by falling ocean freight rates and specialized cargo ships developed for the pulp and paper industry carrying trades but also the development of port facilities and whole logistics systems that ensure efficiency in

---

J. Ojala (✉)
Department of History and Ethnology, University of Jyväskylä, Jyväskylä, Finland
e-mail: jari.ojala@jyu.fi

S. Tenold
Department of Economics, NHH - Norwegian School of Economics, Bergen, Norway
e-mail: stig.tenold@nhh.no

T. Särkkä et al. (eds.), *Technological Transformation in the Global Pulp and Paper Industry 1800–2018*, World Forests 23, https://doi.org/10.1007/978-3-319-94962-8_12

shipping raw materials and finished products to markets (Harlaftis et al. 2012; Hazley 2000; Kurosawa and Hashino 2012; Ojala and Kaukiainen 2012).

Waterways have historically been important for the carriage of the forest industry products, from floating timber to international trade flows. In fact, water transport has particularly large advantages relative to land transport for bulky, high volume goods such as forest products. While they were usually transported in vessels that carried many different cargoes simultaneously, today they are carried in highly specialised ships. These vessels include, for example, specialised ro-ro vessels and open-hatch ships designed to carry paper products, and Japanese wood-chip carriers.

This chapter discusses the overall importance of seaborne trade for the development of the pulp and paper industry. We will first analyse the major shifts in the international trade flows of pulp and paper industry products, thus providing an empirical basis for the development of trade volumes. Thereafter, we will analyse the technological and organizational changes in shipping, with a particular focus on the post Second World War period. It was during these years that seaborne transport of pulp and paper developed from a regional/intracontinental basis to an intercontinental industry. The article uses international databases—such as Sound Toll Registers Online collection (STRO) and the material from the Food and Agriculture Organization (FAO) of the United Nations—as its main sources combined with industry- and company-focused research.

12.2 Trade Flows of Pulp and Paper Products

12.2.1 From Early Developments…

Because forest industry products are very bulky, they have been particularly well suited to seaborne transport, and this trade has been among the key features in economic and maritime history. The Dutch supremacy in the early modern trades was largely dependent on the forest products carried from the Baltic Sea area, as was the case with the rise of the UK as a maritime power and its dependence on (northern) forest products. Initially, trade was centred around round wood, then later on planks of various qualities. While there has been some intercontinental trade, in particular in hardwoods, for centuries, such trade is more recent for pulp and paper—as are the products themselves.

Pulp and paper emerged among the major forest industry products during the mid-nineteenth century when the technology for producing paper from wood was introduced. This also shifted the dominance of the world's forest industries to the Northern Hemisphere, which boasted the suitable forest resources, and thereby increased demand for the transport of raw materials needed by and products of this specific industry.

Trade flows with paper products have changed a number of times over the last 400 years. The Baltic Sea area is a prime example of this change. Before the technology for making paper from wood emerged, paper was produced mainly from rags. The major production areas were in the Dutch Republic, France, Britain, and in the German-speaking area. Thus, the trade flows with paper products went from the North Sea to the Baltic Sea rather than vice versa, which is the situation today. This early modern paper trade to the Baltic is traceable from the Danish Sound Toll registers, which recorded all ships passing the Sound from the late fifteenth century up to the mid-nineteenth century. The ships passing the Sound had to stop at Danish Elsinore to pay the Sound custom dues. These Sound Toll records are well preserved and are available in the Sound Toll Registers Online (STRO) compilation (http://www.soundtoll.nl/index.php/en/).

Forest industry products were among the major items in the early modern trade flows through the Danish Sound, consisting of all sorts of timber products and large quantities of tar and pitch produced in the northeast corner of Europe (Scheltjens 2015; Åström 1988). The volume of paper traded through the Danish Sound was more modest, but it grew alongside the spread of literacy, the expansion of the economy and increasing paper consumption. Early modern paper production was, though, mainly local; thus the records do not show large trade volumes. Nevertheless, the Danish Sound toll captures trade in these products (Fig. 12.1). Altogether c. 26,000 ships passed the Sound from the mid-seventeenth century to the mid-nineteenth century with cargoes of paper on board. Most of these shipments came from the Dutch republic (65%), whereas France, the UK, Spain and Portugal together accounted for 20% of shipments. Paper was not, however, the major commodity on board these vessels passing Elsinore. Rather it was just another good carried by ships designed to carry all sorts of tradeable items. Even the paper grades varied: Sound Toll accounts specify 265 different paper categories passing the Danish officials collecting the custom dues at the Sound.

As Fig. 12.1 shows, there was a growth in shipments of paper during the eighteenth century and a decline thereafter, when calculated by the simple measure: "the number of passages". In reality, the trade expanded as the volumes per passage grew bigger. However, the paper trade declined relative to other products; as a whole, paper made up to less than two per cent of all passages to and from the Baltic in the years 1600–1857. The substantial increase in the early eighteenth century paper trade shown in Fig. 12.1 can mainly be explained by the paper imports to St. Petersburg, which was founded in 1703 and grew rapidly to be among the most important cities in the Baltic. The first paper machines were developed in the early nineteenth century, and the Sound Toll data also capture paper machine deliveries through sea routes; the first one came to the Baltic through the Sound in 1830.

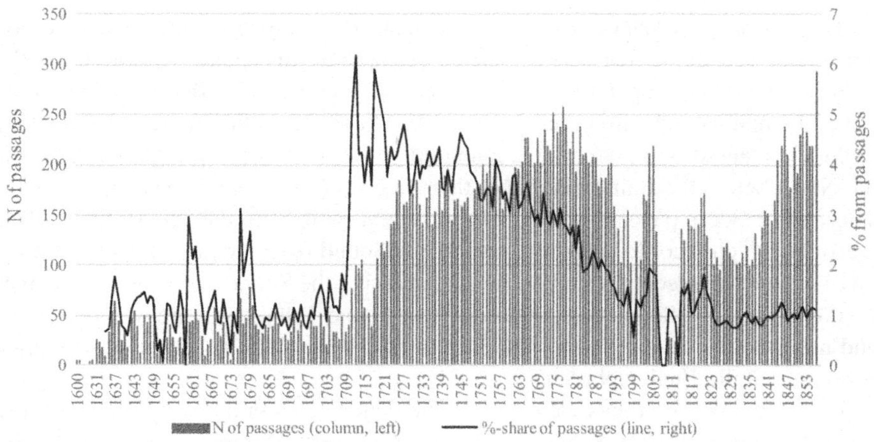

Fig. 12.1 Paper trade through the Danish Sound 1600–1857. Number of passages (left axis) and % share from all passages (right axis) (n = 25,934, N = 1,483,976). *Source* The Danish Sound Toll Account (STRO, http://www.soundtoll.nl/index.php/en/, accessed 19 August 2016) Keyword "pap", N = 25,934, number of passages, 265 products. *Note* only frequency, no volumes nor values. Around 90% of the voyages are ships carrying paper to the Baltic, while around 10% are ships moving westward from the Baltic with paper

12.2.2 ...to Industrialised Paper Production

Above, we used the Baltic paper trade as an example of the early growth of these trades. Unfortunately, we do not have similar data available on a global scale for the period from the early 1860s to the 1960s. Nevertheless, various studies show that the trade in pulp and paper products increased from the mid-nineteenth century after wood emerged as the raw material for manufacturing these products (e.g. Lamberg et al. 2006; Krawany 1910; Rjestoff 1913). There has traditionally been a strong correlation between GDP and paper consumption—thus, the growth of the global economy can be seen in the production of and global trade in pulp and paper (Hazley 2000; Järvinen et al. 2009, 2012). This, however, does not necessarily hold anymore, especially in developed western countries, as there seems to exist a saturation point after which the increasing wealth of nations and individuals does not increase the consumption of paper (Hetemäki and Obersteiner 2001; Järvinen et al. 2012). The existence of a "Kuznets curve" in paper consumption suggests that *per capita* demand for paper outside Western Europe and the Western Offshoots was limited well into the twentieth century. Moreover, the fact that pulp and paper producers primarily served regional markets was reflected in production.

At the beginning of the twentieth century mechanized and industrialized paper-making was almost entirely concentrated in Europe (primarily northern and western Europe) and North America (Table 12.1). As late as the 1970s, almost 80% of the world's paper was still produced in these two regions. However, this share fell to below 60% by early in the new millennium. Today, old western-based companies that

Table 12.1 Global paper and paperboard production 1908 and 1961–2012, average shares (%)

	Eastern Asia	Northern and Western Europe	Northern America	Other regions	Total
1908	1.5	48.2	41.7	8.6	100.0
1961–1969	11.3	22.7	49.7	16.2	100.0
1970–1989	15.7	21.4	42.0	20.9	100.0
1990–2012	26.6	21.0	31.0	21.4	100.0
Total av.	22.9	21.2	34.8	21.2	100.0

Source FAOSTAT faostat.fao.org (accessed 16 August 2016); (Krawany 1910)
Note Averages calculated on the basis of annual production figures

are struggling with over-capacity in their declining domestic and regional markets also own a significant amount of productive capacity on other continents (Hetemäki 1999; Hetemäki and Obersteiner 2001; Järvinen et al. 2012).

The trade flows were still mainly regional from the mid-nineteenth to the mid-twentieth century and concentrated in two areas: European and North American markets. In Europe the Nordic countries met western Europe's (especially the UK, Germany, and France) demand for pulp and paper, while in North America Canadian products were transported to US markets (See especially Särkkä 2012; Kuhlberg 2012; Toivanen 2004, 2012; Lamberg et al. 2012).

The trade in pulp and paper became more global after the Second World War. This was mainly related to two factors. First, the demand for pulp and paper grew in line with rising income levels and the diffusion of economic growth. Second, companies operating in this business internationalized and took advantage of developments in (sea) transport when utilizing their comparative advantages in global trade (Lamberg et al. 2006). For this growth period, from circa 1960s to the turn of the millennium, we also have more comparable data, compiled by the FAO.

As Fig. 12.2 shows, the total volume of global pulp and paper production has increased by a factor of more than five from the early 1960s to 2012, while pulp production grew threefold. The share of paper production that was exported grew from c. one-fifth to nearly one-third, but diminished during the first decades of the twenty-first century. In contrast, exports of pulp have always played a more significant role, representing a roughly one-third share until the early 1990s, subsequently growing to nearly 60%.

Thus, the pulp and paper industry has evolved over centuries from being a local and regional/intracontinental to a global business in which the international transport systems play a crucial role. Exports of both paper and pulp have mainly been transported via sea routes. Moreover, some of the paper industry products have also been transported domestically on water. Trains and trucks are competitive on short distance routes and especially in carrying raw materials.

To get a more coherent view on global trade flows of pulp and paper, we analysed the countries that are the most important players in these trades: eight countries

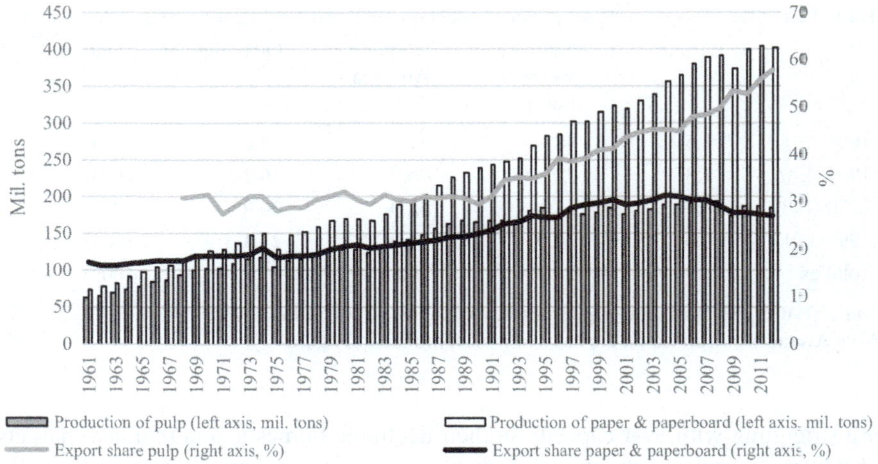

Fig. 12.2 Paper and pulp production and export share of production 1961-2012. Production (left axis, million tons) and export share of production (right axis, %). *Source* FAO (faostat.org, accessed 19 August 2016). *Note* only paper and paperboard, newsprint is not included. Moreover, only pulp for paper is included, not wood pulp

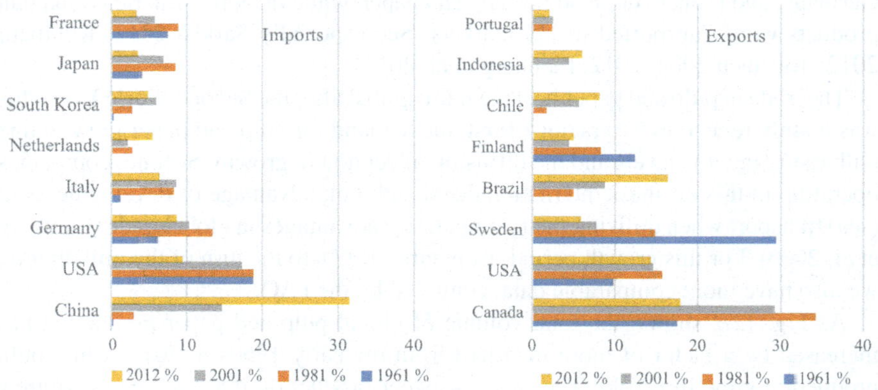

Fig. 12.3 Trade flows of pulp: imports in left, exports in right (% share of global trade flows). *Source* FAO (faostat.org, accessed 19 August 2016)

represent c. 60–70% of imports and 70–80% of exports of global pulp and paper trade in the period 1961–2012 (Figs. 12.3 and 12.4). Furthermore, we selected four cross-cutting years (1961, 1981, 2001, and 2012) to analyse further these trade flows.

From the 1960s to the early 2000s the USA clearly dominated the pulp imports with almost a one-fifth share (Fig. 12.3). In exports, on the contrary, there was no clearly dominant player. First was Sweden's share, which was almost one-third, but it rapidly diminished thereafter as the country's own paper industries used more pulp

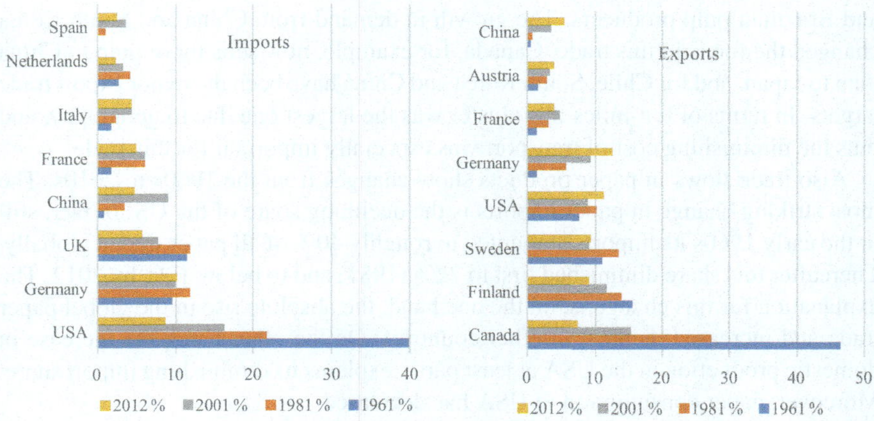

Fig. 12.4 Trade flows of paper and paperboard: imports in left, exports in right (% share of global trade flows). *Source* FAO (faostat.org, accessed 19 August 2016)

produced domestically (Andersson et al. 2016). The trend in Finnish pulp exports, though with smaller figures, were the same as in Sweden, and the reason was also the same: increase in the domestic use of pulp (Jensen-Eriksen and Ojala 2015). Canada, in turn, increased its share of global pulp exports from below one-fifth to one-third. The most striking change occurred from the turn of the millennium, when China's share increased first to c. 15%, and then to roughly one-third of global imports of pulp for paper. At the same time Brazil increased its export share from 5% in 1981 to 16% in 2012. Chile and Indonesia are similarly new entrants in global pulp exports, with both targeting Chinese markets in particular.

Trade flows with pulp to and from the eight most important export and import countries can be divided into three geographical clusters, and the overwhelming majority of this pulp trade was transported by sea. First, the North American cluster has been dominated by Canadian exports to the USA, though some South American counties have also been important in this trade. In 1997, for example, 81% of pulp imports to the US came from Canada. Over the last decades this has gradually been transformed into a "Pacific" cluster, as Canada has also shipped a large volume of pulp to Asia. In 1997, for example, Canadian pulp made up 40% of all pulp imported to Japan. Already by the turn of the millennium, China had emerged as the second most important market for Canadian pulp.

The second cluster is the European one in which Nordic countries (especially Sweden and Finland) have served other European countries. This market has not changed much in terms of trade flows, though occasionally the USA and some South American countries have played some role in this trade. In 1997, roughly a quarter of Germany's total pulp imports came from Sweden and one-fifth from Finland; these figures were roughly the same a decade later.

The third geographic area is the expanding Asian cluster. Until the 1990s Japan was the leading market in this group, which was served by American, Canadian and Brazilian pulp producers. The growth in demand from China and South Korea changed the focus in this trade. Canada, for example, now sells more pulp to China than to Japan, and for Chile, South Korea and China have been the major export trade targets. In terms of ton-miles this cluster was the largest one due to geography, and thus the diminishing cost of transport was especially important for this trade.

Also trade flows in paper products show changes from the 1960s to 2010s. The most striking change in paper imports is the declining share of the US. In fact, still in the early 1960s its imports amounted to roughly 40% of all paper traded globally. Thereafter this share diminished first to 22% (1981) and to below 10% by 2012. The explanation for this change is, on the one hand, the absolute rise in the global paper trade and increased demand in other countries. On the other hand, the increase in domestic production in the USA at least partly explains its diminishing import share. Moreover, paper consumption in USA has decreased.

The growth of paper consumption in China does not show in the figures of the global paper trade as its demand has primarily been satisfied by domestic production from imported pulp. European countries—the UK, Germany, France, Italy, the Netherlands and Spain—have all been important importers throughout the period, although the UK especially has clearly lost its former share.

In terms of paper exports, Canada's decline is even more dramatic than the decline of US imports. Canada suffered especially from the decline of American markets. The USA, Germany, Austria, China, and to a certain extent France have succeeded in increasing their share of global paper exports, while both Finland and Sweden have declined.

The geographic trade flows of paper follow to a certain extent the trends of the pulp trade described above. The trade between Canada and the US dominates the North American cluster. Exports from Sweden and Finland dominate the European cluster, with an increasing role played by Germany. The paper trade flows to Asia do not show clearly in the FAO's statistics, however, the US seems to play the most significant role in Chinese markets.

Figures 12.5 and 12.6 combine the imports of pulp and paper to the clusters described above, namely the USA, Europe, and Asia. These three areas represented between 80 and 90% of global pulp and paper imports in the years 1961–2012. As these figures illustrate, the share of paper imports to the USA has declined rather dramatically from 40 to 9%, while the Asian share increased from 7 to 25%. European paper imports, in turn, first increased during the 1960s when the share grew from one-third to roughly half of the world's paper imports. During the 1990s, Europe's share declined by roughly five percentage points, but it remained well above 40% of global paper imports. Thus, in all, paper imports have remained important in Europe, but less so in the USA.

Pulp imports, in turn, paint a rather different picture. Namely, the share of European imports declined from 70 to 36%, while Asian imports increased from 3 to 47%. US imports, in turn, declined as well. In relative terms the decrease—from roughly 20 to 10%—was as large as the European one, but due to the lower starting point, the effect on the world market was less.

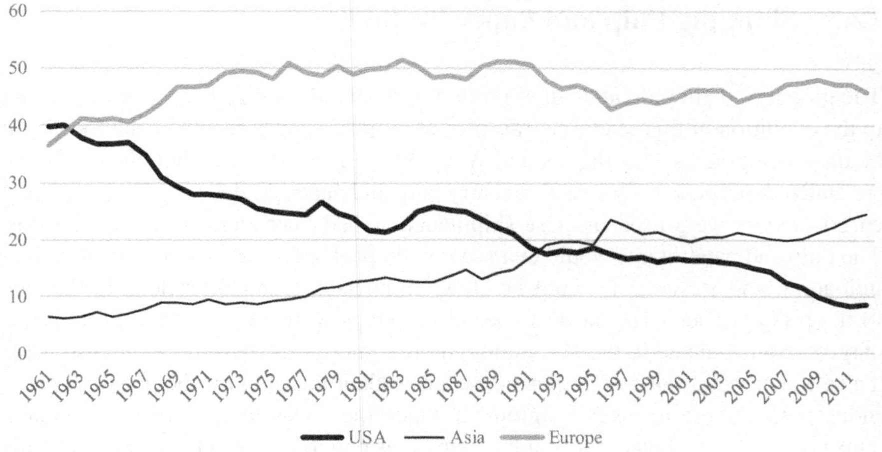

Fig. 12.5 Share of global paper imports in different areas (%). *Source* FAO (Faostat.org, accessed 19 August 2016)

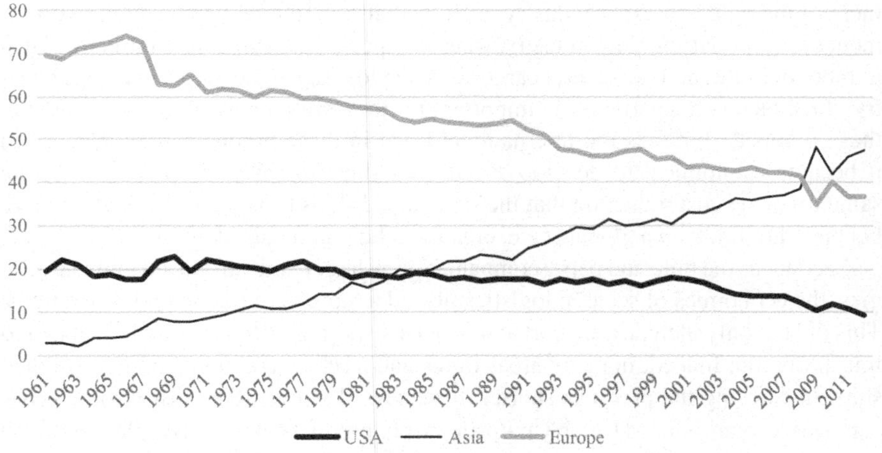

Fig. 12.6 Share of global pulp imports in different areas (%). *Source* FAO (Faostat.org, accessed 19 August 2016)

The absolute tonnage of paper imported, though, increased in the US until the turn of the millennium. Similarly, the volume of European tonnage also increased quite significantly during the same period. Also in absolute terms pulp imports to Europe increased although the share declined. The same trend also occurred in the US.

12.3 Shipping Pulp and Paper Products

The global trade in pulp and paper products has mainly utilized sea routes. Turning to the evolution of this seaborne transport of pulp and paper products, the focus will be on developments after the Second World War. Prior to the conflict, there were no specialized, purpose-built vessels to carry pulp and paper products; ordinary, general cargo carriers were used in these shipments, as had been the case over centuries. The pulp and paper items would be individually packaged and handled, and loading, unloading and stowing it would be slow, cumbersome and often lead to damage to the product itself. The earliest specialized ships were purpose-built to transport cargoes from Canada to the US, although other modes of transport were also used. The efficient solutions in sea transport even made it possible for the pulp and paper industry to emerge in several regions in which there was limited internal demand. This is especially relevant for South America, as pulp (and paper) from Brazil, Chile and Uruguay was shipped in large quantities to South Korea, Japan, China and the US (Tenold 2015, p. 191).

In all, the shipping of pulp and paper products coevolved in line with the development of the pulp and paper industry itself, and also in line with the overall development of the global shipping industry, whereby specialized ships were introduced in a number of trades and for several cargoes. When looking at the pulp and paper industry, three factors are particularly important for the development of the fleet carrying these products. First, as the international trade in these products grew (Fig. 12.2) it became economical to develop specialized ships to carry the products. At the same time, the cost reduction that the specialized ships brought about made it more beneficial to trade on a global scale, creating a benign feedback loop.

Second, as the pulp and paper companies grew larger and introduced global strategies, the economies of scale in logistics played a more vital role in their operations. This did not only include transport at sea, but also port facilities, terminals and inland transportation. Indeed, in many areas trains and trucks were more competitive than ships in carrying the products shorter distances. Moreover, the "just-in-time" concept was also introduced to carry pulp and paper products in the early 1980s with the demand for faster deliveries and lower warehouse costs at ports. The concentration of the industry made the use of economies of scale possible: the ten largest European paper industry companies produced less than a third of the total European sales in 1980, but by the mid-1990s their share was already around one-half. In the US this concentration started earlier. Both in Europe and North America the basis for the increased concentration was mergers and acquisitions.

Third, the value added of traded pulp and paper products increased. This, in turn, meant that there were strong incentives to lower the risk of damaging the cargo during its transport. Nordic pulp and paper producers in particular increased the value added of their products. Whereby they previously exported pulp, now they produced paper (with higher value added grades) to be exported abroad (Jensen-Eriksen and Ojala 2015). Thus, this focus on new and different products also needed new types of carriage capacity.

All these developments in the pulp and paper industry created a situation in which its products were no longer "a fillings cargo for conventional ships", as one major shipping entrepreneur put it in the early 1980s (Lennerfors 2016, p. 52). Now pulp and paper goods were the only, or the main, items to be transported, and they were valuable enough to justify specialized ships to carry them. The main aim was to increase productivity especially in loading and discharging the cargoes, and thus decrease the turnaround time at ports.

The cargo handling revolution since the Second World War has been referred to as the second revolution in shipping—the first one being the change from sail to steam and wood to iron during the nineteenth century. The efficiency gains in ports also enabled yet another substantial change in post war shipping: the enormous growth of the average size of the ships that, in turn, led to an increase in productivity and reduction of the unit costs (Tenold 2006).

The cargo handling revolution in pulp and paper trades relates to two specific ship types: open hatch and ro-ro. Both were used in other trades as well, but were particularly suitable to carrying pulp and paper products. Ships with wide hatch openings and gantry cranes onboard were introduced in the early 1960s; they were first designed to carry paper cargoes on a long-term contract basis and in a fixed trade. The open hatch technology was developed as a cooperative endeavour involving naval architects (Spaulding, Seattle), paper producers (Crown Zellerbach, US/Canada) and ship owners (Østberg, Norway). The two pioneering ships were first used to carry paper from British Columbia to California in the 1960s (Bakka et al. 2009, pp. 50–51).

The advantages with the open-hatch system were manifold. They were unobstructed, completely box-shaped holds without overhang; the speed and security of gantry cranes (faster, labour-saving and with the potential to reduce exponentially the risk of damage); and specialized equipment that was purpose-built to handle a variety of cargoes (both bulk and unitized commodities, including containers). Open hatch ships were especially suitable for forest products, particularly those made by the pulp and paper sector. Thus, open hatch ships were sometimes referred to as "lumber carriers" (Drewry 1978, pp. 5–7; Drewry 1974, pp. 76–77). They would typically be employed in trades where some major customers—for instance paper producers—had entered into long-term Contracts of Affreightment for a certain portion of the capacity, while the remainder of the ships' space would be sold on a voyage-by-voyage basis.

There has been a substantial increase in the size of the world fleet of open hatch vessels since the first ships were delivered in the early 1960s. The world fleet of these type of vessels was smaller than 500,000 dwt in 1970, but by 2010 the world leader (Gearbulk) alone had a fleet of 2.1 million dwt. The maximum size of individual vessels increased from 10,000 to 70,000 dwt and crane capacity from 13.6 to 70 tons over the same period.

According to Stopford (2009), the design of the open hatch bulk vessels, specifically the manner in which space in the holds can be more fully utilized, increases cargo carrying capacity by around 20% compared with traditional bulk ships. As a result of the specialized cranes and the improved operating environment, the cargo handling rates today are around 80% higher than for conventional bulk carriers. In

Table 12.2 Productivity change with open hatch vessels at port

	Previously	Open hatch
Gangs (number)	12	4
Rolls (per hour)	50	300
Rolls/man hour	4.17	75

Source Herbert (1979), pp. 131–141

the 1970s the difference was even larger, and by turning to open-hatch vessels one could cut the turnaround time at port in half and improve both the speed and quality of the cargo handling (Lennerfors 2016, p. 19). Table 12.2 illustrates the productivity gains with open hatch vessels at ports, resulting in a roughly 18-fold productivity increase. A quotation from a presentation of the first purpose-built open hatch ship, *Besseggen*, summarizes this change: 'to handle 9,000 tons of newsprint loading or discharging would normally take three days and employ 60 men, but with these cranes it is expected that 10 men will be able to accomplish the task in ten hours' (Talbot-Booth 1963).

The second important type of vessels used for pulp and paper shipments were ro-ro (roll on–roll off) and sto-ro (including also side-doors to roll on/off cargoes) ships beginning in the 1970s. Besides being efficient in loading and discharging cargoes, ro/sto-ro ships offered shelter from the weather conditions at ports; valuable paper rolls were not exposed to rain and snow. In a similar vein, in the open hatch segment special "totally enclosed bulk carrier" (TEBC) ships were developed by the late 1980s to be "weatherproof" during the loading and discharging.

Three rather expensive TEBC ships, each with a garage-like box on top of the deck with telescopic gantry cranes, were built in order to cater to the "Just-in-time" needs of one major Japanese pulp and paper importer at the late 1980s. Around half of the vessels' capacity was tied up to a joint venture between Canfor and Oji Paper to transport roughly 200,000 tons of newsprint annually. Due to changes in transport demand, the partners wanted to reduce the amount halfway through the contract, and the owners' problems of filling the ships illustrate the disadvantages of ships that are too specialized (Tenold 2015, pp. 200–202, 228–235).

The TEBC ships were developed collaboratively by three parties: the shipping company (Gearbulk), the paper industry (Canfor and Oji Paper) and the shipyard (Mitsui) (Tenold 2015, p. 201). The sto-ro technology was also developed in close co-operation with ship-owners (Sea-Link), pulp and paper companies in Finland and Sweden and their specific shipping companies (Transfennica and Combi Shipping Company) and shipyards (Lennerfors 2016). In some cases even some old ro-ro ships could be later converted to sto-ro ones (simply by cutting side doors into the hull), and this enabled simultaneous loading and discharging from three sides of the ship.

Ro/sto-ro ships were especially suitable for the exports of the Nordic pulp and paper companies. Thus, by the mid-1980s major pulp and paper shipping companies and operators such as Transfennica, Combi Shipping Company, Finnlines, and Sea-Link already used this technology in most of their shipments not only in the European trade but also in voyages to North America and back. In the case of Transfennica, roughly 17% of shipments were transported by ro-ro ships in 1976, while this share was 54% by 1984 (including sto/ro-ro ships) (Lennerfors 2016, p. 62) Ro-ro ships were further developed during the 1980s and 1990s, including special "cassettes" on which the cargoes were already loaded at port before the ship arrived, and then these steel platforms were simply rolled into the ship.

A third type of ship that made inroads into forest product transport was the specialized wood chip carriers, whereby pneumatic loading reduced port time. Again, the specialized ships first appeared in the early 1960s, when Toyo Pulp in 1963 launched the first such vessel. Subsequently, the fleet increased rapidly (Kurosawa and Hashino 2012, 2017). Most of these ships have been owned by Japanese companies, as the use of "wood chips" in production is a phenomenon that is more prevalent in Japan than elsewhere (Fenton 1982; FAO 1976).

As previously mentioned, the relationship between developments in the pulp and paper industry and the shipping industry had the properties of a virtuous circle. By revolutionizing cargo handling and utilizing economies of scale, ship owners improved efficiency to the point that it led to a dramatic reduction in the unit costs of transport. When this occurs, the basis for trade improves, with a concomitant increase in potential volumes. This was particularly important for areas far from market; some would claim that a country such as Chile became competitive as a result of the services offered by shipping companies (Tenold 2015, p. 192). The development of open hatch bulk carriers, wood-chip vessels and sto/ro-ro ships is an important part of the more general post war shipping trend, namely specialization of vessels carrying world trade.

Still, during most of the twentieth century—and during the previous centuries—the ships were not specialized to carry any specific cargo. In fact, in the first decades after the Second World War, most of the merchant tonnage consisted of general cargo carriers. However, their share diminished, especially from the 1960s onwards. Oil tankers, dry bulk ships and container carriers are perhaps the best known examples of this process that includes dozens of other specialized vessels, including LNG and LPG carriers, chemical tankers and car carriers. Specialized vessels were less flexible, but more efficient with purpose built cargo handling and stowage solutions. This specialization revolutionized sea transport, and again particularly in terms of faster turnaround times that declined in many cases from weeks to hours (Michel and Noble 2008, p. 34; Tenold 2015, pp. 55–56).

The specialization of ships was part of the industrial logics of each industry; transport emerged as a more integrated part of the logistics of industries with standardization of cargoes handled both on land and at sea. The container is of course the standardized "box" *par excellence* (Donovan and Bonney 2006; Levinson 2006). The same principles revolutionized the transport for the pulp and paper industries. The specialized ships enabled a rethinking of the twentieth century idea of vertically integrated industrial complexes located in one place. Namely, with declining unit freight costs, just-in-time concepts, specialized vessels, and global pulp and paper companies, this integration could be extended between areas, countries and even continents. Thus, raw materials and semi-finished products (such as wood chips, pulp or recycled paper) could be transported from far-away places to pulp and/or paper factories. The evolution in transport also enabled the true globalisation of the markets for pulp and paper products; still in the 1970s companies were mainly operating in regional/intracontinental domains (e.g. Europe, North America or Asia) rather than on a global scale (Sajasalo 2003; Siitonen 2003). Moreover, the emergence of efficient transport put new areas on the map of the global pulp and paper industry, most importantly South America (Tenold 2015, p. 191; Lima-Toivanen 2012). Pulp and paper companies aimed for long term contracts with ship owners as ships became an integral part of the production chain, streamlining the logistics. This motivated the ship owners, in turn, to invest in expensive, specialised vessels.

The other important change in international shipping had already emerged during the early modern period but has intensified ever since the specialization of shipping enterprises. Shipping agencies and brokerage firms were already known when paper was transported from the Dutch Republic to the Baltic Sea area in the late eighteenth century. However, the business world of shipping was far more complicated 200 years later. This was also the case in carrying pulp and paper industry products. As previously mentioned, there is a close relationship—and often long-term contracts—between ship owners with pulp and paper transport capacity and companies in the industry. Thus, interplay between these two parties is of vital importance, and collaboration with shipyards and charterers became ever more important in order to have the correct type of ship in operation at the right time and place. For example, Transfennica was a Finnish shipping operator that took specialized vessels on time-charters and operated them exclusively for their own cargoes of pulp and paper. Transfennica was founded in 1976 by the Finnish pulp and paper export associations (Finnpap, Finncell, Finnboard and Converta), which in turn were owned by the Finnish pulp and paper companies. After the consolidation of Finnish forest firms and the dissolution of these associations, Transfennica was sold first to the major paper companies (in 1994) and later to the Netherlands.

There was also a clear movement by the shipping companies to concentrate their operations on certain routes. Sea-Link, for example, carried 25% of Swedish and Finnish paper exports to the UK during the mid-1980s (representing 15% of total paper imports to the UK), though its share of total forest industry exports was smaller; four per cent in the Swedish case and six per cent in the Finnish case, respectively (Lennerfors 2016, p. 75). Within open hatch, Gearbulk had a particularly strong position in Latin America, while its compatriot and main competitor, namely Star Shipping, was the market leader in British Columbia.

Along with the changes in shipping organisations the question of domicile has grown in importance. During the Great Shipping Crises of the 1980s the previously dominant shipping nations in Europe and North America lost a substantial share of tonnage in terms of flag of ships. However, the ownership change was not as dramatic as it appears at first glance; the ships were often "flagged out" rather than having undergone a change in their actual ownership. This type of flagging was also rather typical among companies carrying pulp and paper products. Flagging did not necessarily mean changing the flag to low-cost countries or to flags-of-convenience (e.g., in Europe in many cases changing a company's ownership from one country to another could result in taxation benefits), but often the changes were made to receive assistance for constructing new tonnage. The Swedish firm Sea-Link, for example, operated ships under the Swedish flag but also under the flags of Finland, the UK and the Netherlands (Lennerfors 2016).

During the twentieth century the separation of foreign trade and shipping has further evolved both on the corporate and national levels. In recent articles by Tenold and Ojala (2017) and Ojala and Tenold (2017), they created a purpose-built indicator to compare the growth of global trade and global shipping, namely shipping/trade-ratio. It captures the development of countries' and regions' market shares in world shipping over an extended period of time. This share reveals both the coevolution of shipping and trade on a general level and patterns of specialisation in world shipping. However, the shipping/trade ratio-indicator is too rough to capture developments of shipping and trade at an industry level. Thus, it can unfortunately not be used directly to capture the possible changes in the pulp and paper industry and shipping on a detailed level.

However, looking at the three types of technology, a clear pattern emerges. Open hatch bulk shipping has been a Norwegian—or more specifically Bergen—specialty (Tenold 2009; Bakka and Grung 2009). While most of the open hatch fleet has been controlled by a small number of Norwegian companies, it has served customers all over the world, often through Contracts of Affreightment. The shipping companies have invested in terminals to improve efficiency, and have even "developed" and nurtured markets in distant lands such as South America. Ro-ro has been more closely associated with long-term time charters, and a closer relationship with individual pulp and paper producers that bought all the available capacity. The ship owners, like the producers, in Finland and Sweden have favoured this technology, just as the wood chip carriers have been a Japanese phenomenon.

Shipbuilding moved from European countries to low cost Asian yards during the period after the Second World War. This occurred also in the realm of building ships to carry pulp and paper products. Whereas the first open hatch ships were still constructed in "old" western shipbuilding countries (especially Norway and the UK), the specialised ships during the 1970s were mainly constructed in Eastern European countries (e.g., Poland, Yugoslavia etc.) and thereafter in Japan. During the 1980s and 1990s the production moved again, first to South Korea and then to China. Certain ro-ro vessels designed to carry paper products were built in China at roughly half the cost compared to the ones constructed in Europe during the 1990s (Lennerfors 2016, pp. 168–169; Tenold 2015, pp. 101–104, 133, 230; Lorenz 1991).

12.4 Conclusions

Over the last centuries the products of the pulp and paper industry have moved from local markets, via intracontinental exchange, to global markets. The areas of production have been decoupled from the areas of consumption, with distance playing a smaller and smaller role. The basis for this transformation has been the declining cost of sea transport, which has made it easier for producers to utilize their comparative advantage in raw materials. Cheap and efficient transport has been a necessary condition for the growth of the global pulp and paper industries in regions that are remote from the economic centres. Thus, the pulp and paper industries and international shipping have co-evolved since the 1960s, creating global markets and production chains.

In a post war world where political barriers to trade (e.g., tariffs and trade policy measures) declined rapidly, the practical barriers to trade (e.g., transport and transaction costs) became relatively more important. For the pulp and paper industries, high handling costs and expensive damage during transport discouraged intercontinental trade. Consequently, the benefits from trade could not be realized without new technological solutions within transport being developed, and three groups of stakeholders—the pulp and paper producers, industrially-minded ship owners and ship builders (naval architects and shipyards)—worked together to make improvements.

The result was new ship types—open hatch bulk carriers, ro/ro- and sto/ro-vessels and dedicated wood chip carriers—and a change in the land/sea-interface that improved efficiency and dramatically reduced impairment. This created the potential to transform the business models, both on the production side and on the consumption side. "Just-in-time"-production, global operations and integrated logistics systems followed.

Both the dominant technological solution and the type of contracts introduced varied between the main "market clusters". Although the basis for these different strategic choices needs more research, one thing is evident. The global markets for pulp and paper, and the global market for pulp and paper transport, pushed each other forward in a virtuous circle.

References

Andersson LF, Bergquist AK, Eriksson R (2016) Profits, dividends and industry restructuring: the Swedish paper and pulp industry between 1945 and 1977. Scandinavian Econ Hist Rev 64(3):278–296

Åström SE (1988) From tar to timber. Studies in Northeast European forest exploitation and foreign trade 1660–1860. Finnish Society of Sciences and Letters, Helsinki

Bakka D, Grung T (2009) A Star Story. Kremmerhuset, Bergen

Bakka D, Basberg BL, Tenold S (2009) I bølger—Vestfold og Telemark Rederiforening 1909–2009. Vestfold og Telemark Rederiforening, Tønsberg

Donovan A, Bonney J (2006) The box that changed the world: fifty years of container shipping—an illustrated history. Commonwealth business media, East Windsor

FAO (1976) Wood chips production, handling, transport. Food and agriculture organization of the United Nations, Rome

Fenton R (1982) International wood-chip trade (and the South Pacific). Commonw For Rev 61(3):181–194

Harlaftis G, Tenold S, Valdaliso JM (eds) (2012) The world's key industry. History and economics of international shipping. Palgrave Macmillan, Basingtoke

Hazley C (2000) Forest-based and related industries of the European Union: industrial districts, clusters and agglomerations. Taloustieto, Helsinki

Herbert R (1979) The story of designing and building the first open-hatch ship with 'rectangular' holds to carry newsprint. In: Kummerman H, Jacquinet R (eds) Ships' cargo, cargo ships. MacGregor Publications, Hounslow, pp 131–141
Hetemäki L (1999) Information technology and paper demand scenarios. In: Palo M, Uusivuori J (eds) World forests, society and environment. Kluwer, Dordrecht, pp 31–40
Hetemäki L, Obersteiner M (2001) US newsprint demand forecasts to 2020. International institute for applied systems analysis, interim report IR-01–070. Laxenburg, Austria
H.P. Drewry (Shipping Consultants) Ltd. (1974) The market for handy-sized bulk carriers (20/35,000 DWT)—a review of trends in supply/demand. H.P. Drewry, London
H.P. Drewry (Shipping Consultants) Ltd (1978) Specialized Types of Bulk Carrier. H.P. Drewry Economic Studies, no. 59. H.P. Drewry, London
Järvinen J, Lamberg J-A, Murmann JP, Ojala J (2009) Alternative paths to competitive advantage: a fuzzy-set analysis of the origins of large firms. Ind Innov 16(6):545–574
Järvinen J, Lamberg J-A, Nokelainen T, Tikkanen H (2012) Global demand for Paper Products: 2006-2050. In: Lamberg J-A, Ojala J, Peltoniemi M, Särkkä T (eds) The Evolution of Global Paper Industry 1800–2050. A Comparative Analysis. Springer, Dordrecht, pp 307–343
Jensen-Eriksen N, Ojala J (2015) Tackling market failure or building a cartel? Creation of investment regulation system in finnish forest industries. Enterp Soc 16(3):521–555
Krawany F (1910) Internationale Papier-Statistik. Verlag für Fachliteratur, Wien
Kuhlberg M (2012) An accomplished history, an uncertain future: Canada's pulp and paper industry since the early 1800s. In: Lamberg J-A, Ojala J, Peltoniemi M, Särkkä T (eds) The evolution of global paper industry 1800–2050. A comparative analysis. Springer, Dordrecht, pp 101–133
Kurosawa T, Hashino T (2012) From non-European tradition to a variation on the Japanese competitiveness model: The modern Japanese paper industry since the 1870s. In: Lamberg J-A, Ojala J, Peltoniemi M, Särkkä T (eds) The evolution of global paper industry 1800–2050. A comparative analysis. Springer, Dordrecht, pp 135–166
Kurosawa T, Hashino T (2017) Three markets and three types of competitiveness: pulp and paper industry. In: Bouwens B, Donzé P-Y, Kurosawa T (eds) Industries and global competition. A history of business beyond borders. Routledge International Studies of Business History. Routledge, Abingdon
Lamberg J-A, Näsi J, Ojala J, Sajasalo P (eds) (2006) The evolution of competitive strategies in global forestry industries. Comparative perspectives. Springer, Dordrecht
Lamberg J-A, Ojala J, Peltoniemi M, Särkkä T (eds) (2012) The evolution of global paper industry 1800–2050. A comparative analysis. Springer, Dordrecht
Lennerfors TT (2016) Elling Ellingsen. The Shipping Entrepreneur. Medströms Bokförlag, Stockholm
Levinson M (2006) The box: how the shipping container made the world smaller and the world economy bigger. Princeton University Press, Princeton
Lima-Toivanen MB (2012) The South American pulp and paper industry: cases Brazil, Chile and Uruguay. In: Lamberg J-A, Ojala J, Peltoniemi M, Särkkä T (eds) The evolution of global paper industry 1800–2050. A comparative analysis. Springer, Dordrecht, pp 243–283
Lorenz EH (1991) An evolutionary explanation for competitive decline: the British shipbuilding industry, 1890–1970. J Econ Hist 51(4):911–935
Michel K, Noble P (2008) Technological Advances in Maritime Transportation. Bridge 38(2):33–40
Ojala J, Kaukiainen Y (2012) Finnish Shipping—A Nordic Exception? In: Tenold S, Iversen MJ, Lange E (eds) Global Shipping in Small Nations. Nordic Experiences after 1960. Palgrave Macmillan, Basingtoke, pp 129–155
Ojala J, Tenold S (2017) Maritime trade and merchant shipping: the shipping/trade ratio since the 1870s. Int J Marit Hist 29(4):838–854
Ojala J, Lamberg J-A, Ahola A, Melander A (2006) The ephemera of success: strategy, structure and performance in the forestry industries. In: Lamberg J-A, Näsi J, Ojala J, Sajasalo P (eds) The evolution of competitive strategies in global forestry industries. Comparative perspectives. Springer, Dordrecht, pp 257–286

Rjestoff NA (1913) Pappret i Ryssland. Finska Pappersindustriidkarenes förbund, Helsingfors
Sajasalo P (2003) Strategies in Transition. The Internationalization of Finnish Forest Industry Companies. Dissertation, University of Jyväskylä
Särkkä T (2012) The British paper industry, 1800–2000. In: Lamberg J-A, Ojala J, Peltoniemi M, Särkkä T (eds) The evolution of global paper industry 1800–2050. A comparative analysis. Springer, Dordrecht, pp 167–190
Scheltjens W (2015) Dutch deltas: emergence, functions and structure of the low countries. Marit Transp Syst, ca. 1300–1850
Siitonen S (2003) Impact of globalisation and regionalisation strategies on the performance of the world's pulp and paper companies. Dissertation, Helsinki School of Economics
Stopford M (2009) Maritime economics. 3e. Routledge, Abigdon
Talbot-Booth EC (1963) Merchant ships. J. Commer. Shipp. Telegraph, Liverpool
Tenold S (2006) Tankers in trouble: Norwegian shipping and the crisis of the 1970s and 1980s. International Maritime Economic History Association, St. Johns
Tenold S (2009) Vernon's product life-cycle and maritime innovation: Specialised shipping in Bergen, Norway, 1970–1987. Bus Hist 51(5):753–769
Tenold S (2015) Geared for growth: Kristian Gerhard Jebsen and his shipping companies. Bodoni, Bergen
Tenold S, Ojala J (2017) How to sail a sinking ship—adapting to the declining competitiveness of the European shipping industry. In: Bouwens B, Donzé P-Y, Kurosawa T (eds) Industries and global competition. A History of business beyond borders. Routledge international studies of business history. Routledge, Abingdon, pp 269–289
Toivanen H (2004) Learning and corporate strategy: the dynamic evolution of the North American pulp and paper industry, 1860–1960. Dissertation, Georgia Institute of Technology
Toivanen H (2012) Waves of Technological Innovation: The Evolution of the U.S. Pulp and Paper Industry, 1860–2000. In: Lamberg J-A, Ojala J, Peltoniemi M, Särkkä T (eds) The evolution of global paper industry 1800–2050. A comparative analysis. Springer, Dordrecht, pp 49–80

Chapter 13
Technological Transformation in the Global Pulp and Paper Industry: Concluding Remarks

Mark Kuhlberg, Timo Särkkä and Jussi Uusivuori

13.1 Conclusions and New Beginnings

It is axiomatic that the speed and intensity of the industrialization process around the globe have been shaped by a wide variety of factors. For example, variables such as access to natural resources, the cost of and accessibility to transportation, government policy at the regional, national and international levels, demand characteristics and entrepreneurial and corporate strategies have all played a role—to different degrees—in determining both which countries adopted the latest industrial technologies and how they did so. By examining these dynamics, historians have been able to explain, for instance, why countries that shared the same continent or even a border experienced the phenomenon of industrialization in such fundamentally dissimilar ways.

This volume has engaged this broad subject and taken it from the macro to the micro level by focusing on the development of a single industry—pulp and paper—and the role technology played in its evolution in a number of diverse locations literally around the world and over the course of several historical periods. Our study has taken us from ancient China, for example, to assess the earliest methods of making paper, all the way to contemporary Portugal, which was visited in an effort to analyse the most recent steps in the process of nationalising its domestic pulp and paper industry. Similarly, the chapters in this volume have examined their

M. Kuhlberg (✉)
History Department, Laurentian University, Sudbury, ON, Canada
e-mail: mkuhlberg@laurentian.ca

T. Särkkä
Department of History and Ethnology, University of Jyväskylä, Jyväskylä, Finland
e-mail: timo.sarkka@jyu.fi

J. Uusivuori
Natural Resources Institute Finland, Helsinki, Finland
e-mail: jussi.uusivuori@luke.fi

T. Särkkä et al. (eds.), *Technological Transformation in the Global Pulp and Paper Industry 1800–2018*, World Forests 23, https://doi.org/10.1007/978-3-319-94962-8_13

stories in a wide range of environmental frameworks. At one end of the spectrum are countries such as Canada, Finland and Sweden in the Northern Hemisphere, which are well endowed with the fibre, found principally in the form of conifer forests, which emerged in the 1860s as the most important raw material for making pulp and paper. At the other extreme both geographically and environmentally are the lands "down under", Australia and New Zealand, which we examined because their lack of conifers compelled them to search for new fibre sources and develop new pulp and paper technologies upon which they could base their national industries. Technology is the common thread linking all these stories, sometimes because the industry aggressively embraced it in a particular part of the globe and at other times because the opposite was true.

As a result of these analyses, this collected volume has made several important contributions to the literature about the history of the international pulp and paper industry in general and its adoption of technology in particular. Obviously, the departure point for such a study should be the long history of papermaking technology, which reaches back in time nearly two thousand years, and Chap. 2 has provided us with a sterling summary of this subject. The 10 case studies that follow demonstrate, for example, that the industry's establishment and development in a particular location is dependent upon a unique set of factors that defies generalisation on the one hand and lends itself to it on the other. Consider the supply of raw materials available to and the location of a producer relative to markets. Australia (Chap. 10), New Zealand (Chap. 9) and Portugal (Chap. 6) were historically deficient in terms of the industry's traditional coniferous wood fibre resources and they were initially forced to satisfy their domestic demand for pulp and paper products through imports. Although this shortcoming was initially a hurdle to the establishment and growth of their domestic industry, government and industry cooperated to overcome their "fibre challenge" using various strategies at different stages of the twentieth century. In sharp contrast, a producer situated both amidst a bountiful supply of wood and power resources in Canada and within close proximity of an enormous market—the United States—remained surprisingly small because of a host of factors ranging from unfavourable policies implemented by local and foreign governments to the firm's adoption of a corporate culture that embraced monopoly capitalism (Chap. 7).

On the other hand, this volume's chapters have made it clear that a few observations apply to practically all the disparate stories told here. For example, war was both a short term curse but long term blessing for many pulp and paper makers, and the conflicts noted herein certainly highlight how necessity is truly the mother of invention. Chap. 11 deals with British India's pulp and paper industry and underscores how the shortage of raw wood pulp available during the First World War back in the United Kingdom was the impetus behind its producers searching for alternative supplies of raw materials, and how their quest led them to India and its enormous stock of bamboo. This was the precursor to a colonial policy that sought first to foster the development of the colony's native pulp and paper industry and later efforts by both domestic and British firms to develop the technology to process this fibre into marketable paper products. Similarly, Chap. 3 assesses how war dealt producers located in Finland short term pain but long term gain. That country's pulp and paper makers were forced to grapple with the profound challenges wrought by the Second World War and the enormous burden represented by paying reparations to the Soviet Union in its wake. Nevertheless, these trying circumstances precipitated strong cooperation between the Finnish government, the country's universities, and its industrial sector, and the result saw Finland emerge as a global leader in the development and production of the technology used by the pulp and paper industry. Furthermore, Chap. 4 demonstrates how the conflicts in the Middle East during the 1970s—and the consequent oil crises—compelled producers in Sweden to develop biofuels in an effort to wean themselves off fossil fuels. This was the fillip to a period during which players in the country's industry—and the collective industry and government—began working together to develop new technology; cooperation not competition in research and development became their mantra. The capstone achievement in this regard was the development of a greener product to meet a new market demand, a breakthrough that vaulted Sweden's firms to the upper echelon of the world's pulp and paper producers. Finally, Chap. 12 explores international shipping of the industry's raw materials and its products. In doing so, it illustrated how location near both fibre sources and markets decreased in importance during the latter part of the twentieth and early twenty-first centuries as a result of the technological developments in—and subsequent dramatic reduction in cost of—seaborne transportation. Although this observation can be made about many industries during this period, it strikes a particularly resonant cord with the pulp and paper sector.

Nevertheless, the study has raised as many questions as it has answered, and in doing so it has created the opportunity for myriad future studies of the global pulp and paper industry. This volume's focus on technology has underscored the industry's capital-intensive nature and the paramount role that financing has thus played in its development. Although often a difficult subject to unravel, there is fertile ground to be ploughed in the future in terms of examining the different sources of capital—both public and private—that have been tapped to support investment in the industry and the relationship between financial backers and company policy. Similarly, many of the profiles presented here have mentioned the existence of cartels and sales associations in the pulp and paper industry, another subject that has been relatively unexamined in this field. In addition, this investigation has highlighted the enormous role that government policy—on the regional, national and international levels—has played in influencing the establishment and growth of the industry around the world. This, too, is a subject that merits a separate study.

From an economic and policy point of view, it would be interesting to analyse the obstacles that have prevented a more profound division of labour to take place in global pulp and paper industries. It seems that the history tells us that the competitive advantage paradigm has not been completely valid in the international trade of pulp and paper products. We have seen successful cases of paper industries in cultural, economic and natural conditions that defy the likelihood of encountering such examples. A future analysis could look into, for example, the question of what role government policies—in the form of protection or according the industry strategic status—played in developing a domestic pulp and paper industry to replace a country's former reliance on imports. Furthermore, the role of exchange rate regimes and policies in the evolution of the global pulp and paper industry might warrant a closer analysis. It has been heavily influenced by international trade and foreign direct investments and by changes in the currency alignments between economic zones. On the other hand, there are cases in which national exchange rate policies have been influenced by the economic performance of domestic pulp and paper industries; Finland is perhaps the most obvious example of this latter phenomenon. In this instance, it would be interesting to see—as some have suggested—that the value of the Finnish domestic currency prior to it adopting the common Euro-zone currency—was determined to a large extent by the competitiveness of its pulp and paper industry.

To conclude, this study has revealed several obvious reasons for an in-depth study of the political economy of the world's pulp and paper industry. It has also revealed that analysing the history of this industry by using a comparative method is a valuable way of exposing its historical complexities and the challenges it is facing today and will probably confront in the future. Through its comparative analysis, the investigation's scientific results have increased our knowledge and understanding of the global pulp and paper industry both historically and in the present day. Furthermore, an obvious, and perhaps the most important, outcome of this volume is the enhancement of collaboration among the contributors to this project. The research network involved ensures that the research subject is addressed within the on-going international academic discussions and thus allows for the utilisation of the research results in the most efficient way (e.g. in terms of utilisation and interaction with potential end-users and beneficiaries of the research; new collaborative structures and networks; researcher mobility; future publications; potential new, innovative research openings).

Index

T. Särkkä et al. (eds.), *Technological Transformation in the Global Paper Industry 1800–2015: Comparative Perspectives*, World Forests23,
https://doi.org/10.1007/978-3-319-94962-8

T

The manufacturer's authorised representative in the EU is Springer
Nature Customer Service Centre GmbH, Europaplatz 3, 69115 Heidelberg,
Germany. If you have any concerns regarding our products, please
contact ProductSafety@springernature.com

Printed and bound by CPI Group (UK) Ltd, Croydon, CR0 4YY
30/09/2026
02225643-0010